Mobile Technologies and Handheld Devices for Ubiquitous Learning:
Research and Pedagogy

Wan Ng
University of New South Wales, Australia

Information Science REFERENCE
INFORMATION SCIENCE REFERENCE
Hershey · New York

Director of Editorial Content: Kristin Klinger
Director of Book Publications: Julia Mosemann
Acquisitions Editor: Lindsay Johnston
Development Editor: Mike Killian
Publishing Assistant: Milan Vracarich, Jr.
Typesetter: Deanna Jo Zombro
Production Editor: Jamie Snavely
Cover Design: Lisa Tosheff

Published in the United States of America by
Information Science Reference (an imprint of IGI Global)
701 E. Chocolate Avenue
Hershey PA 17033
Tel: 717-533-8845
Fax: 717-533-8661
E-mail: cust@igi-global.com
Web site: http://www.igi-global.com

Library of Congress Cataloging-in-Publication Data

Mobile technologies and handheld devices for ubiquitous learning : research and pedagogy / Wan Ng, editor.
p. cm.
Includes bibliographical references and index.
Summary: "The book provides education and IT stakeholders with an overview of what has been implemented with the different types of mobile technologies, their pedagogical differences, evaluations of the learning outcomes, and new ideas for their uses as both pedagogical and research tools"--Provided by publisher.
ISBN 978-1-61692-849-0 (hardcover) -- ISBN 978-1-61692-851-3 (ebook)
1. Educational technology. 2. Education--Effect of technological innovations on. 3. Mobile communication systems. 4. Wireless communication systems. 5. Mobile computing.
I. Ng, Wan, 1955-
LB1028.3.M59 2010
371.33--dc22

2010018445

British Cataloguing in Publication Data
A Cataloguing in Publication record for this book is available from the British Library.

Table of Contents

Section 3
Research and Innovative Pedagogy

Detailed Table of Contents

ness Map). PERKAM allows the learners to share knowledge, interact, collaborate, and exchange individual experiences. It utilizes ubiquitous technologies to detect the learner's environmental objects and location, then recommends the best matched educational materials and peer helpers in accordance with the detected objects and the current location. This environment provides the learner with Knowledge Awareness Map, which graphically displays the knowledge awareness support. The experiment and the questionnaire results indicated that Knowledge Awareness Map is helpful to understand the learner's situation and it efficiently visualizes the recommendations.

This chapter argues that ubiquitous computing is a necessary, but not sufficient condition for ubiquitous learning. It will propose definitions that clarify some of the contextual features that shape both terms when they are applied in a school context. Using data and experiences derived from a year-long project with primary and secondary schools in Victoria, Australia, examples of classroom activities are analysed to illustrate how the presence of ubiquitous computing did not guarantee ubiquitous learning. These activities are compared to articulate the key features of ubiquitous learning and develop a model that shows how the contribution of ubiquitous computing to ubiquitous learning is constrained by pedagogic frameworks that shape the relationship between handhelds, other elements of the technological suite and learning.

Contemporary research on mobile learning focuses mainly on issues such as the acquisition of knowledge, the development of cognitive skills and the efficiency of these tools with respect to the achievement of specific educational goals. Nonetheless, the consequences of the adoption of a technology within a learning context for educational purposes should not be reduced solely to the cognitive dimension implied in its use, nor should it be measured only in terms of goal achievement. Even if intended as purely educational tools, technologies are complex social objects that redefine the sense of the context, the activity and even the identity of the actors engaged in their use. When educational institutions adopt mobile information technologies they propose more than a supposedly efficient educational instrument or technology-formatted contents. They introduce a form of life. By form of life, the authors mean a repertoire of possible uses, actions, meanings and even intended actors that the users may adopt. A technology is then a condensed social context within which learning takes place. To grasp the richness and the complexity of the learning involved in using mobile information devices, the authors developed a larger and holistic definition of learning that goes beyond simply acquiring knowledge and which involves different aspects of an individual's psychological, cultural and social development.

Questions asked are: how does the use of an iPod affect the students' identity? How does it contribute to the development of social skills and social awareness? Drawing on research involving 123 Canadian university students recruited from different disciplines (on the basis of data coming from diaries and focus groups), this chapter focuses on the multiple consequences of the introduction of this technology as an educational tool in students' academic life.

The practical contribution of this chapter is the understanding of activities in mobile learning environments and how learners use awareness to model their actions for the provision of personalized learning support. The chapter is about the advancement of the human-centric approach to personalized learning through enhanced learner-to-learner interaction – where context and social presence awareness is of vital significance to how learners decide and act on a learning task. It is an expedition towards understanding the phenomenon of mobile learning, where personalized learning and support is a result of social awareness activities of learners as they traverse varied learning contexts. Mobile learning, in this chapter, is signified by mobility of learners regardless of mobile technologies. Activity Theory, which draws attention to mediated activity within a social context, is used to explore how mobile learners use context and social presence awareness to facilitate their ubiquitous social interactions.

The handheld computer as a pedagogical tool has the capacity to enable students to demonstrate understanding through different modes of representations, for example, verbal, text, tables and graph, drawings, writing or written formulas, concept mapping and animations through Flash or Pocket Slides Powerpoint. Its impact as a motivational learning tool has been described in numerous articles. The purpose of this chapter is to describe its use as a research tool for capturing students' thinking processes as they construct representations in science and mathematics, or solve problems in these learning areas on the handheld. By using an avi-screen capture software operating in the background to do this, the research is a non-intrusive method of capturing the verbal and screen-based (visual) elements of students' thinking as they use the handhelds to complete individual or collaborative tasks.

This chapter describes the use of mobile technologies as research tools while discussing both the benefits and limitations of this mode. It maps practical approaches to the management of a mobile, digital

Foreword

Education doesn't have a very good track record with innovative technologies. A few folks reading this will be old enough to remember being banned from using the "new" ballpoint pens for fear that their cursive script handwriting might be ruined. Others, later, may have been prevented from using their slide rules, or in still later years yet had their calculators confiscated, or told that it was the "wrong type" ("we don't allow programmable functions here I'm afraid"). Education has had a track record of first confiscating, and then appropriating, emerging technologies. Even the potential that computers and laptops offered became tamed by a mass of "managed services" and by a stultifying focus on a small suite of tools for office workers, rather than perhaps the freedom of tools used in creative industries, or in play. For many, the power of moving images was reduced to weakly animated text on a PowerPoint slide. Even the remarkable, free, open world offered by the World Wide Web was rapidly closed off by massively filtered internet provision and by parallel "equivalent" services. "We have our own YouTube" authorities would proclaim, literally blind to the 10 hours of video contributed every minute to the real one.

Education's question, faced with most emerging technology, has traditionally been a simple productivity view of "how can this new thing usefully improve what we are already doing?", rather than "what new things might we now do?" The excitement of games became tamed to "spelling space invaders", the art, installation and celebration potential of a computer plus projector was reeled back into an interactive white-board and so on. It has always been relatively simple to achieve this appropriation of new technologies, because the pace of change was rapid but manageable. We had only been on the gentle slope of an exponential curve that is now, finally, beginning to steepen very rapidly indeed. On the gentle part of that slope we had plenty of time to reflect on quite small technological changes - and to head off perhaps the perceived dangers of that potential change. I remember a debate running for several years in the early 80s about whether we should have colour or monochrome computer monitors! People suggested that colour might be damaging for students' eyes! But now that we have reached the steep bit, where the Technology Progress against Time curve leaps skywards, we have but the blink of an eye to make sense of huge leaps in technological capability, and that changes everything.

Mobile technology is the first to appear in our classrooms and to successfully resist education's hunger for appropriation. It is the first technology already on the steep part of the exponential curve. When I put some of the very first "luggable" phones into a classroom in early early 90s the children saw a potential right away: "we could talk to people in France and improve our language skills; it wouldn't matter where they were, or we were, we could talk about what we were doing" they said, eyes alight with excitement. Education didn't give up the fight easily and produced the usual list of hokum to justify the usual bans: texting would ruin children's prose, radiation might fry their brains; when TV used video to show fictional fights and incidents it was "drama", when children did it, it was "happy slapping" and so on. But mobile technology is essentially personal, portable and powerful. As schools were fretting about texting children were already social networking. This time the pace of change meant it was all too easy for the children to remain one step ahead. And they did.

Today of course mobile phone technology is an integral part of very many classrooms: children summarise their understanding, snap images from the board, blog their field trips, bluetooth to their teachers, do day-in-the-life projects through their phones with other schools, sample and exchange data, and - as this book reveals - much more besides. This post-Google generation see a whole new potential for themselves as learners and they see much of it lying beyond the structures and strictures of the old education system of the last century. Mobile learning has already shown itself to be a watershed in learning - the moment when the learners, finally, had an authentic say in what the future might be like. The biggest certainty we have as we move forward is that education won't appropriate this next tidal wave of change. Tomorrow will be a very different place. This excellent book gives us some hints about what kind of place that might be.

Stephen Heppell

***Stephen Heppell** (Professor) is CEO of Heppell.net, Professor of New Media Environments at Bournemouth University, Emeritus Professor Anglia Ruskin University, Visiting Professor University of Wales, Newport and Executive Chairman LP+. Stephen founded and ran Ultralab for a quarter of a century, building it into one of the most respected research centres in e-learning in the world - at one time Ultralab was the largest producer of educational CD-ROMs in Europe - before leaving it in 2004 to found his own global and flourishing policy and learning consultancy Heppell.net which now has an enviable portfolio of international projects all round the world such as Learnometer and HorizonTAL. Stephen pioneered, and was the guiding "father" of, early social networking with seminal projects including Schools OnLine for the UK Department of Trade and Industry in 1995/6, Tesco Schoolnet 2000 from 1999 (the then Guinness Book of Record's largest internet learning project in the world) andThink.com with Oracle from 1999. He was a founder board member for Teachers.TV and sits on the board of the UK's Skillset and on BAFTA's Council. Stephen created in 1998, and then guided for ten, years Notschool.net, at the time a uniquely effective project to re-engage children excluded from school by behaviour or circumstances. His recent awards include: in June 2006 Stephen was awarded the Royal Television Society's Judges Award for Lifelong Services to Educational Broadcasting and in 2008 he received the BETT for Outstanding Achievement in ICT Education.*

Preface

The main aim of this book is to present to education stakeholders – teachers/lecturers, principals, policy makers and researchers in education and information technology, an argument for the inclusion of mobile technologies for ubiquitous learning at all levels of education. The argument is based largely on (i) the capacity of these technologies to make a difference and innovate learning and teaching (ii) research that demonstrates benefits in learning with mobile technologies and (iii) the need for teachers to be sensitive to the upbringing and culture of today's generation of students whose lives are very much influenced by these technologies and the easy access to the Internet.

THE INFLUENCE OF TECHNOLOGY ON SOCIETY

The assumption that computer-mediated learning will occur in the classroom, managed by a teacher, is now being challenged, not by schools and educational software developers, but by the consumer growth of personal technologies. Many children already have access to a wide range of computing and communications devices. (Sharples, 2003, p. 505)

These children are the 'millennial' students who have been born after 1980 (McMahon & Pospisil, 2005; Oblinger, 2003, 2004; Raines, 2002). Chansanchai (2006) describes millennials as young adults who are so attached to their gadgets and that for them, life without their mobile phones, iPods, computers or being online is unimaginable. This view of changes in people is accompanied by changes in understandings of technology. The conceptual shift over the last decade from Information Technology (IT) to Information Communication Technology (ICT) reflects a shift from IT as providing, storing and processing information to a focus on ICT's ability to 'strengthen and multiply communicative relationships – between ideas, especially as a feature of non-linearity and hypertext, as well as between people' (Nixon, Atkinson and Beavis, 2006, p.133). Consistent with this, millennial students are ICT-skilled people who are frequent online users and who embrace technology in their everyday lives and use it as a central plank of their relationship building. They are networked socially through technology and stay in contact with friends, family and peers via email, mobile phones, SMS, MSN, Skype, discussion boards and chatrooms. They become part of online communities and collaborate to create and share information on the Web using Web 2.0 technologies such as wikis, blogs, VoiceThreads and video sharing sites (YouTube) and they socialise through networking sites such as MySpace and FaceBook. Nixon et. al. (2006) have argued that students' activities outside schools embrace ICT in this manner where they are engaged in communicative networks that are not only for recreational purposes, but also for the exchange of skills and knowledge. Mobile devices further enhance Nixon et. al.'s network of a 'multiplicity of communicative relationships' both formal and informal. The mobility of the devices, such as the mobile phones and PDAs with wireless capacity, provides learners with communicative capabilities and access to information and social networking sites anytime and anywhere. Lefoe and Olney (2007) assert that the research (e.g. Sharples, 2006; Seppälä & Alamäki, 2003) in recent years has shifted the focus from the (mobile) technology to the mobility of learners as those learners take advantage of the flexibility in situated learning in environments that are mobile rather than fixed at predetermined locations. In this regard, learning with mobile devices is becoming a normal part of self-directed and lifelong learning. Hence, necessary changes to best facilitate students' learning need to be considered. In addition to content and pedagogy in curriculum design, the view that learning is a cultural activity needs to be recognized. By linking students' learning experiences outside of educational institutions to those within the institutions' contexts, students' learning could be enhanced and institutions should be supportive of this.

Within the range of mobile technologies - IT devices that are portable and personal, are laptops, tablet PCs, mobile phones, iPods, MP3 players, ultramobiles, smartphones, game consoles, global positioning systems (GPS), digital cameras and recorders as well as handheld computers (also called PDAs, pocket PCs and Palm). The emphasis of this book, however, is on portable technologies that fit into the palm of a hand and are pocket sized. Because of this, most authors exclude laptops, tablet PCs and notebooks.

Research (e.g. Hardey 2007) has shown that ownership of handheld technologies, particularly mobile phones, sets the trend that defines the youth culture of today. In education, an understanding and embracing of this culture in the classroom could potentially lead to fewer mismatches between the teacher and students in terms of learning engagement with technology, potentially leading to better learning outcomes. As Kolb (2009, p. 4) said:

Educators dismiss mobile phones, instant messaging and other popular technology communication tools as "distracting" to classroom learning. Yet if educational technology theory, research and pedagogy are reconceptualised to include the tools and knowledge that students already possess, then students will have better opportunities to connect learning inside and outside of school.

The book attempts to reconceptualise theory, research and pedagogy of learning with mobile devices that students are familiar with. The changing nature of people connected with mobile devices and the fast-paced development of ICT and mobile technologies place growing demands on the ICT capabilities of school teachers, tertiary lecturers and teacher educators. The book will provide these and other educational stakeholders (principals, policy makers, researchers and parents) with theoretical frameworks and research-informed pedagogies for learning with mobile technologies across primary, secondary and tertiary education sectors. The chapters discuss the capacity of mobile devices for ubiquitous learning, the appropriate and just-in-time uses of these devices to bring about better learning outcomes independently and collaboratively (including internationally) in both formal and informal situations. The book provides insights into how developing countries are innovatively adopting mobile devices for their children's learning (Chapter 8) as well as how tertiary institutions (Chapters 2, 4, 5 10 and 11) and schools (Chapters 3, 9, 12, 13, 14 and 15) are designing and implementing sound pedagogies into the students' use of mobile and handheld devices for learning. In each case, the challenges and ongoing research are discussed.

The book provides a holistic view of research and pedagogy for using mobile technologies in education. While a number of the chapters focus on mobile devices largely as pedagogical tools, two of the chapters focus on using mobile devices as tools for research. Chapter 6 provides the framework for using video-capture software in PDAs to capture students' 'thinking' as they make use of the devices as pedagogical tools to solve problems. Chapter 7 discusses broadly the use of a suite of mobile devices to capture data and the management of a mobile, digital data collection process and its accompanying challenges in gathering, coding and archiving the research data.

As the focus of this book is on ubiquitous learning with mobile and handheld devices, the authors in the book have made attempts to define ubiquitous and mobile learning within the contexts of their research and work. The term 'mobile learning' tends to be used more commonly by researchers and often interchangeably with 'ubiquitous learning'. My very brief attempt to show the relationship between these two terms in the next section sets the tone for the book.

UBIQUITOUS AND MOBILE LEARNING

Ng, Nicholas, Loke and Torabi (2010, p.43) stated the two dimensions characterizing ubiquitous learning as:

(i) it is not constrained by physical space, plans or timetables but is pervasive and occurs anywhere at anytime and (ii) as a consequence of the distributed nature of the immediate access to a variety of sources of information or means of reflecting on experiences in interaction with others, ubiquitous learning is characterized by the transformation of understanding and the ability to question experiences and information.

A dimension of the "distributed nature of the immediate access to information" is where the learning environment is equipped with technologies that could 'talk' with or support one another to provide the necessary information and help for the learner. In this regard, the authors of Chapters 2 and 3 present slightly varying views of ubiquitous computing environments. One view, as described in Chapter 2, is that the learning environment is fitted with context-aware facilities and objects which are adaptable to the learner's needs in providing resources and peer-assistance. The other view, as described in Chapter 3, is the physical presence of a variety of technologies surrounding the learner which could be drawn on to complement and to enhance learning. This is particularly true at the school level where handheld devices are less sophisticated or are shared due to budgetary constraints. For example, a student with a PDA that does not have camera or video recording functions may need to work with his/her mobile phone's camera or the school's digital camera to capture images for a project. The dispersion of many computing devices and objects throughout the physical environment that could service one person has been described by Weiser (1991, 1996) as the third wave of computing of which we are now part.. Weiser articulated the concept of 'ubiquitous computing' as technologies receding into the background of people's lives such that they are being increasingly used unconsciously for various purposes. In both the context-aware and non-context aware (ubiquitous) computing environments of Chapters 2 and 3 respectively, the technologies are readily available to assist the students with their learning. In this regard, the computing is ubiquitous in that it is so embedded in the students' lives that it "disappears" (O'Malley & Fraser, 2006) and does not occupy their attention all the time. In a ubiquitous computing environment, the student is able to make use of technology seamlessly and effortlessly for learning and to complete tasks. When computing becomes ubiquitous, it has the capacity to support ubiquitous learning.

Ubiquitous learning is able to situate the learner in both the real and virtual world, regardless of time and place, where questions encountered in the real world can be immediately answered individually by accessing and conducting research on the Internet or collaboratively through text messages, twittering or posting online. For these purposes, an important component of ubiquitous learning is mobile learning where mobile devices enable continuity in a person's learning as a consequence of their continual access to learning resources and networks.

The meaning of 'mobile learning' has been defined differently by different people. The complexity involved in its definition and what education around it means have been discussed by Traxler (2009). The more technology-centred definition indicates the portability and capacity of mobile devices to enhance learning. The more learner-centred approach stresses that it is the learner who takes advantage of the opportunities offered by mobile technologies to learn while on the move between areas of life, whenever the need arises regardless of when and where. The book chapters capture mostly one aspect of ubiquitous learning, that is, through student-centred mobile learning in formal and informal settings that are structured around curriculum. The aspect of ubiquitous learning that studies how, when and what students learn informally, whether related to school curriculum or not, is a more difficult methodology and is less pronounced in the book. This is where future research work should focus.

As Stephen Heppell stated in the Foreword of this book, new and emerging technologies present challenges for education. It means that educators embracing mobile technologies in their teaching will have to rethink pedagogy, boundaries and curriculum. This book will assist them in undertaking this 'rethink'.

OVERVIEW OF THE BOOK

The book has 15 research-based chapters that are divided into three sections. In Section I (Theories and Frameworks), various theories and frameworks underpinning mobile and ubiquitous learning are described or developed. These theories and frameworks consider the learning environments of learners using mobile technologies, personal identities of users of handheld devices and the social perspectives of ubiquitous learning. Section II (Design and Implementation) emphasises the need for careful planning and design when implementing new ways of teaching (and learning) with mobile technologies in existing learning frameworks. Section III describes innovative pedagogies and the research associated with them.

The book will provide readers with a rich collection of research-informed ideas for integrating mobile technologies into learning and teaching. Each chapter looks critically at the issues, related benefits and limitations of learning ubiquitously within the context of the research reported. Proposals for ways forward with the research in each chapter will also be made. A brief description of each chapter is given below.

Section 1: Theories and Frameworks

Chapter 1. *The Digital Revolution in Education: Digital Citizenship and Multi-literacy of Mobile Technology*. In this chapter, Ria Hanewald and Wan Ng review the emerging technology landscape by outlining the digital revolution and key developments that have emerged over the past two decades. The chapter contextualises the concept of mobile devices and learning in relation to the various aspects of the digital revolution: technical (online/Internet connection, Web 1.0 and Web 2.0), skills and knowledge (multiliteracies, digital citizenship), pedagogy-associated learning theories (behaviorism, constructivism, situated, collaborative and conversational) and key principles (ubiquitous, mobile and personalized use).

Chapter 2. *Visuslizing Knowledge Awareness Support in Ubiquitous Learning.* Moushir M. El-Bishouty, Hiroaki Ogata and Yoneo Yano describe a ubiquitous computing environment called PERKAM (PERsonalized Knowledge Awareness Map) which supports learners while they are undertaking their tasks and allows the learners to share knowledge, interact, collaborate and exchange individual experiences. The application utilizes ubiquitous technologies to detect the learner's environmental objects and location, then recommends the best matched educational materials and peer helpers in accordance with the detected objects and the current location.

Chapter 3. *Ubiquitous Computing Does Not Guarantee Ubiquitous Learning in Schools: The Case of Handheld Computers.* Howard Nicholas argues in this chapter that ubiquitous computing is a necessary, but not sufficient condition for ubiquitous learning. He articulates key features of ubiquitous learning suitable for school environments to develop a model that shows how the contribution of ubiquitous computing to ubiquitous learning is constrained by pedagogic frameworks that shape the relationship between handhelds, other elements of the technological suite and learning.

Chapter 4. *Beyond Mobile Learning: Identity Construction and the Development of Social Awareness.* In this chapter, André Caron, Letizia Caronia and Pascal Gagné assert that learning through mobile devices is a large and complex process that involves different aspects of an individual's psychological, cultural and social development. Drawing on research involving 123 Canadian university students recruited from different disciplines, they address the two questions of: How does the use of an iPod affect the students' identity? How does it contribute to the development of social skills and social awareness?

Chapter 5. *Activity in a Mobile Learning Environment: Ubiquitous Personalised Learning Using Context and Social Presence Awareness.* Ray Kekwaletswe contributes to the understanding of the phenomenon of mobile learning, where personalized learning and support is a result of social awareness activities of learners as they traverse varied learning contexts. He investigates how learners use awareness to model their actions for the provision of personalized learning support. Ray has based his theoretical framework on Activity Theory and used Contextual Inquiry methods to gather his data to explore how mobile learners use context and social presence awareness to facilitate their ubiquitous social interactions.

Chapter 6. *Insights into Students' Thinking with Handheld Computers.* Wan Ng and Howard Nicholas set up a framework for the exploration of the use of the handheld computer (PDA) as a research tool for capturing students' thinking processes as they construct representations in science and mathematics, or solve problems in these learning areas on their handhelds. By using avi-screen capture software operating in the background to do this, the research documents a non-intrusive method of capturing the verbal and screen-based (visual) elements of students' thinking as they use the handhelds as a pedagogical tool to complete individual or collaborative tasks.

Chapter 7. *Using Mobile Technologies as Research Tools: Pragmatics, Possibilities and Problems.* Ria Hanewald explores the use of mobile technologies as research tools while discussing both the benefits and limitations of this mode. She maps out a practical framework for the management of a mobile, digital data collection process and its accompanying challenges, namely the potential and pitfalls in gathering, coding and archiving research data. The use of mobile technology devices and the amalgamation of hardware, operating systems, networking and software to facilitate and support this process are discussed. The lessons learned from the technical aspects underpinning fieldwork for research purposes will assist others in maximizing the potential of mobile technology to support their research projects.

Section 2: Design and Implementation

Chapter 8. *Mobile Phones, Developing Countries and Learning*. Many of us take for granted the availability of computers and associated technologies. For about half the population in the world, computers and the Internet are almost alien concepts. Elba del Carmen Valderrama Bahamóndez and Albrecht Schmidt share with us an important contribution that discusses the obstacles encountered in developing countries and different ways of overcoming them to address both the scarcity of computers and computer illiteracy in schools in developing countries. They advocate that the accessibility of mobile phones could change this situation and provide the potential to build feasible educational applications that enhance the learning experience of students in these countries to bridge the digital divide.

Chapter 9. *Historic Monument Education: The Impact of a Collaborative Inquiry-based Mobile Learning Strategy on Social Relationship Development*. Ju-Ling Shih, Chien-Wen Chuang and Gwo-Jen Hwang use inquiry-based models and collaborative learning principles to design and implement an inquiry-based mobile learning strategy that integrates the physical environment and digital resources to enhance elementary students' learning about historic monuments during field studies. The use of mobile devices in the strategy enabled students to have more a customized learning pace and process while at the same time being able to get individual attention when they were dispersed in the field. The impact on students' learning performances, learning satisfaction and social relationships are described.

Chapter 10. *Implementation of Mobile Learning at the Open University Malaysia*. In this chapter, Chng Loi Peng, Zoraini Wati Abas, Norlia T. Goolamally, Yuzery Yusoff and Harvinder Kaur Dharam Singh present the introduction of mobile learning at the Open University Malaysia as part of its ongoing effort to offer an effective blend of learning modes. Ensuring that potential students are ready to learn in non traditional ways that embrace new technologies (with a readiness survey) and careful consideration of the pedagogical and technical aspects were crucial for the success for the new style of teaching. The challenges identified associated with creating and delivering learning materials for mobile phones that suit various mobile operating systems and the different screen sizes of mobile phones are described.

Chapter 11. *Educational Podcasts at University Campus Suffolk*. Tim Goodchild and Sam Chenery-Morris present in this chapter, the introduction and development of podcasts at University Campus Suffolk (UCS). The podcasts discussed in the chapter have all been developed in relation to pre-registration health and social care courses within the Faculty of Health at UCS. The chapter describes where podcasts sit in the paradigm of mobile learning, a brief history of podcasting and the development of a model for the educational use of podcasts and their introduction and evaluation at UCS.

Section 3: Research and Innovative Pedagogy

Chapter 12. *Imagine Mobile Learning in your Pocket*. Cecilie Murray examines the experiences of students and teachers in a range of mobile learning projects in the K-12 environment. Four research projects highlight the experiences of students and teachers as they grapple with mobile technologies and the wireless environment, with implementation and technical issues and with learning approaches and pedagogical innovations. The projects focused on Literacy, Mathematics and cross-curricular learning with Australian primary and secondary students as well as students in international collaborative projects. The benefits and challenges are described for each mobile learning project.

Chapter 13. *Using Mobile Phones for Teaching, Learning and Assessing Irish in Ireland: Processes, Benefits and Challenges*. Faced with the issue of decreasing motivation of school children in the learning of the Irish language, Katrina A. Keogh outlines the processes, benefits and challenges of two pilot projects which investigated the integration of mobile phones into the teaching, learning and assessment of Irish in post-primary schools in Ireland. The pedagogy involved is a novel way of motivating students to study Irish in and outside of the classroom. The pedagogy is innovative in terms of the use of a system that sends out text messages with lists of vocabulary words/short questions and is able to record students' oral responses in Irish for teachers to assess and provide feedback at a later time.

Chapter 14. *The Pedagogy of Mobility*. In this chapter, Katherine Stewart and John Hedberg consider the nature and role of field work and discuss pedagogy associated with fieldwork with a view to locating practice within a 21st century learning context. Four innovative case studies are presented to describe a variety of field work events, each employing different mobile devices and activities. General principles around the technologies, the activities and the tools that comprise the pedagogies embodied in these cases are presented.

Chapter 15. *Formal and Informal Use of Handhelds by Australian and British Students: A Comparative Case Study*. Bridging home and school learning is an important aspect of students' education and this chapter explores the use of mobile devices to do this. In the chapter, Wan Ng and Stamatina Anastopoulou present a comparative study of how Year 7 (12–13 year old) students in Australia and UK with personal ownership of handheld computers used them in formal (school) and informal (home) situations and their perceptions of the usefulness of these devices for their learning. The data show that entertainment activities with the handhelds dominate both school and home use for both groups of students and that there is little continuity between the activities carried out in the school and activities in the home. Wan and Stamatina argue that schools have a responsibility to bridge home-school learning and to support these students to become self-directed learners for lifelong learning. Examples of how this could be done are presented.

Wan Ng
University of New South Wales
Australia

REFERENCES

Chansanchai, A. (2006). *Millennials lead the wired life: New generation digital natives in a land of digital immigrants*. Retrieved July 9, 2009, from http://www.msnbc.msn.com/id/14560871/

Hardey, M. (2007). *Going live:Converging mobile technology and the sociability of the iGeneration*. Retrieved November 3, 2008, from http://journal.mediaculture.org.au/0703/09-hardey.php

Kolb. L. (2009). *Toys to tools: Connecting student mobile phones to education*. Moorabbin, Victoria, Australia: Hawker Brownlow Education.

Lefor, G. & Olney, I. (2007). New technologies, new pedagogies: Using scenarios for staff development with mobile technologies. In Norman, A. & Pearce, J. (Eds), Proceedings of the 6th International

Conference on Mobile Learning, (pp. 125-132) Melbourne, Australia: University of Melbourne.

McMahon, M., & Pospisil, R. (2005). Laptops for a digital lifestyle: Millennial students and wireless mobile technologies. In H. Goss (Ed.), ASCILITE 2005 Balance, Fidelity, Mobility: maintaining the momentum? *Proceedings of the 22nd annual conference of the Australasian Society for Computers in Learning in Tertiary Education* (Vol. 2, pp. 421-431). Brisbane, Australia: ASCILITE.

Ng, W., Nicholas, H., Loke, S., & Torabi, T. (2010). Designing effective pedagogical systems for teaching and learning with mobile and ubiquitous devices in Goh, T. T. (Ed). *Multiplatform E-Learning Systems and Technologies: Mobile Devices for Ubiquitous ICT-Based Education.* (pp. 42 – 56). Hershey, PA: IGI Global Publishing.

Nixon, H., Atkinson, S., & Beavis, C. (2006). New media pathways: Navigating the links between home, school and the workplace in Tan, L.W.H. and Subramaniam, R. (Eds). *Handbook of Research on Literacy in Technology at the K-12 Level* (pp 118-136). Hershey, PA: Idea Group Publishing.

Oblinger, D. G. (2003). Boomers & gen-Xers, millennials: Understanding the "new students".*EDUCAUSE Review*, *38*(4), 37–47.

Oblinger, D. G. (2004). The next generation of educational engagement. *Journal of Interactive Media in Education,* (8). Retrieved July 9, 2009, from http://www-jime.open.ac.uk/2004/8/oblinger-2004-8-discpaper.html

O'Malley, C., & Fraser, D. S. (2006). *Literature review in learning with tangible technologies.* Bristol: Futurelab. Retrieved June, 6, 2008 from http://palmsource.palmgear.com/index.cfm?fuseaction=software.category&catid=209&PartnerREF=&siteid=7

Raines, C. (2002). *Managing millennials*. Retrieved June 7, 2009, from http://www.generationsatwork.com/articles/millenials.htm.

Seppälä, P., & Alamäki, H. (2003). Mobile learning in teacher training. *Journal of Computer Assisted Learning, 19*(3), 330 - 335.

Sharples, M. (Ed.). (2006). *Big issues in mobile learning: Report of a workshop by the Kaleidoscope Network of Excellence Mobile Learning Initiative*. University of Nottingham: Kaleidoscope Research Project.

Sharples, M. (2003). Disruptive devices: Mobile technology for conversational learning. *International Journal of Continuing Engineering Education and Lifelong Learning, 12*(5/6), 504-520.

Traxler, J. (2009). Learning in a mobile age. *International Journal of Mobile and Blended Learning, 1*(1), 1-12

Weiser, M. (1991). The Computer for the Twenty-First Century. *Scientific American*, September issue, 94-10.

Weiser, M. (1996, March). *Open house. ITP Review 2.0*. Retrieved March 15, 2008 from http://www.itp.tsoa.nyu.edu/~review.

Section 1
Theories and Frameworks

Chapter 1
The Digital Revolution in Education:
Digital Citizenship and Multi-Literacy of Mobile Technology

Ria Hanewald
University of Melbourne, Australia

Wan Ng
University of New South Wales, Australia

ABSTRACT

This chapter aims to provide an outline of the digital revolution and the way that mobile devices facilitate participation in the Information age. It provides readers with a broad understanding of the key developments that have emerged over the past two decades as well as the current developments in this area. New and emerging practices relating to the use of mobile technologies for learning and their underlying drivers will be explored. The interconnectivity of applications and devices that is closely linked to concepts of multiple literacies and digital citizenship will be discussed. This brief review of the emerging technology landscape allows for greater appreciation and fuller exploitation of the potential that mobile technologies hold and provides a portrayal of its topography to enable conceptualization at a macro-level.

EMERGENCE OF THE DIGITAL REVOLUTION IN EDUCATION

This introduction traces the evolution of the Internet, also known as Web 1.0 followed by Web 2.0 and the emergence of Web 3.0 while describing their range of applications and potential implications for use in educational contexts. It will assist in understanding the development and complexity of these technologies and the flow on effect into mobile devices for educational purposes within major learning theories. Harnessing the power of network communication and information with mobile devices needs a strategic approach to train the next generation of netizens. It will be argued that multiple literacies, digital citizenships (the responsible, ethical and resilient use of digital technology) and fluency with mobile devices will be essential pre-requisites for education, recreation and employment in the future.

DOI: 10.4018/978-1-61692-849-0.ch001

The Internet: Web 1.0 and Web 2.0

From its origins in the 1960's as a military co-ordination system, the Internet has exploded in the 1990's due to commercialization and is now a global data communication system, used by a quarter of the Earth's population (Miniwatts Marketing Group, 2009). In addition to the World Wide Web (www) and electronic mail (email), it also supports file transfer and sharing, gaming, publishing and teleconferencing.

The term Web 1.0 is an often used synonym to describe the Web before 2000/01, and particularly when comparing with Web 2.0. Others argue that there is no distinct cut-off point due to the ongoing development of technology.

A simple way of understanding the main differences between Web 1.0 and Web 2.0 is by comparison. Table 1 shows the contrast in the applications and the limited technical scope of Web 1.0 with the more sophisticated design of Web 2.0. Hinchcliffe (2006) believes that the rise of Web 2.0 was due to technical improvements such as "...broadband, improved browsers, and Ajax, to the rise of Flash application platforms and the mass development of widgetization, such as Flickr and YouTube badges". Those platforms were drivers for change in the field of education, as they offered more pedagogical scope in the use of online environments.

Table 1. Comparison of Web 1.0 with Web 2.0

Comparison	Web 1.0 (one way) is mainly about	Web 2.0 (both ways) is mainly about
Application (Technical scope)	text	video
	reading	writing
	professional	amateur
	dial-up	broadband
	hardware cost	bandwidth coast
	client server	Peer to peer
	wires	wireless
Education (Pedagogical scope) and **Recreation** (Entertainment)	lectures	conversations
	top-down	bottom-up
	information	opinion
	facts	facile
	tool	lifestyle
Implication (Employment & Economy)	companies	communities
	edited and produced	raw and uncensored
	discovery	drivel
	didactic	anarchic
	for few	for many
	owning	sharing

Mobile Devices for Learning

Another emergence of the digital revolution in education is the uptake of mobile devices in learning. The majority of educational institutions have static computers either as a whole class set in computer labs or on trolleys, or in smaller clusters (e.g. in the library or classrooms) to afford their staff and students Internet access. Over the last decade, mobile devices have become sufficiently sophisticated technologically and affordable financially to shift beyond their experimental status. The use of mobile devices in teaching and learning contexts have moved from pilot projects to institution-wide adoption (Becta, 2003; Finn & Vandenham, 2004; Ng & Nicholas, 2007; Palm, 2002; Perry, 2003; van 't Hooft, 2006). Activities using mobile devices are gradually embedded into the curriculum rather than just the occasional "add-ons". This shift has given rise to concepts such as m(mobile)-learning, e(electronic)-learning, p(pervasive)-learning and u(ubiquitous)-learning.

Web 2.0 with its new and emerging applications has gained popularity in classrooms across all education levels due to their potential contribution to innovative teaching and learning and authentic ways of integrating technology into the school curriculum. Web 2.0 technologies cover a range of applications (i.e. blogs, wikis, social networking sites, Google maps), which are claimed to offer not only new and improved ways of interacting for community building and networking, but also for knowledge creation and transfer in collaborative endeavors. There is a fluidity between and hybridism of various Web 1.0 (such as websites, search engines, email) and Web 2.0 (such as Flickr, YouTube, GoogleEarth) tools that link in with static (desktop) and mobile technology users.

Figure 1 provides a broad overview of the main digital technologies and their capacity for education, employment and the economy. It is not an encompassing picture for all digital technologies but rather an overview of the mostly popular and widely used Web 1.0 and Web 2.0 application and the connections. O'Reilly (2009, see website) stated that:

Web 2.0 is all about harnessing collective intelligence. Collective intelligence applications depend on managing, understanding, and responding to massive amounts of user-generated data in real time……… 1990-2004 was the match being struck; 2005-2009 was the fuse; and 2010 will be the explosion.

Berners-Lee (1999) who is credited with inventing the World Wide Web goes even further when he envisaged Web 3.0 or the semantic Web a decade ago. Web 3.0 would be based on semantic programming languages with "smarter" functionalities and able to automatically communicate added meaning (Shannon, 2006). Intelligent agents are a feasibility as they draw on the user's existing social relationships and contexts to extend connections to related events, objects, places or other people. At this point in time, the implication of Web 3.0 in an educational context is speculative but we envisage that personalized learning will be a daily reality, as a learning program can be tailored, tracked and tweaked automatically – using "smarter functions" that caters to the individual digital needs of each mobile user.

CURRENT STATUS OF MOBILE DEVICES FOR LEARNING AND LIVING

Mobile devices are able to access the web and are an integral part of the digital revolution. Commonly, mobile devices (also known as handheld devices or simply handhelds) are regarded as a number of portable, small sized tools that usually have a digital screen with a miniature keyboard

Figure 1. Web 1.0 and Web 2.0 for Education, Employment and Economy

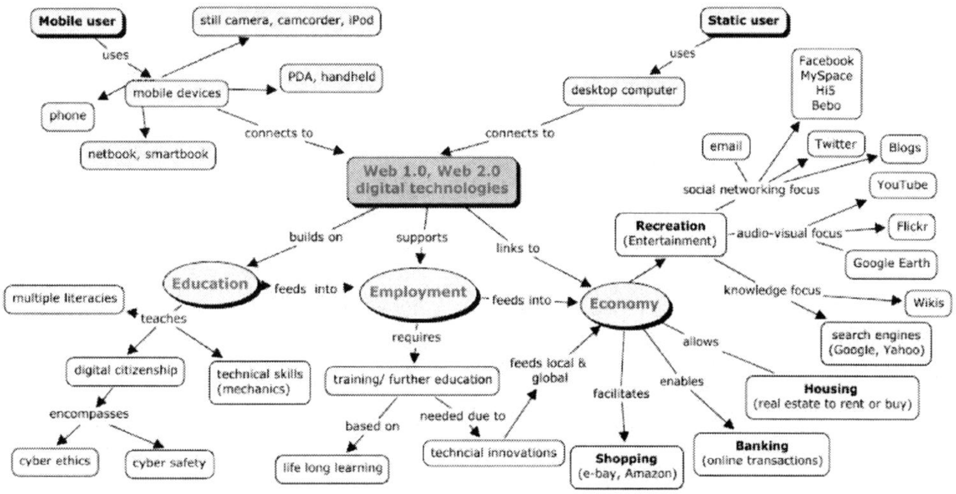

or touch input (with a finger or stylus). The term typically covers personal communication devices such as mobile phones/cell phones and pagers; mobile computers such as PDAs (Personal Digital Assistant) Palmtops, Netbooks, Smartbooks, Tablet PCs; media recorders such as voice recorders, digital still and video cameras (also known as camcorders) and media player/displayers such as the iPod series (Classic, Touch, Nano and Shuffle) and e-book readers. In the broadest sense, mobile technologies may also be moveable and shareable such as Interactive Whiteboards (IWB), video conferencing facilities, kiosks in museums or public spaces.

However, the portability advantage is offset by restricted input methods, limited displays (due to screen size) and often low rates of connectivity. Nevertheless, mobile technologies have become a normal part of every day living, working and learning.

Accessing the web, video, music and information on mobile phones is now a regular and normal part of life in the 21st century as the 2009 Australian Mobile Phone Lifestyle Index revealed. On the whole, a third of Australians check emails on their handsets and more than 70 per cent access mobile entertainment and information services. The survey of 3710 respondents found 36 per cent used email on their mobiles in the past 12 months. Of those, almost half used email daily, which is more than an 80 per cent increase from the year before. At least once a day, 32 percent of respondents accessed social networking sites from their handsets and 21 per cent of respondents visiting websites on their mobile phones. At least once a month, half of the respondents accessed information services such as news, weather and movie times and bought entertainment services (games, ringtones, music downloads) on their mobile phones. The growth in smart phones and the adoption of faster 3G phones has let to a significant increase in mobile Internet use (AIMIA, 2009).

Similarly, in the United States of America, there were 63.2 million people (out of a population of 307 million) who accessed news and information on their mobile devices in January 2009. Of those, 35 percent (22.4 Million) did so daily; which is double the size compared to the previous year (ICT Statistics Newslog).

The capacity to access communication and information nearly anytime and everywhere facilitates usage from home, work, school or the community (via "hot spots".) Learning has thus become mobile in three ways according to Vavoula and Sharples (2002:152), who differentiated between space (learning at home, work, school or in the community via "hot spots"), time (learning during the day, at night, on weekends) and area (related to personal interest, work demands or self-improvement).

Undoubtedly, mobile technologies are now pervasive in every day activities. They support education, recreation and employment in a seamless and unobtrusive manner. Five properties of mobile devices (in this case PDA's) provide the mobility and ubiquity according to Klopfer, Squire, and Jenkins (2002). These are namely portability (due to small seize and weight), social interactivity (affords face-to-face collaboration and data exchange), context sensitivity (capacity to respond to current location, time and environment), connectivity (ability to create a shared network) and individuality (customization for each user).

To fully benefit from the potential that mobile technologies hold for learning and teaching purposes, the focus on individual devices has to move towards embedded classroom practice. A decade ago, Hennessey (1999) envisaged mobile technologies as opportunity for fundamental change in education, replacing sporadic computer lab visits with embedded use in the classroom and beyond. But to make any difference in the classroom at all, Soloway, Norris, Blumenfeld, Fishman and Marx (2001) make the case that mobile devices must be 'ready at hand', provide individualized learning experiences and give increased ownership and thus responsibility over their work. Ideally, those instances would be collaborative to promote learning through social interaction and situated to promote learning within an authentic context and culture. Mobile devices are perfectly suited to support learning and teaching, particularly in coordinating learners and resources for learning activities (e.g. school excursion or camps, swimming carnivals, for distance education or during prolonged absences such as hospital stays). Learning experiences outside the classroom or a formal curriculum are increasingly common in a contemporary society that values informal and lifelong learning.

In reality, many schools ban mobile phones, still cameras, camcorders, MP3 players to prevent recording behavior inside their grounds and during school events due to privacy concerns for students and their families. Schools (for example Gladstone State High School, 2009:3) argue that these "…technological devices can cause significant disruption to the learning and teaching rights of others." Rather than seeing them as disruptive devices, Sharples (2003) suggests that teachers should see them as opportunities to benefit learning and employ their students' devices for educational activities.

Mobile learning was seen previously as segregated and isolated tasks according to Naismith, Lonsdale, Vavoula and Sharples (2006). More recently, the view has shifted to seeing it as opportunities for integrated, collaborative and conversational experiences within and outside the boundaries of a dedicated physical learning environment. Observable is a further shift from small-scale pilots to institution-wide implementations based on a growing interest in taking advantage of mobile devices' capacity for learning. These shifts have implications for teachers, students and schools. Naismith et al. (2006) emphasizes that a number of issues need to be considered to make implementation of mobile technologies a success. For example, the power to gather and utilize contextual information may conflict with the learner's desire for privacy and anonymity. Likewise, increased mobility allows learners to link activities from outside the classroom, in effect 'escaping' adult supervision and possibly engaging in deeds that are not appropriate. Another issue is ownership, which may create tension as students want to control their own personal mobile device but may be asked to share it within a classroom environment. Informal use of mobiles for learning may encroach on students' social networks and this may become an issue if they see this as a threat and abandon technologies. Yet another issue is the informality of ad hoc learning situations. Rogers (1997) termed it "natural" learning although it is also known as incidental learning, unplanned learning, unstructured learning and unintentional learning. The situated, contextual social interaction with others triggers natural learning, with characteristics that are considered as the most effective in formal learning situations. However, this spontaneous, random learning may clash with the need for structured, continued, predictable learning and accountability of those experiences in lesson plans and evaluations. This need for transparency and reporting may require effective tools to record, organize and retrieve naturally, impromptu occurring (mobile) learning experiences. These and other more practical concerns (i.e. cost, usability, technical and institutional support) can be addressed by research-informed guidelines, which support a blended approach to enable learning with mobile technologies (Naismith et al., 2006). Regardless of these challenges, mobile technologies are slowly penetrating schools with an impact on teaching and learning. It will take place in the users' environments, making learning more personal, collaborative, situated and lifelong as mobile technologies increasingly become more embedded, ubiquitous, and networked. According to Naismith et al. (2006) these enhanced capabilities will afford abundant social interactions, context awareness and networked access to information and communication, making learning an integral, constant and permanent part of daily life.

MOBILE DEVICES AND LEARNING THEORIES

This section situates mobile learning devices around major learning theories: Behaviourist in the 1960's and 1970's, where learning was a reinforcement of stimulus – response activities; Constructivist in the 1970's and 1980's, where learning was constructed based on both previous and current knowledge; Situated in the 1990's encourages learning within authentic context and culture whereas Collaborative and conversational - also developed in the 1990's - promotes learning through social interaction, communication and team work between learners.

Behaviourist learning draws on Skinner's work (1968), where he asserts that learning is best achieved through reinforcement of a specific stimulus and a response. In this paradigm, the mobile device is used to offer learning materials, then seek a response from the learner and provide feedback. In the 1960's, the earliest computer assisted learning (CAL) utilized this approach, which become known in the vernacular as "drill and kill". This approach is useful in the gathering of data to track the progress of each student and to tailor content and feedback to a particular subject matter. Rochelle (2003) sees the benefits of a whole class responding to a specific question in sharing misconceptions while remaining unidentified and thus avoiding embarrassment if a wrong answer is given. This basic function for mobile devices provides anonymity but also speed in aggregating and displaying collective understandings.

Constructivist learning offers more direct ways for learners to use mobile devices in authentic learning contexts. An extension of constructivism is constructionism, which Papert (1980) is associated with. He saw that the computer is no longer just a presenter of information but as a tool for active manipulation of that content. His programming language Logo and the well known "logo turtle" enabled learners to actively construct their own knowledge by building interactive models and becoming part of a microworld. For mobile technologies, constructivist principles are exemplified in participatory simulations. There, students carry networked devices which makes them part of the simulation itself, with the effects of their actions immediately observable on the whole system. For example, the spread of a disease on a population as part of a study on biology or "becoming" a human in various ages of mankind (i.e. prehistoric, ancient Greece, medieval times) to study history (http://www.human-age.com/). However, while this may provide enjoyment and engagement for students, it is questionable whether learning within simulations remains in that contained environment or transfers across other situations to become general knowledge. A student engaged in ancient Egypt may elevate to the status of a Pharaoh within the simulation but may not have any general knowledge of that historic period or insight into its significance.

Situated learning has learners participate within a community of practice, with knowledge presented in authentic contexts. Lave and Wenger (1991) argue that learning is not just acquisition of information and knowledge but the active process of social participation. This approach to learning is based on Vygotsky's (1978) socio-cultural psychology.

Mobile devices can be used in the three strands within the situated learning paradigm: problem-based learning, case-based learning and context-aware learning. Problem-based learning (PBL) develops learners' thinking skills by presenting a problem, which they might encounter in their professional practice. Mobile devices were employed in nursing PBL (Higgins 1994) and business administration PBL (Merchant 1995). Case-based learning (CBL) is similar but more flexible as it can be used as a catalyst for class discussions and/ or assessment exercise. Context-aware learning with mobile devices draws on the information in the user's immediate environment. Their portability makes them available in different contexts (i.e. in galleries or museum) and increases the learning (i.e. additional information about exhibits and displays) based on the users' location within those spaces.

Collaborative learning is well suited to mobile devices sharing files, messages and other data through networks, which enables communication and teamwork. This approach is based on the notion that individuals learn through conversations with each other, by sharing understandings of ideas and concepts of the world around them. In this regard, Ng, Nicholas, Loke and Torabi (2010), citing Sharples, Taylor and Vavoula (2007) and Rochelle, Patton and Tatar (2007), have discussed the conversational and social collaborative perspectives of mobile technologies in bridging formal and informal learning. Sharples et. al's (2007) theory of mobile learning focuses on the communicative interactions between technology and people and see learning as conversational processes taking place across multiple contexts within systems where people and technology are in 'continual flux'. According to them, conversation requires the learner to describe him/herself and his/her actions, explore, extend, converse and externalize that description or understanding to subsequent activities. Technology can be involved as a conversational partner mediating conversations bridging formal and informal learning. Rochelle, Patton and Tatar (2007) view handheld computers as learning technologies that are representational tools for augmenting cognition and a mediational tool for social and cultural participation. They see learning with handhelds and within networked communities as the merging of cognitive augmentation and social mediation perspectives that link formal and informal learning, and the linking of representations mediated by mental operations and social practice. Ng et.al (2010) added the requirement that the learner using the handheld needs to be acknowledged as a 'partner' and collaborator in the conversation and learning rather than just a 'recipient' of information. In other words, the learner is a shaper and presenter of information and values.

CAPACITIES OF MOBILE DEVICES FOR LEARNING

To gain a better understanding of how mobile devices such as mobile phones, handheld computers and iPod Touch or iPhones can bring about ubiquitous learning, it is necessary to look at what these devices are capable of doing and how those capacities contribute to learning. These include (Nicholas & Ng, 2009): (i) collaborating with peers using Bluetooth in class or emailing outside of class (ii) taking notes, audio recording and drawing figures (iii) writing short text paragraphs (iv) researching on the Web with wireless (v) reading e-documents (vi) undertaking interactive exercises e.g. multiple choice questions (vii) interacting with simulations and multimedia educational material (viii) watching animations and movies (ix) entering data, drawing graphs, calculating mathematical problems using Excel Mobile and the calculator (x) collecting data using handheld sensors and probeware with PDAs (xi) organising daily activities using the calendar function (xii) keeping reflective logs (xiii) taking photos and images (xiv) video recording (xv) listening to downloaded podcasts, lectures or interviews (xvi) learning from software packages e.g. foreign languages, dictionary and encyclopedia (e.g. Lexipedia), Periodic Tables and other subject-related software (xvii) creating PowerPoint presentations using PocketSlide software in PDAs and (xviii) creating concept maps, for example with iThink for iPods and Inspiration for PDAs. This list of capabilities suggest that information and learning materials could be delivered in a variety of format in mobile devices – visual, text, audio and multimedia. In this regard, meaning making – the understanding of existing information and synthesizing new ones, for learners is not just text-based only but entails multiple representations of concepts under study as well as the ability to conceptually link the different modes of representations (multimodal) in order to demonstrate their reasoning processes and understandings of the concepts (Ainsworth, 1999; Prain & Waldrip, 2006; Waldrip, Prain & Carolan, 2006). If we take science learning as an example, the modalities of representations afforded by mobile technology include written, verbal, visual, gestural/embodied and spatial as shown in Table 2 (Ng, in press).

Table 2. Modes of representation in science enabled by mobile devices and software

Mode of representation	Demonstrated in	Examples of hardware/software
Written	Text	Word
	Worksheets; puzzles	Word; Hot Potatoes
	Assignments; projects	Word; Web search for information
Verbal	Oral presentations	Audio/video recording; podcasts
Visual	Drawings/diagrams	Drawing software; digital camera to capture diagrams
	Concept maps	Inspiration; Kidspiration; iThink
	Tables and graphs	Excel
	Animations/simulations	Flash, Java applets; simulated experimental work
	Presentations (PowerPoint)	PocketSlide
Gestural/Embodied	Role play; science drama	Digital video recorder; digital camera
	Experimental work & report	Word for text; Excel for graphs; digital camera to capture results; video recording a process
	Modelling (hands-on)	Digital camera and or video recorder
Spatial	3-dimensional models	Modelling software e.g. Rasmol for molecular structures

MULTILITERACIES: PLURALITY OF LITERACIES WITH MOBILE TECHNOLOGIES

The influence of digital technologies on multimodal meaning-making is one aspect of 'multiliteracies', a term first proposed by *The New London Group* in 1996. The term is intended to highlight two arguments. The first argument is that as today's society becomes more linguistically and culturally diverse, meaning making differs according to cultural, social and professional contexts. The Group argues that to be effective citizens in a locally diverse and globally connected society, there is a need to be able to interact with "multiple languages, multiple Englishes" and deal with communication patterns that often cross cultural and national boundaries. In this regard, mobile users make up a community that uses an abbreviated language in texting mobile messages, commonly called SMS texting or textese, which uses a combination of single letters, pictures and/or numbers to convey whole word meanings. The language ignores grammar, conventional spelling and punctuation from a pragmatic perspective to address the constraint in size of keyboard and screen and increase in cost for long messages. While quick to type, *textese* could take longer to read (Fahmy, 2008). The second argument is that the advent of media and communication technologies is enabling meaning making to be increasingly multimodal where the written-linguistic modes are integral of visual, audio, gestural and spatial patterns of meaning. Considering this, the way language is used to provide meanings and for expressing understanding is influenced by the (rapidly changing) new communications media. The London Group argues that no matter how well taught, there cannot be just one set of standards or skills that constitutes literacy learning. Similarly, Unsworth (2001, p.8) has argued for the plurality of literacies saying that:

In the twenty-first century the notion of literacy needs to be reconceived as a plurality of literacies and being literate must be seen as anachronistic. As emerging technologies continue to impact on the social construction of these multiple literacies, becoming literate is the more apposite description.

Williamson (2005) describes multimodality, multisemiotics and multiliteracies as the different ways and contexts in which people experience communication and develop understandings. The forms of communication that individuals encounter in their daily life are diverse and it is necessary for the individual to be able to identify the dependency of the multiple modes of images, gestures/actions, text and speech on each other in enabling him or her to 'design' (a term initiated by the London Group, 1996) his or her own understanding. Williamson argues that with the diverse visual and aural forms of public communication (movie, music, Internet and computer games) gaining in popularity and cultural importance, multiliteracies mean the ability to read all these media and the different modes associated with them to construct meanings.

In education, it has been shown that multimodal learning is more effective than unimodal learning and that the most effective designs for learning should include a combinations of modalities and a variety of media. (CISCO, 2008).

Users of mobile devices are able to access multiple modes of information and communication on a daily basis. For example, e-books with text and images, use GPS software for navigation and undertake spatial-based activities, trawl through loads of static and multimedia-based information on the World Wide Web including interactive simulations, listen to podcasts, download and watch movies, communicate via emails and chat spaces, and interact with Web 2.0 software (Wikis, Blogs, YouTube videos) and participate with the global community via text, images, visual and aural recordings. Educating young people to be able to identify meanings from the multimodal forms of communication presented by media, to critically analyse the meanings and synthesise their own understandings of the topic being learned, or to make decisions in regards to participation in social networked communities. Being digitally literate, at both physical and cognitive levels, is critical in supporting multiliteracies within the digital revolution that this chapter is about.

DIGITAL CITIZENSHIPS

In a society saturated with digital tools and media, digital skills are becoming pre-requisites for employment, a source of enjoyment for recreation and a necessary part of education to participate in the technologies that afford new opportunities. Ribble and Bailey (2007) argue that the exponential increase in digital interaction within society and in the classroom makes digital citizenship an integral and important part of schools. Digitally literate people, also called digital citizens, communicate and collaborate effectively with other members of a digital society. Digital citizens - have the opportunity to use technology for education, employment and social interaction. Digital citizens need to be familiar with their responsibilities within a digital society and the accepted norms and rules to use technology appropriately (Ribble & Bailey, 2007). In this regard, digital citizens are also called netizens, a word first used by Michael Hauben, to describe a person who uses the Internet needing to possess the same sense of civic responsibility in a virtual community as (s) he would in a physical community.

Digital Literate Skills

Digital citizenship is a concept that helps to understand what technology users should know in order to use technology appropriately and what they need to learn in order to be prepared for a society full of technology. The actual skills are term digital literacy, although similar terms such as computer literacy, ICT literacy, information technology literacy, web literacy, multi media literacy, mediacy, online literacy and transliteracy (the ability to interact across a range of platforms, tools and media). Digital literacy refers to the skills needed in locating, organizing, processing and creating information through and with digital technology. It includes computer hardware, software, the Internet and various mobile devices. Existing literature (Gilster, 1997; Inoue, Naito, & Koshizuka, 1997; Lenham, 1995) describes digital literacy as a range of technical, cognitive and sociological skills necessary for successful interactions in digital environments. "Digital literacy can be defined as survival skills in the digital era." (Eshet-Alkalai, 2004: 102). He distinguished five different types of digital skills under the umbrella of digital literacy. The subcategories are a) photo-visual literacy; b) reproduction literacy; c) information literacy; d) branching literacy and e) socio-emotional literacy. Photo-visual literacy is defined as the art of reading visual representations, reproduction literacy is identified as the art of creative recycling of existing materials, information literacy is described as the art of skepticism, branching literacy is explained as hypermedia and non-linear thinking and socio-emotional literacy is the most complex subset of skills as users need to be. "...willing to share data and knowledge with others, capable of information evaluation and abstract thinking, and able to collaboratively construct knowledge." (Eshet-Alkalai, 2004: 102).

Cyber Safety

At another level, and as participating in online environment is fast becoming a lifestyle of the younger generations where millions of young people 'meet' online to chat socially, exchange ideas and collaborate on project using Web 2.0 technologies, the issue of cyber safety and potential risks in online participation is increasing (Conroy, 2007; Hanewald, 2008). Cyber safety is about keeping young people safe online. The potential risks that children face when online include being bullied, stalked and harassed and exposed to identity fraud and inappropriate materials such as pornographic and illegal materials. The impact of cyber bullying on young people could lead to both short and long-term psychological harm which is manifested in depression and anti-social behaviour. This in turn would have negative implications on family lives and schooling. Part of being digitally literate is an awareness of cyber safety (Becta, 2006) and the need to keep personal information as private as possible. A digitally literate person should be able to recognise when (s) he is being threatened and have the skills to deal with the situation, for example whether to ignore, report or respond to the threat. The fostering of respect and responsibility when communicating through digital technologies and the understanding of ethical and cultural issues associated with digitally-based communities should be a part of the digital literacy education of students (Berson & Berson, 2003).

CONCLUSION

This chapter has attempted to contextualise the concept of mobile devices and learning into the various aspects of the digital revolution: technical (online/ Internet connection, Web 1.0 and Web 2.0), skills and knowledge (multi-literacies, digital citizenship), pedagogy associated learning theories (behaviorism, constructivism, situated, collaborative and conversational) and key principles (ubiquitous, mobile and personalized use).

As mobile devices become more and more inextricably linked to learning and everyday living, more research is needed to provide a greater understanding of the issues and nature of the ubiquitous learning with mobile technologies, including online learning and the social landscape inhabited by users with mobile devices. In *Building Learning Future,* Heppell, Chapman, Millwood, Constable and Furness' (2004) have argued that "new technologies have provided both a catalyst for change and a means of changing" (p. 3) and have provided several scenarios of future schools. Mobile and ubiquitous learning and skills associated with the learning, as described in this chapter, would no doubt be components of their notion of 'dissolved schools' in the 21st century.

REFERENCES

AIMIUA. (2009). *Australian Mobile Phone Lifestyle Index*. 5th Edition – Special topic: Mobile commerce, September 2009. Retrieved October 21, 2009, from http://www.aimia.com.au/i-cms?page=1.32.1491.2841

Ainsworth, S. (1999). The functions of multiple representations. *Computers & Education*, *33*(2/3), 131–152. doi:10.1016/S0360-1315(99)00029-9

BECTA. (2006). *Safeguarding children in a digital world: Developing a strategic approach to e-safety*. Retrieved January 5, 2009, from http://publications.becta.org.uk/display.cfm?resID=25933.

Becta, (Britian Educational Communications and Technology Agency) (2003). *Becta ICT Research Online*. Retrieved July 30, 2009, from http://www.becta.org.uk

Berners-Lee, T., & Fischetti, M. (1999). *Weaving the Web: The Past, Present and Future of the World Wide Web by its Inventor*. Britain: Orion Business.

Berson, I. R., & Berson, M. J. (2003). Digital literacy for effective citizenship. *Social Education*, *67*(3), 164–167.

CISCO. (2008). *Multimodal Learning Through Media: What the Research Says*. Retrieved October 31, 2009, from http://www.cisco.com/web/strategy/docs/education/Multimodal-Learning-Through-Media.pdf

Conroy, S. (2007). Labor's plan for cyber-safety (fact sheet). Retrieved January 5, 2009 from http://www.alp.org.au/download/now/labors_plan_for_cyber_safety.pdf.

Eshet-Alkalai, Y. (2004). Digital literacy: A conceptual framework for survival skills in the digital era. *Journal of Educational Multimedia and Hypermedia*, *13*(1), 93–106.

Fahmy, M. (2008). *If u cn rd this quickly, gd 4 u*. Retrieved October 31, 2009 from http://www.reuters.com/article/technologyNews/idUSTRE4B90YA20081210?rpc=64

Finn, M., & Vandenham, N. (2004). The handheld classroom: Educational implications of mobile computing. *Australian Journal of Emerging Technologies and Society*, *2*(1), 1–15.

Gilster, P. (1997). *Digital Literacy*. New York: Wiley.

Gladstone State High School. (2009). *Recording Device Policy*. Retrieved October 21, 2009, from www.gladstonshs.eq.edu.au/.../School%20Information/Recording%20Device%20Policy.pdf

Hanewald, R. (2008). Confronting the Pedagogical Challenge of Cyber Safety. *Australian Journal of Teacher Education*, *33*(3), 1–16.

Hennessy, S. (1999). *The potential of portable technologies for supporting graphing investigations. British Journal of Educational Technology, 30*(1), 57-60. (hybrid article full version retrieved October 20, 2009, from www.education.leeds.ac.uk/research/mathseducation/gcalc_hennessy.pdf).

Heppell, S., Chapman, C., Millwood, R., Constable, M., & Furness, J. (2004) *Building learning futures*. Retrieved November 26, 2009, from http://rubble.heppell.net/cabe/final_report.pdf

Higgins, L. (1994). Integrating background nursing experience and study at the postgraduate level: an application of problem-based learning. *Higher Education Research & Development*, *13*, 23–33. doi:10.1080/0729436940130103

Hinchcliffe, D. (2006). *All We Got Was Web 1.0, when Tim Berners-Lee Actually Gave Us Web 2.0.* Retrieved October 21, 2009, from http://web2.socialcomputingmagazine.com/all_we_got_was_web_10_when_tim_bernerslee_actually_gave_us_w.htm

Inoue, H., Naito, E., & Koshizuka, M. (1997). Mediacy: What it is? Where to go? *The International Information & Library Review*, *29*(3/4), 403–413. doi:10.1006/iilr.1997.0059

Keamy, R., & Nicholas, H. (2007). Personalised learning: can governments guarantee diversity for individuals? *International Journal on Diversity in Organisations. Communities & Nations*, *7*, 137–146.

Klopfer, E., Squire, K., & Jenkins, H. (2002). Environmental Detectives: PDAs as a window into a virtual simulated world. *Proceedings of IEEE International Workshop on Wireless and Mobile Technologies in Education* (pp95-98). Vaxjo, Sweden: IEEE Computer Society.

Lave, J., & Wenger, E. (1991). *Situated Learning: Legitimate Peripheral Participation*. Cambridge, England: Cambridge University Press.

Lenham, R. (1995). Digital literacy. *Scientific American*, *273*(3), 253–255.

Merchand, J. E. (1995). Problem-based learning in the business curriculum: An alterative to traditional approaches. In Gijselaers, W., Tempelaar, D., Keizer, P., Bernard, E., & Kasper, H. (Eds.), *Educational innovation in economics and business administration: The case of problem-based learning* (pp. 261–267). Dordrecht: Kluwer.

Miniwatts Marketing Group. (2009). *Internet World Statistics. Usage and Population* Statistics. Retrieved October 21, 2009, from http://www.internetworldstats.com/stats.htm

Naismith, L., Lonsdale, P., Vavoula, G., & Sharples, M. (2006). *Futurelab Report 11: Literature review in mobile technologies and learning.* University of Birmingham, UK. Retrieved October 21, 2009, from www.futurelab.org.uk/research/lit_reviews.htm

Ng, W. (in press, accepted 4 March 2009). Empowering students to be scientifically literate through digital literacy in Rodrigues, S. (Ed). *Multiple Literacy and Science Education: ICTs in Formal and Informal Learning Environments.* Hershey, PA: IGI Global Publishing.

Ng, W., Nicholas, H., Loke, S., & Torabi, T. (2010). Designing effective pedagogical systems for teaching and learning with mobile and ubiquitous devices in Goh, T. T. (Ed). *Multiplatform E-Learning Systems and Technologies: Mobile Devices for Ubiquitous ICT-Based Education.* (pp. 42 - 56). Hershey, PA: IGI Global Publishing.

Nicholas, H., & Ng, W. (2009). Ubiquitous learning and handhelds: An overview of theory and pedagogy. In Rogers, P., Berg, G., Boettcher, J., Howard, C., Justice, L., & Schenk, K. (Eds.), *Encyclopedia for online and distance learning* (*Vol. IV*, pp. 2171–2176). Hershey, PA: IGI Global Publishers.

O'Reilly, T. (2009). *Web Squared: Web 2.0 Five Years On*. Retrieved October 28, 2009, from http://www.web2summit.com/web2009/public/schedule/detail/10194

Palm Inc. (2002). *Palm Education Pioneers Program Final Evaluation Report*. Retrieved October 21, 2009, from http://www.palmgrants.sri.com/PEP_Final_Report.pdf

Papert, S. (1980). *Mindstorms: Children, Computers, and Powerful Ideas*. Brighton: Harvester Press.

Perry, D. (2003). *Handheld computers (PDAs) in schools*. Coventry: Becta (British Educational Communications and Technology Agency).

Piaget, J. (1955). *The construction of reality in the child*. London: Routledge & Keegan Paul.

Piaget, J. (1972). *Psychology and epistemology: Towards a theory of knowledge*. London: Penguin University Books.

Prain, V., & Waldrip, B. (2006). An exploratory study of teachers' and students' use of multi-modal representation of concepts in primary science. *International Journal of Science Education*, *28*(15), 1843–1866. doi:10.1080/09500690600718294

Ribble, M., & Bailey, G. (2007). *Digital citizenship in schools*. Washington, DC: ISTE.

Rogers, A. (1997). Learning: Can we change the discourse? *Adults Learning*, *8*(5), 116–117.

Roschelle, J. (2003). Unlocking the learning value of wireless mobile devices. *Journal of Computer Assisted Learning*, *19*(3), 260–272. doi:10.1046/j.0266-4909.2003.00028.x

Roschelle, J., Patton, C., & Tatar, D. (2007). Designing networked handheld devices to enhance school learning. *Advances in Computers*, *70*, 1–60.

Shannon, V. (2006). *A 'more revolutionary' Web*. International Herald Tribune. Retrieved October 22, 2009 from http://www.iht.com/articles/2006/05/23/business/web.php.

Sharples, M. (2003). Disruptive devices: mobile technology for conversational learning. *International Journal of Continuing Engineering Education and Lifelong Learning*, *12*(5/6), 504–520. doi:10.1504/IJCEELL.2002.002148

Sharples, M., Taylor, J., & Vavoula, G. N. (2007). A Theory of Learning for the Mobile Age. In Andrews, R., & Haythornthwaite, C. (Eds.), *The SAGE Handbook of E-learning Research* (pp. 221–247). London: Sage.

Soloway, E., Norris, C., Blumenfeld, P., Fishman, B. J., & Marx, R. (2001). Hand-held devices are ready at hand. *Communications of the ACM*, *44*(6), 15–20. doi:10.1145/376134.376140

Statistics Newslog, I. C. T. *Mobile Internet Becoming A Daily Activity For Man*. Wednesday, March 18, 2009. Retrieved October 21, 2009 from http://www.itu.int/ITUD/ict/newslog/Mobile+Internet+Becoming+A+Daily+Activity+For+Man.aspx

The New London Group. (1996). A Pedagogy of Multiliteracies: Designing Social Futures. *Harvard Educational Review*, *66*(1), 60–92.

Unsworth, L. (2001). *Teaching multiliteracies across the curriculum: Changing contexts of text and image in classroom practice*. Buckingham, UK: Open University Press. van 't Hooft, M. (2006). Tapping into digital literacy: Handheld computers in the K-12 classroom. In Tan, L. W. H., & Subramaniam, R. (Eds.), *Handbook of Research on Literacy in Technology at the K-12 Level* (pp. 287–307). Hershey, PA: Idea Group Publishing.

Vavoula, G., & Sharples, M. (2002). KLeOS: A personal, mobile, knowledge and learning organisation system. In Milrad, M, Hoppe, U and Kinshuk (eds) *Proceedings of the IEEE International Workshop on Mobile and Wireless Technologies in Education* (WMTE2002) (pp.152-156), Aug 29-30, Vaxjo, Sweden.

Vygotsky, L. S. (1962). *Thought and language*. Cambridge, MA: MIT Press. doi:10.1037/11193-000

Vygotsky, L. S. (1978). *Mind in Society: the Development of Higher Psychological Processes*. Cambridge Mass, London: Harvard University Press.

Waldrip, B., Prain, V., & Carolan, J. (2006). Learning junior secondary science through multi-modal representations. *Electronic Journal of Science Education, 11*(1), 87–107.

Weiser, M. (1991). The Computer for the 21st Century. *Scientific American Ubicomp Paper*. Retreived November 2, 2009, from http://nano.xerox.com/hypertext/weiser/SciAmDraft3.html

Williamson, B. (2005). *What are multimodal, multisemiotics and multiliteracies?* Retrieved November 15, 2009, from http://www.futurelab.org.uk/resources/publications_reports_articles/web_articles/Web_Article532

Chapter 2
Visualizing Knowledge Awareness Support in Ubiquitous Learning

Moushir M. El-Bishouty
The University of Tokushima, Japan

Hiroaki Ogata
The University of Tokushima, Japan

Yoneo Yano
The University of Tokushima, Japan

ABSTRACT

The aim of this research is to provide the learner with knowledge awareness support according to his current need and location. Therefore, we presented a ubiquitous learning environment in order to support the learners while doing tasks; this environment is called PERKAM (PERsonalized Knowledge Awareness Map). PERKAM allows the learners to share knowledge, interact, collaborate, and to exchange individual experiences. It utilizes ubiquitous technologies to detect the learner's environmental objects and location, and then recommends the best matched educational materials and peer helpers in accordance with the detected objects and the current location. This environment provides the learner with Knowledge Awareness Map, which graphically displays the knowledge awareness support. The experiment and the questionnaire results indicated that Knowledge Awareness Map is helpful to understand the learner's situation and it efficiently visualizes the recommendations.

INTRODUCTION

Ubiquitous computing is a new information and communication technology that utilizes a large number of cooperative small nodes with computing and/or communication capabilities such as handheld terminals, smart mobile phones, sensor network nodes, contact-less smart cards, and RFID (Radio Frequency Identification)…etc (Sakamura & Koshizuka, 2005). The RFID system (Klaus & Rachel, 2000) consists of a tag, which is made up of a microchip with an antenna, and an interrogator or reader with an antenna. The reader sends out electromagnetic waves. The tag antenna is tuned to receive these waves. A passive RFID tag draws

DOI: 10.4018/978-1-61692-849-0.ch002

power from a field created by the reader and uses it to power the microchip's circuits. The chip then modulates the waves that the tag sends back to the reader and the reader converts the new waves into digital data.

Computing becomes ubiquitous. The knowledge workers use, in computing and communication services, is less limited to solitary moments at an office desk. Instead, it is extended in multifaceted ways to all aspects of the life, both public and private (Yoo & Lyytinen, 2005). The challenge of the future computer systems is therefore not to provide information at anytime and at anywhere but to say the right thing at the right time in the right way to the right person (Fischer, 2001; Fischer & Konomi, 2005). Ubiquitous computing evolution has recently been accelerated by improved wireless telecommunications capabilities, open networks, continuous increasing in computing power, improved battery technology, and the emergence of flexible software architectures. With those technologies, an individual's learning environment could be embedded in his or her daily life (Ogata & Yano, 2004). Almost on a daily basis, handheld devices are becoming cheaper for a wide range of people. Meanwhile, mobile devices are now being introduced as learning devices and some researches use them as collaborative tools (Cole and Stanton, 2003). The main characteristics of a Computer Supported Ubiquitous Learning (CSUL) environment are shown as follows (Chen et al., 2002; Curtis et al., 2002):

- Permanency: Learners never lose their work unless it is purposefully deleted. In addition, all the learning processes are recorded continuously every day.
- Accessibility: Learners have access to their documents, data, or videos from anywhere. The information is provided based on their requests.
- Immediacy: Wherever learners are, they can get any information immediately. Thus, learners can solve problems quickly. Otherwise, the learner can record the questions and look for the answers later.
- Interactivity: Learners can interact with experts, teachers, or peers in the form of synchronous or asynchronous communication. Hence, the experts are more reachable and the knowledge becomes more available.
- Situating of instructional activities: The learning could be embedded in our daily life. The problems encountered as well as the knowledge required are all presented in their natural and authentic forms.

Moreover, a computer supported ubiquitous learning environment could be a Computer Supported Collaborative Learning (CSCL) environment (O'Malley, 1994) that focuses on the socio-cognitive process of social knowledge building and sharing (Yoo & Lyytinen, 2005). Ubiquitous computing environments strongly support collaborative and situated learning approaches. The use of ubiquitous computing tools within a situated learning approach is recommended to facilitate the students' attainment of curricular content, technology skills, and collaboration skills (Lin et al., 2005).

Many teachers and learners believe that *learning by doing* (Schank, 1995) is the best way for learning. In *learning by doing* model, the teachers identify a specific set of skills to teach, embed that skills in a task, an activity, or a goal that the student will find interesting or motivational, then the teachers can evaluate the learner's understanding and skills according to how much the learner succeed to reach to the goal. While the learner is practicing to reach to the goal, he usually looks for some knowledge. There are two logical ways to obtain the desired knowledge; one way is to refer to one or more of the educational materials that match the learner's needs such as books, journals, or video lectures. The other way is to ask for an aid from other learners who have enough knowledge about the learner's request. In the latter case, the learners can interact and exchange their knowledge in collaborative ways. In a ubiquitous learning environment, it is difficult for a learner to know who has this knowledge even though they are at the same place. In this case, the learner needs to be aware of the other learners' interests that match his request.

The aim of this research is to support the learner with *Knowledge Awareness Map,* which is personalized according to his current need and location. The proposed system can recognize the environmental objects that surround the learner-those that he uses during his practice study. It uses RFID tags to detect the surrounding physical space objects then it generates the learner's digital space where each object in the physical space has its corresponding one in the digital space. The system matches the educational materials/topics with the learner's current task then recommends and visualizes the best matched materials according to how much they match the learner's current task and how near their physical locations are to the learner's current location. The system also matches the current learner's need with the other learners' interests and locations, recommends the best peer helpers, and visualizes the relative distances between the learner's current need and the peer helpers' interests and locations. The learner can then refer to some of the educational materials or contact peer helpers, forward to them his *Environmental Object Map*, interact, and collaborate.

The outline of this chapter is as follows: the main concept of the personalized *Knowledge Awareness Map* is illustrated, the system architecture and implementation are explained, and followed by the system evaluation and discussion. Finally, the conclusion and future work are presented.

PERSONALIZED KNOWLEDGE AWARENESS MAP

In order to get help from another learner, you have to be aware of his interests and past actions. Therefore it is very difficult to find suitable partners at the beginning of the collaboration. Dourish and Bellotti (1992) defined *Awareness* as the understanding of the activities of others, which provides a context for your own activity. *Knowledge Awareness* (KA) is defined as awareness of the use of the knowledge (Ogata, Matsuura & Yano, 1996). KA has a close relation to the learner's curiosity (Ogata & Yano, 2000). Collaborative awareness is frequently achieved by means of lightweight messaging tools and dynamic information displays that function as notification systems (Carroll et al., 2003). *Knowledge Awareness Map* graphically displays KA information. It displays the surrounding environmental objects, the matched educational materials, and the recommended peer helpers. This map plays a very important role in finding peer helpers and inducing collaboration (Ogata et al., 1999). Hatano and Inagaki (1973) identified two types of curiosity: particular curiosity and extensive curiosity. Extensive curiosity occurs when there is a desire for learning that makes the learner's stock of knowledge well balanced by widening the learner's interests. Particular curiosity is generated by the lack of sufficient knowledge, and it is very useful because the learner can acquire detailed knowledge. *Knowledge Awareness Map* excites both types of curiosity.

In PERKAM, Knowledge Awareness Map is personalized according to the learner's need and the physical location (El-Bishouty et al., 2007). Personalization can be defined as the way in which information and services can be tailored in a specific way to match the unique and specific needs of an individual user (Renda & Straccia, 2005). Personalized applications lay more emphasis on specific needs, and location of individual learners in designing the learning content on handheld devices (Cui & Bull, 2005). *PERKAM* satisfies the conditions that Cui and Bull (2005) presented for the personalized learning system:

1. Individualized according to the learner's knowledge.

2. Individualized according to the learner's location and needs in that location.
3. Mobility.

This type of learning systems supports the learners to make their own decisions based on the changed context or sensor information (Mitchell & Race, 2005). Therefore, it encourages *decision-making* and reflection (El-Bishouty et al., 2010; Song, 2006).

We assume that the knowledge space consists of a number of unique keywords. These keywords define the learners' interests, the environmental objects and the content of the educational materials. Therefore, the system model consists of the items described below.

Environmental Object

It represents the available real objects that may surround the learner. It may be computers, electronic parts, chemicals…etc. Each object has its own keywords that specify its configuration. One object may share one or more keywords with one or more other objects.

Learner

Each learner has his own profile. The learner's profile represents his interests and experiences, it is translated into a number of keywords, and it is determined from the following sources:

1. Learner's explicit registration: The learner introduces his personal information and his interesting topics.
2. Learner's academic level: The system detects the knowledge that the learner gained from his past academic records.
3. Learner's actions: The system records the learners' actions while using it.
4. Learner's folder: It contains the educational materials that the learner is aware of.

Location

The location of the learners is represented by one dimension in the knowledge space. It is supposed that, everywhere there is a RFID tag to identify the location. The system can detect the learner's physical location using RFID tags. However, the relative physical distances between the different locations are predefined in the system.

Educational Material

This is the available educational material related to the education process. It includes materials from books, videos and/or audios. Each material has its own keywords that specify its content. One material may share one or more keywords with one or more other materials.

Environmental Objects Map

While a learner is interacting with other learners remotely trying to explain to them his current environment and situation, it may be difficult or at least may need a long time to describe exactly the available objects that he uses during his study. The role of *Environmental Objects Map* (Figure 4) is to map the physical space into the digital space, where each object in the physical space is detected, recognized and presented graphically by this system. The learner can forward this map to the peer helper in order to facilitate easy understanding of his environment, which augments the collaboration between them. For example, consider a learner (learner1 in Figure 1) doing an experiment at a chemistry lab, there are many objects surrounding him, each object is recognized, mapped into the digital space and forwarded to the peer helper. According to the *Environmental Objects Map* information, the peer helper can understand the learner's situation and he can efficiently collaborate with him.

Peer Helpers Map

This map displays a two dimensions knowledge space of the recommended learners who are using the system and have enough knowledge about the learner's request (Figure 5). It represents the level of recommendation of the peer learners based on their interests in one dimension, and how far their physical locations are from the learner's location in the other dimension. The closer the recommended helper to the learner's request in the horizontal (x) dimension, the more matched the peer helper's interests with the learner's request.

On the other hand, the closer the recommended helper to the learner's request in the vertical (y) dimension, the nearer his physical location to the learner.

Educational Materials Map

This map displays a two dimensional knowledge space of the recommended materials (Figure 6). It represents the strength of the relation between the recommended materials and the learner's request in one dimension, and how far their physical locations are from the learner's location in the other dimension. Therefore the learner can get information about the appropriate material that satisfies his need. The closer the material to the learner's request in the horizontal (x) dimension, the more matched the material with the learner's request. On the other hand,

The closer the material to the learner's request in the vertical (y) dimension, the nearer the material to the learner's physical location. In case that many educational materials are matched, the system recommends the best three of them according to the recommendation score function (El-Bishouty et al., 2008).

SYSTEM IMPLEMENTATION

In *PERKAM*, it is supposed that each learner has a PDA (Personal Digital Assistant) handheld device connected to the Internet through wireless connection, each device is equipped with a RF (Radio Frequency) reader, and for each object there is RFID (OMRON - I.Code V720S-HMF01, 13.56MHz) tag attached to identify it. The learner can access the system 'anywhere and anytime', face his PDA to all objects that surround him, and then the system will recommend the related educational materials and the appropriate helpers who are aware of the requested knowledge. According to the level of recommendation, the learner can refer to an educational material or select the helper whom will interact with, forward

Figure 1. Environmental objects

a) Physical space

b) Ubiquitous environment

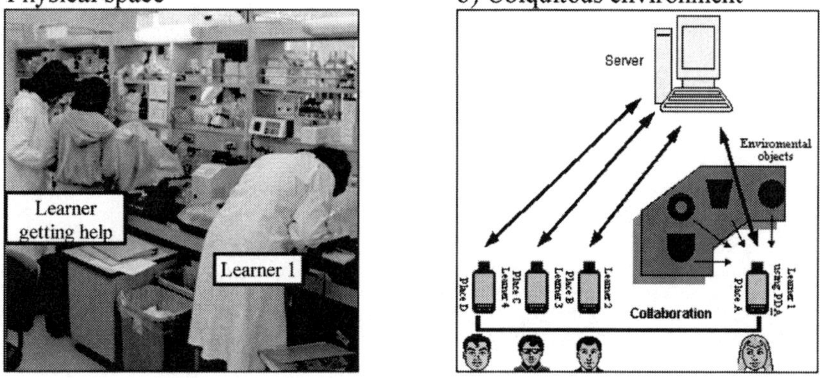

his *Environmental Objects Map* and exchange knowledge with him.

System Architecture

PERKAM system consists of the modules shown in Figure 2.

- Learner model: It represents the learner's information such as personal data, past actions, interests, folder…etc.
- Object model: It represents information about the physical educational objects that may be used by the learner.
- Location model: It represents information about the places (buildings and rooms) and the distances between them.
- Material model: It represents information about the educational material such as its title, author, type and keywords.
- Message system: It provides the learner with an easy tool to exchange messages with other learners.
- Detection manager: It detects the location and the objects that surround the learner.
- Search engine: It matches between the learner's request, and the available educational materials and peer helpers.
- Map generator: It represents the surrounding environmental objects, and the recommended educational materials and peer helpers.
- Map visualization: It visualizes the KA maps.

Software Prototype

During the implementation of *PERKAM* prototype, the limited CPU speed and memory capacity of PDA devices is put into consideration. In order to get high performance software, most of the computing processes are done on the server side.

Figure 2. System architecture

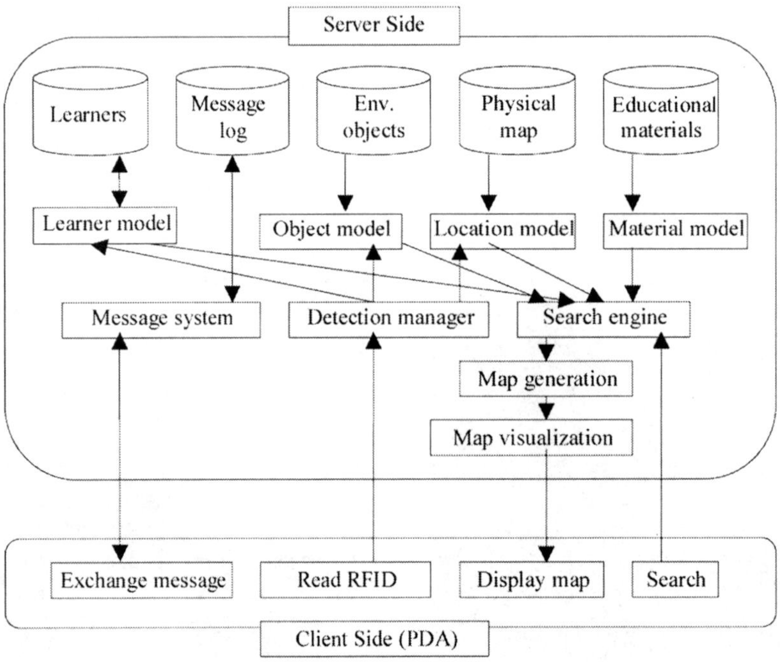

Therefore, the PDA is mainly used to send and receive data. Database schema is designed and implemented using Microsoft SQL server. The software consists of the following applications:

- Detect-Tag: It is a client-server application, the client side application is developed using embedded C++ language. The target of this application is to read the stored data in the RFID tag (which represents the objects and the locations) and send it to the server. The server side application is developed using visual C++ language. The function of this application is to listen on the network to the incoming client packets, receive client data (a particular learner's location or environmental object) and to connect to the database server in order to update the learner's profile.
- Search-Collaborate: This is the core of our system. It is a web based client-server application developed using ASP.Net and C# languages, where the learner can use anywhere using any Internet browser. A learner can login the system using his ID and password, browse his own folder, search about some knowledge, view *Knowledge Awareness Maps*, and exchange messages.
- Visualize-KA-Map: It is an embedded flash object. Here, *Knowledge Awareness Map* is dynamically designed and displayed. The application is developed using Macromedia Flash ActionScript.

System Interface

After logging-in successfully, the system will redirect to *User Page* (Figure 3-a), where the learner can check the incoming messages. *New Messages Alert* list provides the learner with the new unread messages. He can click on any message to read it then the read message will be moved to *Message History* list. The message title takes the sign *(R)* when the learner replies to this message.

Also, the learner can click on an educational material item in his own folder to get more information about it (such as its type, keywords, location). *User Page* is refreshed automatically every 30 seconds but the learner can refresh it any time by pressing *Refresh* button.

When the learner needs a help, he can press *Search* button where the system will redirect to *Search* page (Figure3-b). In *Search Page*, the learner can enter some keywords that identify his request but he does not have to do that because the system can detect the environmental objects

Figure 3. System Interface

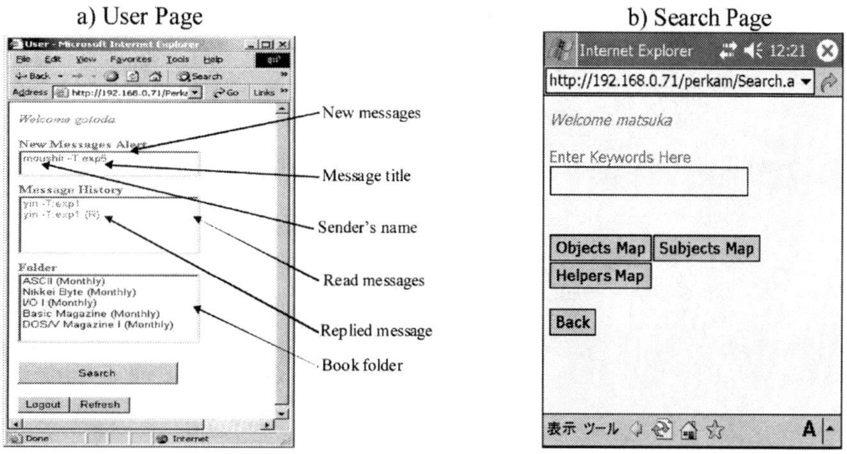

Figure 4. Environmental objects map

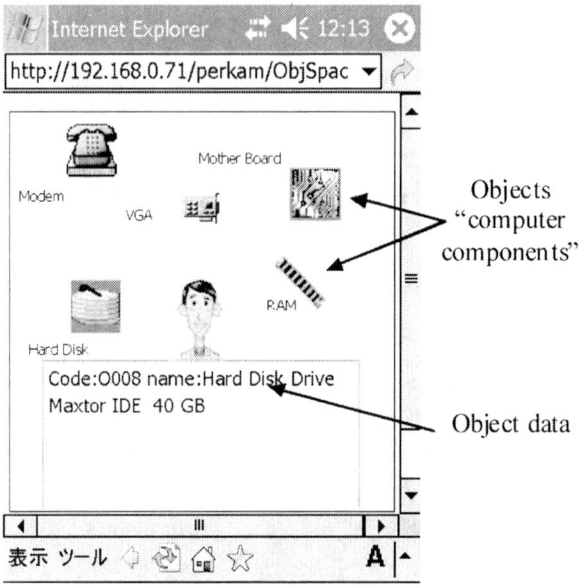

that he is using and translate it into a number of keywords that represent his query. In case of entering some keywords, these keywords will be combined with the detected environmental objects keywords to build the learner's query.

If the learner presses on *Objects Map* button, he will see his *Environmental Objects Map* (Figure 4) where each object in the physical space is detected, recognized and presented graphically as an icon. The learner can recognize the object type from its icon shape. When the learner moves the pointer over any of these objects, the object information will appear in the text box located at the bottom of the map. The person's character

Figure 5. Peer helpers map

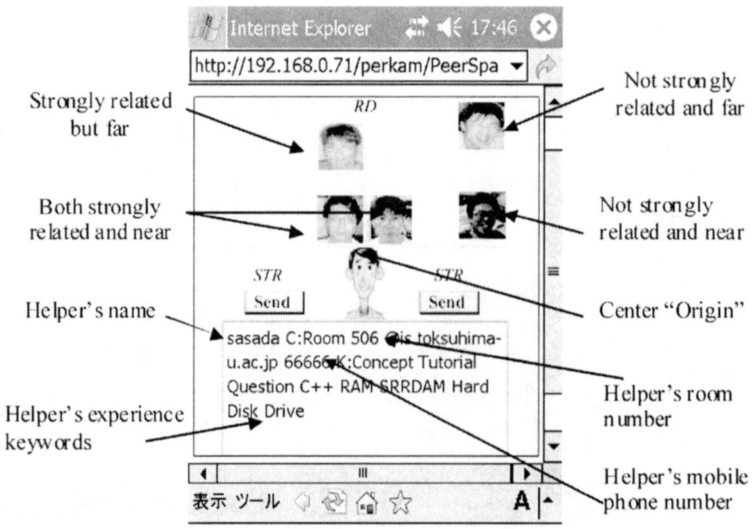

Figure 6. Educational materials map

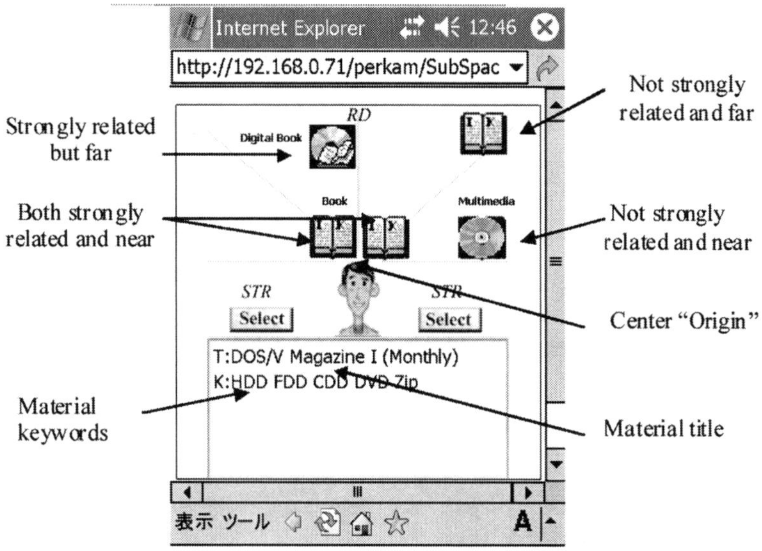

represents the current learner who is doing a task and it indicates the learner's request.

Returning to *Search Page*, if the learner presses *Helpers Map* button, the recommended *Peer Helpers Map* will be displayed (Figure 5), where each photo represents one of the recommended peer helpers. The peer helper's name, contact information, and his matched interest keywords are displayed in the text box at the bottom of the map when the learner moves the cursor over his photo. The map shows how strong the relation is (*STR*) between the helper's knowledge and the learner's request in the horizontal dimension and how near his physical location is to the learner's location in the vertical dimension The best matched and nearest peer helper is the closest to learner's head (origin). The learner may select a suitable helper and press *Send* button then the system will redirect to *Mail Page* where he can send a message to this helper, attach his *Environmental Objects Map*, and ask the helper to reply to his question or to come to help him.

Returning again to the *Search Page*, if the learner presses on *Subjects Map* button, then the *Educational Materials Map* will be displayed (Figure 6). He can click over any material icon to get its information in the text box at the bottom of the map like, its title, keywords, and location. The learner can recognize the material type from its icon shape. The map shows how strong the relation (*STR*) is between the material content and his task in the horizontal dimension and how near its physical location is to his location in the vertical dimension. -The best matched and nearest material is the closest to the learner's head (origin). The learner may select a suitable educational material, then the system will redirect to the *Material Page* to get more information about it. It may be very close to his current location, so he can refer to it and complete his task.

SYSTEM EVALUATION

An experiment was done in order to evaluate the system usage, how much it can improve the learners' skills, and the learners' satisfactions. The proposed experiment is based on *learn by doing model*, where the teachers identify a specific set of skills to teach, embed that skill learning in a task, an activity, or a goal that the student will find interesting or motivational. Finally the teacher can evaluate the learner's understanding and skill according to how much the learner succeeds to reach to the goal.

In this experiment, the teacher of a computer hardware course is asking students to do tasks regarding personal computer (PC) assembling - examples of these tasks are illustrated in Table 1. Each student is given many hardware components. The teacher defines a fixed time to complete the task and to reach the final goal of getting the computer to work properly. The teacher supplies each learner with a PDA connected to the wireless LAN and equipped with RF reader. While the learner tries to assemble the PC, he is often lacking in some knowledge; for example, how to adjust a certain Hard Disk Drive (*HDD*) jumper to *Slave* mode. In this case, he picks up his PDA and directs it to the location tag and all hardware components, and then *PERKAM* system detects his location and all the environmental objects that surround him and builds an *Environmental Objects Map*. This map represents each object with a small photo of the same object type in order to make it easier for the learner to understand and recognize the physical objects. After building the *Environmental Objects Map*, the learner can ask the system to recommend a number of related educational materials to refer to, and/or peer helpers to collaborate with in order to complete the task (Figure 7).

The Experimental Procedure

The main target of this experiment was to evaluate the learners' skills in PC assembling before and after using the system, and to measure how much they have learnt during using the system. This experiment consisted of five phases where 16 students were involved. In its first phase the teacher asked each student to fill in a questionnaire to evaluate his knowledge and experience about PC assembling, it included the followings:

- Learner's theoretical knowledge.
- Learner's practical experiences.
- Learner's references and his own educational materials.

Depending on the responses from the questionnaire, the students were divided into two groups:

1. Experts group: 6 students who were experts and had a strong knowledge about PC assembling.
2. Learners group: 10 students who were beginners.

In the second phase, the teacher asked the experts group to undertake some tasks individually in order to be sure about their strong experiences in

Table 1. Examples of the tasks

Task1	Plug a Hard Disk drive 40 GB as a Master device and a CD ROM as a Slave drive using one IDE cable.
Task2	Plug a Hard Disk drive 30.7 GB as a Master device and a CD ROM as a Master drive.
Task3	Plug an AGP VGA card 32 MB and 2x128 MB RAM.
Task4	Plug an AGP VGA card 16 MB and 1x256 MB RAM.

Figure 7. Using PERKAM system during the experiment

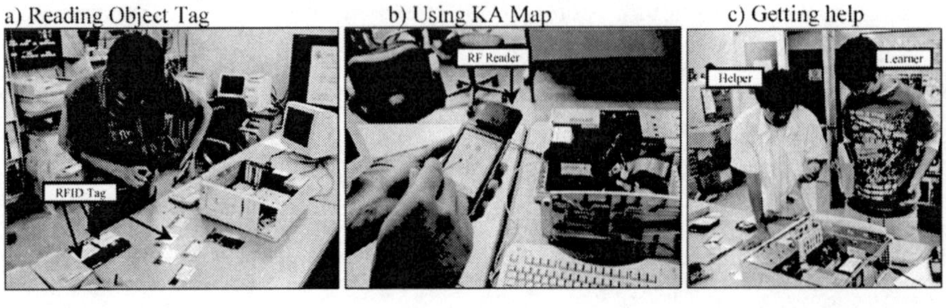

PC assembling. In the third phase, 5 students from the learners group were asked to do some tasks individually without using the developed system, and they were allowed to use an Internet facility to search for appropriate materials that might help them to complete the task in the fixed time (30 minutes). In the fourth phase, 9 students from the learners group (three of them were involved in the third phase but failed to do their tasks) were asked to do the tasks using *PERKAM* system, get recommendations and collaborate with the peer helpers to complete the tasks during the defined time; while the 6 expert students were asked to be online using *PERKAM* at different locations in the campus during their normal daily work, and to be ready to receive any messages from the other learners. After completing the tasks, all students from both groups (experts and learners) were asked to fill in a questionnaire in order to measure the learner's satisfaction.

In the fifth phase (after two months), all students who used the system in the fourth phase and successfully completed their tasks were asked to do their tasks without getting any help to measure how much they have learnt while using the system.

Results and Discussion

In the third phase of the experiment, 40% of the involved students (2 out of 5) hardly completed the task in time (30 minutes) while getting help from related websites. They used search engines like *Google, Yahoo and AltaVista.* They entered some keywords to identify their requests such as, Primary, Slave, or SDDR RAM. Sometimes they entered wrong keywords, and at other times the search results returned as unrelated websites.

In the fourth phase (while using *PERKAM* system), when the learners stopped doing their tasks because of a shortage of knowledge, they used the *Environmental Objects Map* to recognize the computer parts and tried to complete their tasks. If they still could not finish their tasks, they used *Peer Helpers Map* to get the system recommendation of the peer helpers. About 33% (3 out of 9) of the learners depended only on the *Environmental Objects Map*, while 67% (6 out of 9) of the students asked for a help from the recommended helpers in order to complete their tasks; however 66% (4 out of 6) of them asked the peer helpers to come and interact face-to-face. They did not refer to any educational materials; they preferred to collaborate with the peer helpers instead of referring to the recommended education materials. They sent messages using *PERKAM* message system, asked the peer helpers to recognize the computer components that they were using from their *Environmental Objects Map,* proposed some questions, and then tried to use the helpers' answers to finish their tasks. Finally in case they failed to finish the tasks, they invited the helpers to come and to interact face-to-face in order to show them how to do the tasks. An example of a message dialog between learner "A" who was doing Task1, and helper "B" is shown in Table 2.

In the fifth phase, 89% of the students (8 out of 9) completed their tasks during the allowed time. They did not receive any kind of help. They depended only on what they have learnt while using *PERKAM* system before. It implies that there was an obvious improvement in the students' skills after using the system, and they have learnt and gained new experiences while using the system.

Table 3 illustrates the questionnaire results, from the answers of questions 1 and 2, it is clear that, all students agree that the system is easy to use and its performance is acceptable. From questions 3 to 6, the students indicate excellent understanding and very effective use of *Knowledge Awareness Maps.* Questions 7 and 8 declare that most students prefer to collaborate with other learners to complete their tasks rather than to refer to related educational materials. Questions 9 and 10 indicate the efficiency of the system to recommend the peer helpers. Question 11 refers to the importance of the location dimension to recommend the peer helpers, where 50% of the

Table 2. Sample of a message dialog

Message	Student
I cannot understand. What does it mean "Primary"?	A
"Primary" is a plugging setting. What is the capacity of your task HDD?	B
The HDD capacity is 40GB.	A
Connect it to the motherboard in a place where "pri" is written, and use one cable to connect both CDD and HDD.	B
Use the correct HDD; check its specification from the *Environmental Objects Map.*	B
I have plugged the two devices, what shall I do next?	A
After plugging in, you can find a teeny part called a "Jumper" adjust it as a "Primary" for the HDD and "Slave" for the CDD.	B
I cannot understand, could you please come now and show me how to do it?	A
Ok…	B

students prefer to ask the helper to come and interact face-to-face, while the other 50% prefer to communicate using the message system or the mobile phone. Questions 12 and 13 reflect the efficacy of using the system as a learning tool.

After completing the fifth phase, the learners were asked to score the improvement of their skills in PC assembling. The average score was 91%. It indicates strong satisfaction of the learners for using the system as a learning tool.

Also the peer helpers were asked about the main reason for offering their help and supporting the learners' requests with their experiences; 50% of them said that "they are looking forward

Table 3. The questionnaire results

Question	For (Learner/ Helper)	Agree Percentage	Disagree Percentage
1-Can you easily use the system?	Both	100%	0%
2-Was the system response time fast?	Both	100%	0%
3-Can you easily use *Knowledge Awareness Maps*?	Both	100%	0%
4-Can you easily get information about the computer components from the *Environmental Objects Map*?	Learner	100%	0%
5-Can you easily get information about the learner's surrounding computer components from the *Environmental Objects Map?*	Helper	100%	0%
6-Do you think that the *Environmental Objects Map* is useful to recognize the learner's problem?	Helper	100%	0%
7-Did you use only the *Environmental Objects Map* to complete your task?	Learner	33%	67%
8-Did you collaborate with a recommended peer helper to complete your task?	Learner	67%	33%
9- Did you get the suitable helpers according to your task from the *Peer Helpers Map*?	Learner	100%	0%
10-Do you think that the incoming message questions are suitable for your experiences?	Helper	100%	0%
11-Do you prefer to collaborate with other learners face-to-face, or using the message system?	Both	50%	50%
12-Do you think that it is useful to use this system for learning?	Both	100%	0%
13-Have you learnt and got new experience while using the system?	Learner	100%	0%

to getting help from the learners in the future" while the others defined the main reason as "they like to offer the help to everybody". Both reasons guarantee keeping the collaboration between the students going. Finally, some comments are given by the students to improve the system interface. The learners suggest allowing them to take and transfer photos to the helpers, while the helpers recommend getting sound alerts when receiving new messages.

CONCLUSION AND FUTURE WORK

While a learner is doing a task to practice what he has learnt, he usually needs some help and instructions because of a lack of some knowledge, but the problem is the difficulty for the learner to find the required knowledge in time even if the source of this knowledge is very near to his physical location. Therefore, a ubiquitous computing environment is presented to support the learner while doing a task; this environment is called *PERKAM.* It utilizes the RFID ubiquities technology to detect the learner's environmental objects and location, then recommends the best matched educational materials and peer helpers in accordance with the detected objects and the current location. This environment provides the learners with *Knowledge Awareness Maps,* which graphically display the knowledge awareness support. An experiment was done in order to evaluate the system usage, efficacy and the learner's satisfaction. The proposed experiment is based on *learn by doing model.* The experiment and the questionnaire results indicate that *Knowledge Awareness Map* is helpful to understand the learner's situation and efficiently visualizes the recommendations. Also, the results showed that the map plays a very important role in finding peer helpers and inducing collaborations

The proposed future work is to empower the peer helper's map to provide social knowledge awareness support, and to improve the peer helper's recommendation algorithm in accordance with the similarity between the learner's current situation and the peer helper's previous situations.

REFERENCES

Carroll, J., Neale, D., Isenhour, P., Rosson, M., & McCrickard, D. (2003). Notification and awareness: synchronizing task-oriented collaborative activity. *Journal of Human-Computer Studies, 58*(5), 605–632. doi:10.1016/S1071-5819(03)00024-7

Chen, Y. S., Kao, T. C., Sheu, J. P., & Chiang, C. Y. (2002). A Mobile Scaffolding-Aid-Based Bird Watching Learning System. In *Proceeding of International Workshop on Wireless and Mobile Technologies in Education,* IEEE Computer Society, pp.15-22.

Cole, H., & Stanton, D. (2003). Designing mobile technologies to support co-present collaboration. *Personal and Ubiquitous Computing, 7*(6), 365–371. doi:10.1007/s00779-003-0249-4

Cui, Y., & Bull, S. (2005). Context and learner modeling for the mobile foreign language learner. *System, 33*(2), 353–367. doi:10.1016/j.system.2004.12.008

Curtis, M., Luchini, K., Bobrowsky, W., Quintana, C., & Soloway, E. (2002). Handheld Use in K-12: A

Descriptive Account. *Proceeding of the International Workshop on Wireless and Mobile Technologies in Education,* IEEE Computer Society, pp. 32-30.

Dourish, P., & Bellotti, V. (1992). *Awareness and coordination in shared workspaces. Proceedings of Computer Supported Cooperative Work '92* (pp. 107–114). ACM Press.

El-Bishouty, M. M., Ogata, H., Ayala, G., & Yano, Y. (2010). Context-aware support for self-directed ubiquitous-learning. *Int. J. Mobile Learning and Organisation*, *4*(3), 317–331. doi:10.1504/IJMLO.2010.033558

El-Bishouty, M. M., Ogata, H., & Yano, Y. (2007). PERKAM: Personalized Knowledge Awareness Map for Computer Supported Ubiquitous Learning. *Journal of Educational Technology & Society*, *10*(3), 122–134.

El-Bishouty, M. M., Ogata, H., & Yano, Y. (2008). A Model of Personalized Collaborative Computer Supported Ubiquitous Learning Environment. *Proceedings of the 8th International Conference on Advanced Learning Technologies (ICALT08)*, Spain, IEEE Computer Society, pp. 97-101.

Fischer, G. (2001). User Modeling in Human-Computer Interaction. [UMUAI]. *Journal of User Modeling and User Adapted Interaction*, *11*(1/2), 65–86. doi:10.1023/A:1011145532042

Fischer, G., & Konomi, S. (2005). Innovative Media in Support of Distributed Intelligence and Lifelong Learning. In *Proceeding of the International Workshop on Wireless and Mobile Technologies in Education*, Japan, IEEE Computer Society, pp. 3-10.

Hatano, G., & Inagaki, K. (1973). *Intellectual curiosity*. Chuo Koronsya Press.

Klaus, F., & Rachel, W. (2000). *RFID Handbook: Radio-Frequency Identification Fundamentals and Applications*. NY: John Wiley & Sons.

Lin, Y., Swan, K., & Kratcoski, A. (2005). Situated Learning in a Ubiquitous Computing Classroom, *Journal of the Research Center for Educational Technology,* Fall 2005, from http://www.rcetj.org/?type=art&id=3530&.

Mitchell, K., & Race, N. J. P. (2005). uLearn: Facilitating ubiquitous learning through camera equipped mobile phones. In H. Ogata, M. Sharples, Kinshuk & Y. Yano (Eds.). In *Proceedings of the Third IEEE International Workshop on Wireless and Mobile Technologies in Education 2005*, Tokushima, Japan, pp.274-281.

O'Malley, C. (1994). *Computer supported collaborative learning* (*Vol. 128*). NATO ASI Series, Computer and Systems Sciences.

Ogata, H., Imai, K., Matsuura, K., & Yano, Y. (1999). Knowledge Awareness Map for Open-ended and Collaborative Learning on World Wide Web. In *Proceeding of the International Conference on Computers in Education ICCE99*, pp.319-326.

Ogata, H., Matsuura, K., & Yano, Y. (1996). Knowledge awareness: Bridging between shared knowledge space and collaboration in Sharlok. *Proceeding of Educational Telecommunications*, *96*, 232–237.

Ogata, H., & Yano, Y. (2000). Combining Knowledge Awareness and Information Filtering in an Open ended Collaborative Learning Environment. *International Journal of Artificial Intelligence in Education*, *11*, 33–46.

Ogata, H., & Yano, Y. (2004). Knowledge Awareness Map for Computer-Supported Ubiquitous Language-Learning. *Proceeding of the International Workshop on Wireless and Mobile Technologies in Education*, Taiwan, IEEE Computer Society, pp. 19-26.

Renda, M., & Straccia, U. (2005). A personalized collaborative Digital Library environment: a model and an application. *Journal of Information Processing and Management*, *41*(1), 5–21. doi:10.1016/j.ipm.2004.04.007

Sakamura, K., & Koshizuka, N. (2005). Ubiquitous Computing Technologies for Ubiquitous Learning. *Proceeding of the International Workshop on Wireless and Mobile Technologies in Education*, Japan, pp.11-18.

Schank, C. (1995). What We Learn When We Learn by Doing. *(Technical Report No. 60),* Northwestern University, Institute for Learning Sciences.

Song, Y. (2006). What consequences have handheld educational uses brought about. *Proceeding of the International workshop on Mobile and Ubiquitous Learning Environments MULE in conjunction with ICCE2006*, China, pp. 1-7.

Yoo, Y., & Lyytinen, K. (2005). Social Impacts of Ubiquitous Computing: Exploring Critical Interactions Between Mobility, Context and Technology. *Journal of Information and Organization, 15*(2), 91–94. doi:10.1016/j.infoandorg.2005.02.006

Chapter 3
Ubiquitous Computing Does Not Guarantee Ubiquitous Learning in Schools:
The Case of Handheld Computers

Howard Nicholas
La Trobe University, Australia

ABSTRACT

In this chapter, I will argue that ubiquitous computing is a necessary, but not sufficient condition for ubiquitous learning. I will propose definitions that clarify some of the contextual features that shape both terms when they are applied in a school context. Using data and experiences derived from a year-long project with primary and secondary schools in Victoria, Australia, examples of classroom activities are analysed to illustrate how the presence of ubiquitous computing did not guarantee ubiquitous learning. These activities are compared to articulate the key features of ubiquitous learning and develop a model that shows how the contribution of ubiquitous computing to ubiquitous learning is constrained by pedagogic frameworks that shape the relationship between handhelds, other elements of the technological suite and learning.

INTRODUCTION

Handheld technology and its frequent synonym, mobile technology include portable computer-based devices such as mobile phones, tablet PCs, personal digital assistants (PDAs), iPods, MP3 players, smartphones, graphics calculators, probewares, digital cameras, digital video recorders and games consoles. This multitude of portable devices can be used to create ubiquitous computing, a term first used by Weiser in 1988 (Weiser, Gold & Brown, 1999). Weiser asserted that the most profound technologies associated with ubiquitous computing are those that disappear as they weave themselves into the framework of our everyday lives. Weiser described computing in three waves. The first wave of computing saw the use of one mainframe computer by many people; the second wave saw a 1:1 computer to human ratio where individuals were connected to desktops or laptops. We are now in the third wave of computing where an individual possesses or is served by many computers dispersed throughout

DOI: 10.4018/978-1-61692-849-0.ch003

the physical environment. As these technologies recede into the background of people's lives, they are being used unconsciously for various tasks (Weiser, 1991; 1996). As they do this, they become 'calm technology' (Weiser and Brown, 1996). When technology becomes 'calm', it does not occupy the learners' attention all the time but is able to move seamlessly and effortlessly between the learners' periphery and centre of attention.

More recent discussions of ubiquitous computing (Hwang, Tsai and Yang, 2008, p.82) have identified ubiquitous computing with access to a range of context-aware technologies.

From the system designer's point of view, physical integration and spontaneous interoperation are the two main characteristics of ubiquitous computing systems (Kindberg & Fox, 2002). Physical integration means that a ubiquitous computing system involves some integration between computing nodes and the physical world. For example a smart coffee cup, such as Media-Cup (Beigl et al, 2001), serves as a coffee cup in the usual way, but also contains sensing, processing and networking elements that let it communicate its state (full or empty, held or put down) ...

In the general school context, this identification demands too much since tools such as Radio Frequency Identification chips and sensors cannot be universally implanted or accessed within the resources of most schools. So, in this article, I restrict the concept of ubiquitous computing to one that is more realistic for students and schools– universal access to a suite of technologies that includes mobile and internet connected devices.

The general claim within the literature is that once the computing becomes ubiquitous, it begins to have the capacity to support ubiquitous learning. Hwang et al (2008) argue:

... by employing this new technology in education, the learning system is not only adapted to the individual's needs, but is also actively involved in his or her learning activity (p. 82) ... [so that there can be] ... a special definition of u-learning that employs mobile devices, wireless communications and sensor technologies in learning activities, called "context-aware u-learning" ...

The above is a view of learning that is personal and able to occur regardless of time or place. However, context-aware u-learning focuses on the technology rather than on the nature of the learning process that occurs with the technology. A similar gap is in the more general sense of ubiquitous learning that is regarded as 'anywhere at any time'. I will argue that for the purposes of looking at beneficial learning in schools, we need also a view of how the learner's relationship with the object of learning changes such as it is not only personal, but also (at least potentially) transformative (Keamy & Nicholas, 2007).

Ubiquitous computing in the less technologically-intense sense defined above is, therefore, a necessary but not sufficient condition for ubiquitous learning since ubiquitous learning requires two additional features: the systematic presence of and interaction between handheld and other forms of computing and systematic pedagogic design that creates opportunities for transformative experiences i.e. pedagogies that acknowledge the learner (aided by technology) as a source of knowledge with the capacity and space to question and reframe the knowledge, behaviour or attitudes/values presented by the teacher.

The relationship between ubiquitous computing and ubiquitous learning with school students using handheld computers will be explored. Using the result of the exploration, a framework for ubiquitous learning that encompasses and positions ubiquitous computing will be outlined. The particular handheld computer alluded to in this paper is the PDA, sometimes also referred to as a palmtop computer or pocket PC. Within the context of Victorian schools in Australia, where handheld computers in schools are still a new phenomenon, a ubiquitous learning framework needs to consider both schools where students have personal handhelds and schools where students only make sporadic use of class sets of handhelds. This chapter focuses on the latter as ownership of personal handheld computers by individual students is still relatively rare and the latter condition presents a more challenging situation for the support of ubiquitous learning. The context within which this paper is based is a year-long investigative study with five schools (three primary and two secondary) implementing handheld computer programs of various kinds. The research identified differences between how the primary and secondary schools made use of pocket PCs in their classes and the successes, problems and pedagogical changes associated with them (Ng & Nicholas, 2009). The focus of this paper, based on detailed observations of three primary classes using handheld computers and informal interviews with teachers and students during the observations is to seek evidence of ubiquitous learning supported by the handheld computers in each of the classes by identifying (i) how teachers structure this kind of teaching and learning in handheld computer classes and (ii) other requirements that are necessary to make ubiquitous learning accessible and real within the classroom environment. The aim is to define a view of ubiquitous learning that is workable within the constraints of a school environment. The question for this paper is: What are the factors that schools and teachers need to consider in order to achieve ubiquitous learning with handheld computers for school students?

THE HANDHELD COMPUTER AS A UBIQUITOUS LEARNING TOOL

Handheld computers have the potential to be physically ubiquitous as a result of their small size and light weight. They are not as powerful as laptops or tablet PCs but the range of uses is evolving quickly as more software applications are produced and new devices become available. Therefore, to meet the requirements of ubiquitous computing, schools using handhelds must develop means by which the use of the handheld is part of the use of a suite of technologies – each chosen for its functional appropriateness. Different models of handheld computers come with different functionalities. The most basic models have capabilities for word processing, spreadsheet and graphing functions, Bluetooth, calendar, calculator, notepad, voice recording and presentation (but not necessarily editing) software. A more powerful (and more costly) handheld computer would have in addition a camera, audio and video recording tools and Wi-fi to access the Internet, increasing the number of supports for ubiquitous and seamless learning and varying the nature of the relationship needed with other technologies.

Roschelle, Patton and Tatar (2007) identified three historical large scale successes with handhelds in schools as the graphics calculator, probeware and networked response systems for quick feedback. The capabilities of handhelds enable students to carry out a variety of contextualized learning activities that are not possible with a desktop or laptop computer. They provide students with the capacity to interact with other computers and between themselves as well as to work both independently and in groups on activities that include: collaborating with peers using Bluetooth in class or emailing outside of class, working collaboratively on multimedia materials, taking notes, audio recording and drawing figures using the notepad function, researching via wireless access to the Web, reading and writing e-documents, writing e-books/articles, undertaking interactive worksheet exercises e.g. those created with Hot Potatoes software; interacting with simulations and multimedia educational materials, watching animations and movies with Media Player, entering data, drawing graphs, calculating mathematical problems using Excel Mobile and the calculator, collecting data using handheld sensors and probeware, organising learners' daily activities via the calendar, making images and video recordings, listening to downloaded podcasts, lectures or interviews, learning from software packages e.g. foreign languages, or accessing reference material such as in a dictionary or encyclopedia, Periodic Tables and other subject-related software and creating and displaying presentations (Curtis, Luchini, Bobrowsky, Quintana, & Soloway, 2002; Garland, 2006; Nicholas & Ng, 2009; Shin, Norris & Soloway, 2007; van 't Hooft, 2006).

Soloway, Norris, Blumenfeld, Fishman, Krajick, & Marx (2001) argued that if computers are not 'ready-at-hand' and readily available to students, they will have little impact on the students' learning because they will not be used routinely and in a 'day-in, day-out' manner. If all of the potential resources are realized, having access to a handheld computer is like having the pens, text books or other written resources, cameras, calculators, voice recorders, clocks and Internet access all in one device to support (i) self-directed, independent and personalised learning, (ii) portability and access to the Internet allowing learning and data collection anywhere and anytime, (iii) social interactivity where students collaborate and exchange data in both online and offline environments, (iv) situated learning where real-time or simulated data can be collected in real-time working environments and (v) connectivity to learning communities online, or networks of connected handhelds in shared learning environments (Kimber & Wyatt-Smith, 2006; Naismith, Londsdale, Vavoula & Sharples, 2006; Perry, 2003; 2006; Sharples, 2003, 2006; Waycott, Jones, & Scanlon, 2005). These affordances are aspects of ubiquitous computing and offer the potential for seamless learning.

UBIQUITOUS LEARNING WITH HANDHELD COMPUTERS IN THE CLASSROOM

Theories that support learning with handheld computers (for example Rochelle, Patton & Tatar, 2007; Sharples, Taylor & Vavoula, 2007) commonly embrace aspects of the socio-constructivist learning theories of Piaget, Bruner and Vygotsky. Socio-constructivism asserts that the learner is an active participant in the construction of his/her own knowledge and that existing knowledge (Piaget, 1972; Bruner, 1990) and a socially interactive environment are built into the learning experience as factors that affect the construction of learners' knowledge (Vygotsky, 1978). Sharples, Taylor & Vavoula's (2007) theory of learning for a mobile age focuses on the communicative interactions between technology and people for the advancement of knowledge. They see learning as conversational processes (citing Pask, 1976) taking place across multiple contexts within systems where people and technology are in 'continual flux'. According to them, conversation requires the learner to describe him/herself and his/her actions, as well as to explore, extend, converse and externalize that description or understanding to subsequent activities. Technology can be involved as a conversational partner, mediating conversations that bridge formal and informal learning. Rochelle, Patton and Tatar (2007) view handheld computers as learning technologies that are representational tools for augmenting cognition and a mediational tool for social and cultural participation. They see learning with handhelds and within networked communities as the merging of cognitive augmentation and social mediation perspectives that link formal and informal learning, as well as a means to link representations mediated by mental operations and social practice. To this should be added the requirement that the learner using the handheld needs to be acknowledged as a 'partner' in the networked conversations rather than just a 'recipient' of information. In other words, the learner needs also to be seen as a shaper and presenter of information and values.

Ubiquitous learning has two dimensions. First, it is not constrained by physical space, plans or timetables but is pervasive and occurs anywhere at any time. Second, as a consequence of the distributed nature of the immediate access to a variety of sources of information or means of reflecting on experiences in interaction with others, ubiquitous learning is characterized by the transformation of understanding and the ability to question experiences and information. Extending the formulation of Schenker, Kratcoski, Lin, Swan and van 't Hooft (2007:172), learning is therefore 'the processing of encountered information [extended to include experiences, values or representations] that leads to changes in knowledge, beliefs, abilities, and behaviours'. The notion of 'ubiquitous' learning builds on this to emphasise how this processing can occur through a variety of modes and modalities unconstrained by time and location. Because of their capacity to situate the experiences and transforming reflections in both immediate and more removed contexts, handheld technologies offer a powerful resource for ubiquitous learning in being able to provide a portable, interactive learning environment capable of both multimedia functions and Internet access and supporting both self-directed, independent learning and interactivity with others. Consequently, handheld technology should be able to foster active and creative learning as it is capable of creating opportunities for students to collaborate with peers in project work (Bluetooth, beaming) and to undertake independent research and communication over distance (wireless) as well as to be engaged in problem solving in real-life contexts. The portability of the devices allows for context-based data collection and 'just-in-time' learning.

At a broad level, ubiquitous learning (u-learning) encompasses the multiple forms of electronic learning (e-learning) usually associated with mobile learning (m-learning). At the level of the learners, there are more specific requirements placed on technologies if they are to be used for ubiquitous learning. Ogata & Yano (2003), referencing the work of Chen (2002) and Curtis et. al. (2002) identified these requirements as (i) *permanency* where learners never lose their work unless it is deleted on purpose, (ii) *accessibility* where learners are able to access their files, documents and data from anywhere, (iii) *immediacy* where learners are able to obtain information immediately and (iv) *interactivity* where learners are able to interact with teachers, peers or experts through synchronous or asynchronous communication, enabling knowledge development and transformation to occur more quickly and readily. There is also a crucial requirement of ubiquitous learning to make explicit the mutual engagement of multiple learners through (v) *situating of instructional activities* where learning is embedded in the learners' daily lives and (vi) *adaptability* where learners are able to get the right information at the right place in accessible ways. To these six requirements a seventh needs to be added –(vii) *externalising learning* where learners are able to represent their (evolving) understanding. These characteristics are enabled by the educational affordances that handheld computers can offer, as described in the next section.

THE RESEARCH STUDY

Method

As mentioned above, the focus of this study comes from detailed observations of three primary classes in action with handheld computers. At the time this study took place, the classes had used handhelds for about 9 months. Teachers had had about 12 months of developing an understanding of the pedagogy associated with handhelds (the initial three months were for teachers' personal exploration). The teachers chose the time for the researchers to visit their classes and an assumption could be made that they were ready to 'display' good lessons using handhelds for the visits.

The handhelds were not personal devices as each school had purchased student sets of 20-30. Their usage was targeted at specific year levels or specific groups in the schools. The primary schools had a historical practice of working together in a collegial manner and for the purpose of the handheld project, they shared an IT support person who provided professional development and who helped developed small multimedia tasks for the handhelds called eBytes. The IT support person was based in one of the primary schools and support for the other two schools waned after the initial 3-4 month period.

During the observations in the classes, informal interviews were conducted with students to gain a better understanding of motives and reasons for using the handhelds in specific manners. Observations were focused on the nature of the activities, how the handhelds were used by the students, whether the handhelds were used in conjunction with other resources or technology, the collaboration undertaken by the students and the degree of students' cognitive engagement in undertaking the handheld tasks. These parameters were chosen because they reflect the characteristics of ubiquitous learning outlined above.

Results

In this section, a description of the activities observed in each class will be provided, as well as an evaluation of the use of the handhelds for the activities and an assessment of the ubiquity of learning during the lesson. A summary of the observations of the three primary school classes is presented in Table 1. One task from each of the three primary schools is described in order to document the variable relationships between ubiquitous computing and ubiquitous learning.

Primary School 1. The Size, Colour, Use, Material, Parts, Shape (SCUMPS) task undertaken in School 1 with a composite class of years 4 and 5 students was a fairly straightforward task. Students were asked to select an object to examine and to describe it in relation to the above aspects. They had then to design a new version of the object focusing on those features. The objects selected for description were outside, so that using the PDAs for recording and storing information while outside revealed one aspect of ubiquitous computing.

The 'copying' of instructions onto the students' handheld computers in the initial part of the activity (in the computer room) took the students about 15 minutes to complete. The students could then work outdoors, looking for interesting objects and typing or scribing their findings into Word on the handheld. The collaboration between partners through conversation was high but did not involve collaborative uses of handhelds. The conversations reflected a medium degree of cognitive load (Maker, 1991; 1993) as students made connections between the size, colour, shape, materials and parts and ways in which the object could be used. However since the task was essentially descriptive, there was not a lot of space for transformations in thinking.

In this lesson, little ubiquitous learning was demonstrated. The 'copying' of instructions from the desktops involved re-typing the instructions and/or voice recording the instructions onto the students' handhelds. Each of these acts was manual and involved only low level thinking skills. A direct connection between the handheld cradle and desktop to permit synching would have increased the time available for more cognitively challenging activities.

Even though there were digital cameras available and students had used them in previous lessons that had been observed, students did not ask for nor use them. Activities that are constructed or interpreted as closed do not encourage ubiquitous learning. Frequently, such activities are designed around the equipment rather than the learner. To encourage ubiquitous learning, the inclusion of two open components in the task such as (i) a presentation component using photos or short video recordings would enhance the communicative aspect or (ii) an extension component where students are encouraged to think beyond the object of interest and its SCUMPS would enhance the reflective component. For example, if the object chosen for SCUMPS is a see-saw in the school

Table 1. Summary of observations at the three schools: activities, availability of resources and levels of collaboration and cognitive challenge associated with the activities.

Activity	Handheld computer use	Availability of other technology	Use of other technology	Collaboration High (H), Medium (M), Low (L)	Degree of cognitive challenge High (H), Medium (M), Low (L)
School 1, years 3/4, 27 students, lesson on mathematics eByte activity Size, Colour, Use, Materials, Parts, Shape in computer lab (20 desktops)					
Getting the instructions from desktops for SCUMPS activity	'Copying' instructions from desktop to handheld by typing, scribing or voice recording	Set of 30 handhelds stored in the cupboard. Desktop computers Digital cameras Data projector	Desktop	L (individual)	L (routine)
SCUMPS activity outdoors	Portable Entering attributes of target of study by typing, scribing or voice recording			H (Discussing how selected object rated against the attributes)	M (making connections)
End of SCUMPS activity	Synching document with attributes to student folder on desktop and naming files		Desktop	L (individual)	L (routine)
School 2, years 5/6 composite, 15 students, mathematics eByte activity on Shapes in computer lab (25 desktops)					
Getting task from desktop which was presented as a movie on shapes. Instructions given at the end of the short movie	Downloading task from desktop to handheld	Set of 20 handhelds, each connected to a desktop computer via cradle. Colour printer Interactive white board Digital cameras	Desktop Printer	L (Individual)	L (routine)
'Shapes' activity outdoor	Describing objects in Word Mobile Drawing shapes of objects in Notes			H (Discussing responses to questions)	L (Ignored all irregular shapes e.g. bark)
End of 'Shapes' activity	Synching work file on 'Shapes' from handheld to desktop		Desktop		
School 3, years 4 & 5 composite, 27 students, literacy activity on Sea Creatures					
Activity on sea creatures. A movie on whales was loaded in all the handhelds. Instructions on task were provided at the end of the movie. Task consisted of set questions and an open research project	Open tasks – students do handheld tasks on a rotational basis. Using Word Mobile and Notes to answer questions Using Internet to search for information	Set of 20 handhelds available to use at will 6 desktop computers Interactive whiteboard Data projector Internet access ports	Internet	M (set questions involved discussions with partners) L (open research task was more individual)	M (set questions) H (open task to find out and write about a sea creature)
Concept map of sea creatures	Inspiration software installed in the handhelds		Inspiration software	M (mainly technical)	H (make connections between different concepts)

yard, the students could take a photo of the object and voice record a description and explanation of the object by referring to the features, dimensions and uses the object serves. Questions such as 'In what other situations could you see the object that you are investigating being used?', 'How can the object be improved or modified to make it better or to make it useful?' could encourage students to relate the see-saw to levers useful for lifting heavy objects. The students will be most unlikely to be able to discuss principles of levers, but providing them with an opportunity to link everyday experiences to their learning and making connections to other objects that they have seen helps promote higher order thinking by providing a basis for enquiry-based engagement with others.

At the end of the SCUMPS lesson the teacher said that the students were comfortable with and proficient in their use of the PDA and that there was a need for more 'educational software' for the handhelds that could promote 'genuine learning activities'. The reference to 'genuine learning activities' implies ones with transformational character, but the dominant focus of the teacher was on the technological issues. She commented on the 'hassle' in managing the hardware. She and the other teachers using the set of handhelds had to ensure that the handhelds were charged by ensuring that the students placed them back in their cradles in a lockable cupboard where all the cradles were connected to a set of powerboards. When some units were not charged, it could take about ten minutes to trace back among the tangle of cords to find the chargers that were not plugged in, or where plugs were not seated properly. So although the technology was ubiquitous, the focus of the activity was not consistent with ubiquitous learning as outlined above.

Primary School 2. In this school years 5 and 6 students in a composite class undertook a 'Shapes' activity. The task was produced by the teacher in the form of a movie. The movie contained images of shapes, mostly in real environments, accompanied by audio and text, which included instructions for the tasks to be carried out. Essentially these were to (a) comment on where the students saw shapes when in the yard and (b) to list all the shapes observed in the playground. They were to voice record their responses to the shapes that they observed (a) and to write in Word and draw in Notes their list that would be their responses to (b).

As each desktop in the computer lab had a handheld cradle connected to it, the students were able to quickly download the movie with the instructions onto their handhelds. There were two students who decided to type the questions in Word on the desktops and then printed a hard copy. These girls still preferred hard copies despite about 7 months of handheld usage.

The task appeared to be specifically designed to make the most use of the technical capabilities of the handhelds through specifying use of voice recording, writing and drawing. However, the task itself was not challenging. The first question, 'Where do you see shapes?' was regarded by some students as an odd question since they said that 'you see shapes everywhere' and began to discuss spaces in which there were 'no shapes' and said that such spaces would be 'weird'. In the students' minds, the second (listing) task appeared to assume observations and listing of regular shapes. When presented with a piece of bark and asked about its shape, they were unable to comment. All students ignored irregular shapes and none presented the teacher with shapes they could not name. The impression that they gave was that they just wanted to get on with the handheld task and complete it. For more challenging learning to have occurred, the students could have used the digital cameras and taken pictures of shapes of irregular objects such as a tree or stone and taken them back into the classroom for discussion.

Some of the students drafted out their speeches on paper before voice recording them. When asked why, one of the responses was 'We don't like to speak our thoughts into the voice recorder. We might make a mistake and it would sound silly'. Learning is hindered when the handheld is used for the sake of using it rather than as a tool that supports reflection on different and novel perspectives or if, as often is the case with work written on paper, the act of recording is seen as creating an object that is difficult to change.

Overall, the learning in this lesson focused on the handhelds and how to get the best use out of them. The teacher was enthusiastic about the use of handheld computers. She searched for useful software to download, for example a drawing program with ruler and protractor and stopwatch programs. Like the principal and the other two handheld teachers in the school, she frequently described the purpose of her work with handhelds as promoting higher order thinking and not using them for the sake of the technology itself, although that did not appear to be the case in this class. There were other technologies available in the classroom to promote ubiquity in learning. The teacher was proficient with Movie Maker on the desktops as a result of experience with the format for the eBytes, which were a standard tool created and used within the school. However she realized the awkwardness of having students reading instructions and doing the work on the handhelds simultaneously and was beginning to provide printed cards to accompany eByte tasks. She exclaimed enthusiastically about discovering the usefulness of the 'switcher bar' function in the handhelds, which is similar to the minimization/maximization functions of desktops and laptops. This alleviated the frustrations of students having to close and open applications all the time. On return to the classroom, the students could have used the interactive whiteboard to share their drawings of the different shapes they had made and drawings or images of irregular shaped objects could have been displayed for discussions. These opportunities were not used.

Beaming was an issue. Both the teacher and the students cited an experience where the students had been taken on an excursion and encouraged to beam messages to each other in the bus as a 'disaster because of the confusion with too many machines in use'. Without teasing out and addressing the issues surrounding this 'disaster', it was difficult to see Bluetooth-based collaborative tasks being incorporated into the ubiquitous learning of this group of students. No such analysis appeared to have taken place.

Primary School 3. This composite class of years 4 and 5 students undertook a Sea Creature task in a literacy segment. The class set of handhelds was available to the students whenever they needed to use one – either individually or collectively. There were also specific handheld tasks assigned for the class. This particular task was to view a presentation pre-loaded on the handhelds on the subject of whales and then to respond to questions at the end of the presentation. The presentation involved still images, video and text. The questions at the end ranged from finding definitions of words to creating short written passages. After starting the whale presentation on their handhelds, the students were largely focused on their individual handheld units. There was some pausing of the presentation to make comments to a neighbour such as "look at this" or "what's this about?", but generally the discussion was sporadic. There were variations in the choice of written data entry for this task. Several of the students said that having to get the questions at the end of the presentation from the handheld was a factor against also writing on the handheld. These students wrote in their paper-based notebooks instead. Other students swapped back and forth on the handheld from the presentation to a Word document – a process that they described as tedious due to the absence of a minimisation/maximisation button that they were all used to with desktop versions of Windows. Occasionally, these students swapped applications. One had made brief jottings on paper about the questions while another said that he had tried hard to remember the questions to avoid the hassle of swapping applications. One student innovatively used an additional handheld – one to view the presentation and questions, the second to write his responses. Two of the groups voice-recorded their responses. This was done after brief discussions with their partners, a bit like instant script writing followed by performance. At the end of the literacy segment, one of these recordings was played to the class over speakers connected to the teacher's laptop computer.

Students used a combination of book-based and Internet-based resources and dictionaries to complete the task. Students using book-based resources had mentioned the technical interruptions to browsing with the handhelds (a wireless connectivity problem) as being frustrating. Two also mentioned the small screen size as being something that they found annoying when browsing with the handhelds. One student finished writing and editing on the handheld then transferred the material to the desktop computer for file storage and printing.

The task this class undertook was semi-structured with guided questions at the beginning followed by an open research task. The level of cognitive engagement required ranged from medium (set questions) to high (research project).

The teacher of this class thought that a handheld on its own and in combination with other technologies was useful in assisting student learning, *as tools*. She would have liked more applications and educationally useful software for the handhelds. There were two default games left on the handhelds. She was not worried about students getting off task as these games could be used in 'free' moments, and she saw them as having value in developing logical thinking.

DISCUSSIONS

The study of the three primary classes indicated that teachers from schools 1 and 2 structured activities for their students' use of handhelds by providing specific instructions, for example to voice record or draw in Notes. The SCUMPS and 'shapes' tasks were fairly low level thinking tasks requiring only listing and description of shapes or objects. There was little flexibility for students to direct their own learning and the students' learning was not shared with the class, for example through the interactive whiteboards. Activities that are structured around the technology in this way could impede learning and stifle students' development of higher order thinking. They lack some of the crucial features of ubiquitous learning described by Ogata & Yano (2003): (iv) *interactivity,* (v) *situating of instructional activities* and (vi) *adaptability*, as well as the extension, (vii) *externalising*. The teacher from school 3 approached the use of handhelds slightly differently. The task that she designed included both structured and open elements so that there was a requirement for higher levels of thinking. Unlike the teachers from Schools 1 and 2, this teacher did not constrain the students to work in set ways. Instead the students had freedom to work in ways that best suited their needs. This classroom demonstrated a more ubiquitous learning environment and a contributing factor was the task design of the teacher based on the relationship of trust and rapport that she had built with her students. The task involved both situatedness and adaptability, with opportunities for both externalizing and interactivity.

Regardless of their school, all teachers in the study indicated that a personal handheld for each student would be ideal for supporting seamless learning. Since it was unlikely that parents would be able to afford such an arrangement, especially in schools 2 and 3, the next best thing for schools to provide opportunities for ubiquitous learning is the availability of the handhelds in the classroom on an 'as needs' rather than routinely scheduled basis. However, a difference between school 3 and schools 1 and 2 was the location and availability of the handheld computers. The first two schools had the handhelds located in the schools' computer laboratories – either connected to a desktop or stored in a cupboard with cradles ready for charging. School 3 had the set of handhelds in a tub that was brought to class from the computer laboratory where they were charged. The handhelds were free for the students to use whenever they were available.

Handheld computers are not powerful enough to make use of the full range of software applications and they cannot function independent of other learning or technological tools. Ubiquitous learning in schools with handhelds is, therefore, blended learning. It requires multiple technologies to be in place – Internet access points, connection to printers and interactive white boards and software such as concept mapping and graphing software. Handhelds can be used in interactions with a wide variety of technological and human partners and in a wide variety of contexts so that learning can take place both formally and informally. These kinds of uses require sustained links between the student and the handheld, which means that these forms of blended learning begin with personal relationships with the equipment. All teachers interviewed were aware that more challenging tasks have to be designed for the students. But when the handheld is another piece of equipment in the school that is not readily accessible, for example, when students can only access it for an hour or two a week, there is a tendency for teachers to feel that the students have to get maximum value out of the handheld when they are taken to the computer room to use the equipment and therefore to make a more 'forced' use of the equipment rather than to go to their espoused principle of a natural embedding in the curriculum. The result is that forced efforts are made to design curriculum around the handheld, that is ***what can the students do with the handheld for this session*** (focus on the equipment), rather than ***how can the students use the handheld for learning in real time*** (focus on student).

You start from bottom, use it for a purpose not for the sake of using it…money spent does not drive it to work. It has to come from the classroom and kids'needs. (staff#19) (Ng & Nicholas, 2007, p.13)

These contextual and other financial issues work against the principles of 'calm' technology since they create a focus on the technology itself – even if the only issue is 'when' and 'how' the technology can be made available. Consequently, the more opportunity students have to hold the handhelds, the better will be the result of the blending of students' general learning with learning with the handheld in unforced and powerful ways. As one teacher said: "students will be more connected if the pocket PC is personal" – meaning something that they could carry around with them and use in ways that they determined to be appropriate. Students having access to their own handhelds all the time enables the handhelds to be blended in as a part of their education so that the students can find different ways of using them. Email access can also be made available anytime and anywhere but this use requires more investment on the part of the school to have the appropriate infrastructure installed. If these options are realised, the handheld becomes a natural and immersed part of the students' learning. It creates ubiquitous computing with the possibility of the ubiquitous learning mentioned above.

In addition, teachers' IT and pedagogical skills with handheld computer usage need to be constantly updated. The former provides an understanding of the capacities and limitations of the handhelds necessary for teachers to design pedagogically sound tasks that foster learning by encouraging higher levels of thinking.

Finally, learning requires students to be able to externalize it via representations. In addition to the six requirements of ubiquitous learning: permanency, accessibility, immediacy, interactivity, situating of instructional activities and adaptability (Chen, 2002; Curtis et al., 2002; Ogata & Yano, 2003) a seventh component is needed - *externalizing learning* where learners are able to represent their learning to open it to transformation of understanding via different representations on handheld or other technology.

POSITIONING UBIQUITOUS COMPUTING IN RELATION TO UBIQUITOUS LEARNING

The handheld computer is just one element in a suite of technology that students may have access to. Good access enables issues that are being explored to be immediately and flexibly pursued. There is no single method of doing this, though use of each particular technology or application has features that need to be practised.

The figure below identifies the various components in a framework for ubiquitous learning and shows how ubiquitous computing forms one component within the framework although it is a vitally important mediator of experiences. The figure shows how handheld computing overlaps with the 'student' to offer one means of mediating between the wider technological network, the various members of the student's life communities and the wider world. The figure does not show all possible interactions to avoid clutter.

As outlined in the figure, each student is embedded in a series of face-to-face and virtual networks and communities. Students with handheld access have the capacity to engage with others through either means. How students do this depends on two elements: the technical options available, but more importantly the pedagogic framework that shapes relationships with other people and knowledge/experiences/values. Handhelds can be used to bring insights from the wider world into the classroom to be used in interactions with teacher(s) and peer(s). Whether students have access to this kind of resource reflects the social dimension of the pedagogic framework. How students and their peer(s) or teacher(s) engage with this knowledge reflects both the cognitive and the social dimensions of the framework.

Implicit in the diagram is that there are frequent opportunities for externalisation (presentation) of learning or thinking in a variety of forms and to a variety of audiences in an evolving series of interactions. It is the presence of these features (and particularly the understanding that presentations can be changed) that makes the ubiquitous technology part of a ubiquitous learning experience. But vital to the support of ubiquitous learning is

Figure 1. Framework for ubiquitous learning

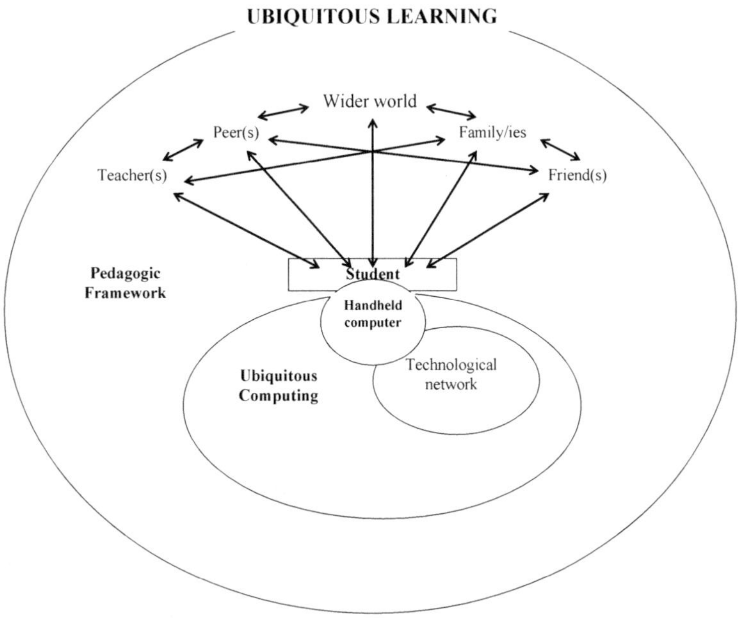

how the pedagogic framework makes central the questioning interaction of the student through the various forms of technology and with schooled and wider community/world.

CONCLUSION

The use of mobile technology in schools is about generational changes in pedagogy. Current students are characterized as products of the online world, who are 'technologically-tuned' and keen to explore new innovations such as SMS, ring tones and mobile phone icons. To these students, the Internet and mobile devices are an integral part of their world. However, as this study has demonstrated, the potentially powerful learning uses of technology are not an embedded part of the students experiences. They need to be modelled, practised and embedded in contexts where the learning benefits can be experienced. While a large percentage of school students, particularly the adolescents, have mobile phones, their educational usage in the classroom is currently limited. Laptop computers are less widespread. Handheld computers on the other hand have the advantages of mobility and capabilities to do many of the things computers are capable of doing. Thus, the achievement of ubiquitous learning requires thoughts and learning by both students and teachers. These devices offer the potential for ubiquitous learning in ways that are independent and self-paced for the learners, providing that there is thoughtful planning to support both personalization of the mobile devices and sustained interactions between mobile and stable technologies – providing the thoughtful and consistent integration with other devices and the active pursuit of good pedagogy are also present.

REFERENCES

Bruner, J. (1990). *Acts of meaning*. Cambridge, MA: Harvard University Press.

Chen, Y. S., Kao, T. C., Sheu, J. P., & Chiang, C. Y. (2002). A Mobile Scaffolding-Aid-Based Bird Watching Learning System. In *Proceedings of IEEE International Workshop on Wireless and Mobile Technologies in Education* (eds. U. Hoppe, M. Milrad, & Kinshuk), pp. 15-22. New York: IEEE Computer Society Press.

Curtis, M., Luchini, K., Bobrowsky, W., Quintana, C., & Soloway, E. (2002). Handheld Use in K-12: A Descriptive Account. In U. Hoppe, M. Milrad, & Kinshuk (Eds.), *Proceedings of IEEE International Workshop on Wireless and Mobile Technologies in Education* (pp. 23-30). New York: IEEE Computer Society Press.

Garland, V. (2006). Digital Literacy and the use of wireless portable computers, planners and cell phones for k-13 education. In Tan, L. W. H., & Subramaniam, R. (Eds.), *Handbook of Research on Literacy in Technology at the K-12 Level* (pp. 309–321). Hershey, USA: Idea Group Publishing.

Hwang, G. J., Tsai, C. C., & Yang, S. J. H. (2008). Criteria, strategies and research issues of context-aware ubiquitous learning. *Journal of Educational Technology & Society*, *11*(2), 81–91.

Keamy, K., & Nicholas, H. (2007). Personalised learning: can governments guarantee diversity for individuals? *The International Journal of Diversity in Organisations. Communities and Nations*, *7*, 137–146.

Kimber, K., & Wyatt-Smith, C. (2006). Using and creating knowledge with new technologies: a case for students as designers. *Learning, Media and Technology*, *31*, 19–34. doi:10.1080/17439880500515440

Maker, C. J. (1993). Creativity, intelligence, and problem solving: A definition and design for cross-cultural research and measurement related to giftedness. *Gifted Education International*, *9*(2), 68–77.

Maker, C. J., & Schiever, S. (1991). Enrichment and acceleration: An overview and new directions. In Davis, G., & Colangelo, N. (Eds.), *Handbook of Gifted Education* (pp. 99–110). Boston: Allyn & Bacon.

Naismith, L., Lonsdale, P., Vavoula, G., & Sharples, M. (2006). *Literature review in mobile technologies and learning*. Bristol, UK: Futurelab.

Ng, W., & Nicholas, H. (2007). *Pocket PCs for mGeneration learners*. Unpublished research report. Bundoora: La Trobe University.

Ng, W., & Nicholas, H. (2009). Introduction of pocket PCs in schools: attitudes and beliefs in the first year. *Computers & Education*, *52*(2), 470–480. doi:10.1016/j.compedu.2008.10.001

Nicholas, H., & Ng, W. (2009). Ubiquitous learning and handhelds: An overview of theory and pedagogy. In Rogers, P., Berg, G., Boettcher, J., Howard, J., Justice, C., & Schenk, K. (Eds.), *Encyclopedia of Distance Learning* (2nd ed., *Vol. IV*, pp. 2171–2176). Hershey, USA: IGI Global Publishers.

Ogata, H., & Yano, Y. (2004). Context-aware support for computer supported ubiquitous learning. In J. Roschelle, T. W. Chan, Kinshuk & S. Yang (Eds.), *Proceedings of the 2nd* IEEE (pp. 27-34). New York: IEEE Press.

Pask, G. (1976). *Conversation Theory: Applications in Education and Epistemology*. Amsterdam, New York: Elsevier.

Perry, D. (2003). *Handheld computers (PDAs) in schools*. Coventry: Becta.

Perry, D. (2006). *Learning2Go Wolverhampton Mobile Learning: Report - end of Phase II*. Retrieved March 15, 2009 from http://www.learning2go.org/media/download_gallery/dpa%20L2G%20Ph2%20FINAL%20Report.pdf

Piaget, J. (1972). *Psychology and epistemology: Towards a theory of knowledge*. London: Penguin University Books.

Roschelle, J., Patton, C., & Tatar, D. (2007). Designing networked handheld devices to enhance school learning. In *Advances in Computers, 70,* 1-60.

Schenker, J., Kratcoski, A., Lin, Y., Swan, K., & van 't Hooft, M. (2007). Researching ubiquity: ways to capture it all. In van 't Hooft, M., & Swan, M. (Eds.), *Ubiquitous computing in education: invisible technology, visible impact* (pp. 167–186). Mahwah, NJ: Lawrence Erlbaum Associates.

Sharples, M. (2003). Disruptive Devices: Mobile Technology for Conversational Learning. *International Journal of Continuing Engineering Education and Lifelong Learning*, *12*, 504–520. doi:10.1504/IJCEELL.2002.002148

Sharples, M. (2006). *Big issues in mobile learning: report of a workshop by the Kaleidoscope Network of Excellence Mobile Learning Initiative*. Nottingham: University of Nottingham.

Sharples, M., Taylor, J., & Vavoula, G. N. (2007). A Theory of Learning for the Mobile Age. In Andrews, R., & Haythornthwaite, C. (Eds.), *The SAGE Handbook of E-learning Research* (pp. 221–247). London: Sage.

Shin, N., Norris, C., & Soloway, E. (2007). Findings from early research on one-to-one handheld use in K-12 education. In van 't Hooft, M., & Swan, M. (Eds.), *Ubiquitous computing in education: invisible technology, visible impact* (pp. 19–39). Mahwah, NJ: Lawrence Erlbaum Associates.

Soloway, E., Norris, C., Blumenfeld, P., Fishman, B., Krajick, J., & Marx, R. (2001). Handheld Devices are Ready-at-Hand. *Communications of the ACM, 44*, 15–20. doi:10.1145/376134.376140

van 't Hooft, M. (2006). Tapping into digital literacy: Handheld computers in the k-12 classroom. In L.W.H. Tan & R. Subramaniam (Eds.), *Handbook of Research on Literacy in Technology at the K-12 Level* (eds.), 287-307. Hershey, PA: Idea Group Publishing.

Vygotsky, L. S. (1978). *Mind in society: The development of higher psychological processes*. Cambridge: Harvard University Press.

Waycott, J., Jones, A., & Scanlon, E. (2005). PDAs as lifelong learning tools: an activity theory based analysis. *Learning, Media and Technology, 30*, 107–130. doi:10.1080/17439880500093513

Weiser, M. (1991). The Computer for the Twenty-First Century. *Scientific American*, (September issue), 94–10. doi:10.1038/scientificamerican0991-94

Weiser, M. (1996, March). *Open house. ITP Review 2.0*. Retrieved March 15, 2008 from http://www.itp.tsoa.nyu.edu/~review.

Weiser, M., & Brown, J. S. (1996). The coming of age of calm technology. Retrieved March 15, 2008 from http://www.ubiq.com/hypertext/weiser/acmfuture2endnote.htm.

Weiser, M., Gold, R., & Brown, J. S. (1999). The origins of ubiquitous computing research at PARC in the late 1980s. *IBM Systems Journal, 38*, 693–696. doi:10.1147/sj.384.0693

Chapter 4
Beyond Mobile Learning:
Identity Construction and the Development of Social Awareness

André H. Caron
Université de Montréal, Canada

Letizia Caronia
Università di Bologna, Italy

Pascal Gagné
Université de Montréal, Canada

ABSTRACT

Contemporary research on mobile learning focuses mainly on issues such as the acquisition of knowledge, the development of cognitive skills and the efficiency of these tools with respect to the achievement of specific educational goals. Nonetheless, the consequences of the adoption of a technology within a learning context for educational purposes should not be reduced solely to the cognitive dimension implied in its use, nor should it be measured only in terms of goal achievement. Even if intended as purely educational tools, technologies are complex social objects that redefine the sense of the context, the activity and even the identity of the actors engaged in their use. When educational institutions adopt mobile information technologies they propose more than a supposedly efficient educational instrument or technology-formatted contents. They introduce a form of life. By form of life, we mean a repertoire of possible uses, actions, meanings and even intended actors that the users may adopt. A technology is then a condensed social context within which learning takes place. We might then ask, what kinds of learning are at stake? To grasp the richness and the complexity of the learning involved in using mobile information devices, we need a larger and holistic definition of learning that goes beyond simply acquiring knowledge on particular topics, or processing information for some formal educational purpose. Learning through mobile devices is a larger and complex process that involves different aspects of an individual's psychological, cultural and social development. How does the use of an iPod affect the students' identity? How does it contribute to the development of social skills and social awareness? Drawing on research involving 123 Canadian university students recruited from different disciplines (on the basis of data coming from diaries and focus groups), this chapter focuses on the multiple consequences of the introduction of this technology as an educational tool in students' academic life.

DOI: 10.4018/978-1-61692-849-0.ch004

INTRODUCTION

A number of authors see a variety of aspects related to *mobile learning*. Naismith, Lonsdale, Vavoula and Sharples (2004) identify six major approaches to mobile learning: behaviourism (reinforcement of data by several repetitions), constructivism (immersive experiences supported by the device), situated learning (learning environments), collaborative learning (interactions such as gathering and sharing information), coordination (access to resources) and informal/lifelong learning (everyday experience). Each pedagogical perspective represents technology in a different way. Hence, each orientation has different learning objectives presumed in their design of learning activities with mobile devices.

The underlying assumption by many is that this new generation often referred to as "*digital natives*" or "Y generation" will spontaneously and eagerly adopt an iPod or another type of mobile device for formal learning. However, many authors see this as to be partly unfounded and question these assumptions (Kennedy et al, 2006; Bachfischer et al, 2008; Caron Caronia, 2008). To some students, mobile learning devices are "just annoying", "invasive", "provides too much potential for distraction", may weaken "the social aspects of talking and interacting with others" and cause "disengagement from social situations" (Bachfisher et al, 2006: 47).

But undeniably, the major change inherent in the mobile shift is linked to the fusion of boundaries between the private (or personal) sphere and the public (or social) sphere. This phenomenon goes well beyond the simple transfer of traditional tasks specific to a domain in the time and space specific to the "Other". Much more radically, communication and mobile information technologies have transformed every public space into a stage where individual behavior becomes a social performance (Katz, & Aakhus, 2002; Caronia & Caron, 2004). The audience is undeniably there, and participates - willingly or not - in the unfolding of the play. Obligatory captives, they seem to be drawn into a performance which they did not necessarily decide to attend, and yet they constantly condition it. Whatever one might believe, individual behavior is, from the outset, designed to be performed on the public stage. It is constantly adjusted to the involuntary audience and keeps traces of this script-writing process for the benefit of a third party that is always included. Whether it involves using earpieces of a digital device as acoustic screens (Gumpert & Drucker, 2007) or removing them to indicate availability for interaction, the attitude of the actors reflects their awareness of the Other. Even by ignoring their presence, wanting to be positioned as a solitary individual or deliberately unconcerned from Others' standpoint is a behavior that requires a simulation effort.

In short, the use of new technologies is a communicative behavior in itself that can be compared with the enactment of oneself on the public stage (Goffman, 1967).

It is not surprising that the symbolic nature of the use of mobile technologies has been grasped by young users almost immediately following the introduction of the technologies. Although not all have the interest or expertise, most young people have been the best interpreters of this change and have very rapidly created behaviors, lifestyles and codes required to situate themselves in this mobile shift in contemporary society (Colombo, & Scifo, 2005; Kasesniemi & Rautianen, 2002; Ling, 1999; Lobet-Maris, 2003; Kasesniemi, 2003). Everything from their social ties, to the organization of their daily lives, the generation gap and cultural consumption is mediated by these devices, whose use is imperative to feel and be perceived as a member of a community of practices that shares "*moving cultures*" (Caron & Caronia, 2007). However, beyond the practical and utilitarian functions, beyond the advantages, the new possibilities of social networking, communications, academic learning, cultural sharing and community construction, the uses of technologies are a sensitive domain for young people. When it comes to owning and using these tools, identity issues ("Who am I and how do I want to appear?") fuse with social issues ("To which community do I belong?") and moral issues ("What are the social consequences of my individual behavior?"). The use of these technologies thus becomes a symbolic mechanism that constantly states the identity of the user, his or her multiple social attachments, integration in the context and sensitivity to the ethical dimension of group life. It may be on the symbolic front that this phenomenon is most interesting to examine learning environments.

When educational institutions promote 'mobile learning' they are not only proposing a technology, contents and new ways of learning, but also introducing ways and forms of life for the students.

Learning, as we know, is more than acquiring knowledge on some specific topic in a formal educational setting. Learning is a larger and complex process that involves different aspects of an individual's psychological, cultural and social development. We might then ask ourselves how do mobile learning devices such as *iPod* and *Mp3* players contribute to such a process in higher education contexts. How does the use of an iPod affect a student's identity? How does it contribute to the development of the student's social skills and social awareness?

A QUASI-EXPERIMENTAL RESEARCH: DESIGN AND METHODOLOGY

The present chapter reports on data collected in 2007 in a larger study undertaken by the *Centre for Research on Emerging Technologies (CITE)* and designed to acquire an understanding of students' technologically mediated environment and habits as well as their interests and opinions regarding mobile learning.

The study was intended to answer an apparently very simple and general question: What happens when "iPods"[1] are introduced into otherwise traditional university courses? It was intended as a description of practices, particularly related to a mobile learning innovation in an educational context. Focusing on innovation in educational settings implies introducing changes in the ordinary context in ways that are meaningful for the participants (i.e. teachers and students). By consequence, we adopted a quasi experimental approach. A quasi-experimental research design in education is oriented to explore what process or changes occur under which conditions or circumstances (Pontecorvo, 1991; Caronia, 2002). While an ethnographic research is mainly interested in describing "what happens in natural contexts", a quasi experimental one group posttest-only design is intended to describe "what may occur in a natural context if" (Cook, Campbell, 1979). It thereby introduces alterations in natural contexts while maximizing the ecological validity of the results (Cicourel, 1982). That is to say, it tries to maximize the probability that the phenomena observed during the research process are similar to what happens or would happen in the same context if analogous alterations were introduced.

To attain a high level of ecological validity we submitted and negotiated the research project with five university professors who were identified by the CEFES (Centre for Faculty Teaching and Learning) as particularly interested in technological innovations in education. They accepted to participate in the study and to adopt mobile digital devices as learning teaching tools in their courses.

The study involved 123 students of three different faculties at the *University of Montréal*—Architecture, Arts and Sciences and Pharmacy—attending courses of Industrial Design, Italian Language, English Language, Pharmacy and Communication. All the participants (professors and students) were provided with a mobile learning device (iPod by Apple or The *Zen Vision* by Creative and their accessories) for the entire research period that spanned a four months semester. Although more then half the participants already owned an MP3 type device, all were provided with the latest model that allowed for both audio and/or visual *podcasts* viewing. The time length of the study also allowed for the control for novelty factor. The research team created a specific and private website to allow the participants to prepare, download, share and exchange podcasts. The researchers also prepared instructional materials (both audiovisual and printed) to help participants format their iPods and Zens and to facilitate their social and educational use. Professors were asked to prepare podcasts in both audio and audio-visual mode according to their own pedagogical criteria and needs: content, format, length and their degree of consistency with respect to the course would be those considered appropriate by each teacher. Although the material was relevant and complementary to the other academic course material, the one common criterion shared by the professors involved, was that listening to the course podcasts was optional with respect to student evaluation. This optional nature of podcasts would free students from any formal obligation and allow them to decide whether or not they really wanted to use the academic podcasts. For further discussion of this issue see (Caronia, Caron 2009). There was one exception, an industrial design class that was mostly composed of weekly two-hour podcasts. Obviously, students needed to consume them but not necessarily on their "iPods".

Students were asked to use their mobile device in their everyday activities as they usually do (if they had owned one), to consider listening/viewing podcasts prepared by the teacher and to consider the possibility of constructing educational podcasts by themselves, to upload and share them with the teacher and classmates (through the common website), thus participating into a knowledge building process.

The purpose of the study was thus not to evaluate the academic value of the podcasts per se but their mobile consumption and the interest of students in freely participating in the construction of academic knowledge through podcasting. Professors participated in individual interviews both at the beginning and at the end of the study. The first interview was aimed at gathering some background information on their own expectations, perceptions, and approaches regarding mobile learning. The second interview was intended to gather the professors' evaluation of the pedagogical experience of using podcasts in their teaching.

The methodological tools used with the students were: a pre-experience survey, an online log book and a post-experience focus group. The online log book covered two one week periods: one at the mid-session and the second at the end of the semester. Therefore, it allowed us to see how young people's uses evolved over time. We gathered data on contents, locations and times of the "iPods" use in students' everyday life as well as in their academic life. In focus groups, students basically discussed and exchanged information about their everyday use of "iPods", podcasts and the mobile learning experience they had participated in.

Data from the survey and the logbooks were quantitatively analyzed. Discursive data from open-ended questions in logbooks and focus groups were qualitatively analyzed. Both quantitative and qualitative data were analyzed according to two distinctive (albeit strongly interrelated) dimensions:

- mobile information technologies in young people's everyday social life and peer culture;
- mobile information technologies in young people's academic life and school culture.

For the purposes of this chapter, and mostly bearing on discursive data, we will focus on the relationship between the uses of mobile learning devices in the development of a particular sense of the "Self". Further, we will investigate the particular role played by these devices in making students reflect on the issues of sharing, etiquette and performance according to the moral aspects of their social life.

DISCOURSE AS DATA

At the heart of our research, we have seen that beyond and within the uses of technologies, it is the discourse *on* technologies that best illustrates the symbolic value that students confer on the usages of technologies. To update these non-functional dimensions of mobile culture among young university students, we therefore analyzed their discourse on the iPod. The choice of discourse as a primary data source arises from its modeling character (Wetherell, Taylor & Yates, 2001). Through discourse on a given subject, the actors are situated and defined relative to what they talk about and the interlocutor. It is indeed this process of "self staging" that allows us to trace the backdrop of the cultural and theoretical models and points of view of the subjects, based on discursive occurrences that are more concrete and idiosyncratic.

As soon as the use of mobile technologies in public space becomes a "subject of discourse," the elementary categories of identity, social belonging and ethical thought surface and structure the discourse. The rationality of the technology relative to its goals, functionality and efficacy are criteria subject to an order of reflection that concerns mainly and above all "Who am I?" "What community am I part of?" "What is right? What is wrong? And why?" Young people constantly define and redefine these categories, and analyze the conditions of their application.

Obviously, before addressing those questions, we wished to verify if educational podcasts were being listened to in the first place, and in what context. Findings indicated that few educational podcasts were listened too - only ten minutes on average a day out of close to three hours, and almost always on computers[2]. Students felt they needed an adequate setting to concentrate and listen to them; therefore, there was little *"mobile" consumption*. Similar results were found in other studies (Deal, 2007; Huntsberger & Stavitsky, 2007; Lane, 2006; Malan 2007). Even if apparently unused mobility wise for traditional knowledge acquisition, mobile learning devices such as iPods appeared to challenge students in their everyday cultural practices in an academic setting.

TO BE OR NOT TO BE AN IPODER: DEFINING ONE'S SELF THROUGH MOBILE LEARNING DEVICES

In modern society, where people undeniably place a lot of importance on image and physical appearance, they are usually very quick to judge others, and first impressions are always decisive. New types of interactions become established, new forms of behavior develop, and new prejudices appear. Earpiece:

Well I wouldn't necessarily want to join the iPod clique. It might seem a bit ridiculous, but you know an iPod says 'I'm young and stylish', and there are iPod battles. (Student in Design)

I would say that — not a long time ago —, I was thinking just like you; now that I'm in the gang, I find it really cool to have an iPod. Rather than being an 'has been' with a fake one, I am a 'has been' with a real one. (Student in Design)

White earpieces, you know immediately that it is someone who has one [an iPod]. But, if I had the choice between showing it or not, I think I would prefer not. (Student in Communication)

I almost forgot, but you know this session I had a class in Philosophy in the other main building... and you know philosophy students are a little this type... you know brown socks, woolen sweater and notebook ...So when I came to class with my iPod, it's kind of stylish... it would really be strange to see a philosophy professor with a really great PowerPoint. You know often they only use chalk and the chalkboard so I felt a little strange to be taking out my iPod in a philosophy class. It seemed a little antagonistic with what they do and what they think... their philosophy. (Student in Design)

I had a few prejudices at first about iPod and the only iPod attitude. When I saw that (Zen), I was a little disappointed not to have an iPod, but in the end I like that one way better and if I were to buy one, I'd buy a Zen... (Student in Italian language)

The fact of owning a mobile learning device can also generate a feeling of belonging to a specific group. By having such a device, the participants realized that they were part of a large community of users, and discussed their experiences with their peers.

Belonging to the group can seem like an exclusive experience; some students try especially hard to maintain this image, particularly with respect to the community of Apple users.

Me, I was sold on Apple, I have the computer and everything... I know that's what I'm going to buy first and nothing else... for me, it's important. I don't know why... in my family, we're all like that. (Student in English language)

Some prestige accompanies the use of the latest technology. Some brands of mobile learning devices, as we have seen, are objects of desire in stylish communities. They are fashion accessories. Like cell phones, there are many models of players and they come in countless colors. Students choose them not only for their technical capabilities but also in relation to their personality and according to the most crucial drive in young people's attitude: "coolness". For many the mobile device has become an extension of their identity, culture and membership.

THE WALTZ OF CONTENT: SHARING AND REINFORCING FORMS OF LIFE

Because it contains the user's cultural landmarks and archives, because it is "a little bit of your life baggage," not only is the iPod itself an identity symbol, but its content also makes it a depository of identity icons whose access is subject to a very selective social filter.

As noted in our previous analysis (Caron & Caronia 2008; Caronia & Caron 2009) "mobile" consumption of academic podcasts did not appear very popular overall, as students preferred to view these podcasts in more traditional environments. For the most part, academic podcasts were consumed on their computers at home. Students considered that their computers provide greater comfort for watching and listening to the content available.

Nonetheless they underlined the importance of the mobile devices in sharing and interacting with their fellow students; for example, sharing recordings of classes and academic podcasts. More fulfilling interactions also occurred in the course of group work, when every member of the group was able to consult the parts of the work done by other team members without having to have a computer. Others recorded courses and gave them to friends who had been unable to attend class.

Well I found it useful; you know sometimes when I recorded classes for my friends who hadn't come to class. And I sent them the file afterwards. They could just listen to it and then take notes like that, instead of borrowing written notes. (Student in Pharmacy)

Students were delighted with being able to record; it often became a reason or even a pretext for interaction.

There are no longer specific spaces or places designated for media consumption. Students combine a number of activities while traveling in public transport or by car, or even on foot. They carry their schoolwork home and their personal content to school, and thus redefine the notion of being "there," by no longer distinguishing an activity from where it is performed.

I showed my pictures from my holiday in Punta Cana to my friends at school and my friend used the player to listen to music in a café while we were studying [...] I used the Zen to save some work I'd done at school and be able to work on it at home. (Student in Communication)

I went on the site to download my professor's podcast, but I couldn't enter my username and password correctly. Anyway, when I talked about it with a friend, she showed me what I was doing wrong! I'm going to try again tomorrow! (Student in English language)

Knowledge exchange practices and sharing are important dimensions of the use of mobile learning device both for academic purposes and leisure. New users also need expert users to help them understand the device's functionalities and configurations.

The use of the mobile learning device becomes a catalyst for social interactions and expectations. Sharing permits socialization and the creation and strengthening of social networks. Use of the player thus creates new communication times and spaces, along with special social practices and interactions.

THE WALTZ OF EARPIECES: SHARING AND CONSTRUCTING SOCIAL TIMES

The intimate link between the use of technology and the relation to the Other generates a subtle coding of social significance of linked behaviors and consequently the need for rules of courtesy. Putting on your earpieces, removing one or two, sharing the content or the device become letters of a social alphabet through which young people state and constantly weave their social links and their perception of the Other.

Well I think it's the contrary, sometimes I am more likely to share an earpiece when I am with someone. [...] but it's true if you are with someone you know for a long stretch, listening with both earpieces on without having any conversations it could create a little 'malaise,' I imagine. (Student in Communication)

...maybe in my case I don't find this as depressing to have all the people around me listening to their iPods because it has happened to me to do this but at least I was sharing an earpiece with my friend and we were really having fun listening to this comedian. (Student in Communication)

Lending an earpiece to someone and thus sharing the content is a sign of a social tie, just as keeping *both* earpieces on while being with someone one knows is a negation of the link, and thus a behavior that can generate social malaise. The codification of the social significance of sharing also extends to the lending of the device itself. Sharing of an iPod as such is perceived as a gesture of inclusion in the recipient's "own world." As the excerpts below show, the loan of a mobile learning device, almost more than a computer or a cell phone, signifies and certifies the nature of the social relationship.

To get back to privacy...sharing...everything like cell phones, computers, laptops... if someone asks me for my cell phone, no problem, I lend it... on the other hand, if he asked to borrow my MP3... I'd be much more reluctant... it sounds stupid but... I think not... Because the computer it doesn't matter, he might need it, but... for my MP3 it's no. It's my world... and not everyone has access to my world. (Student in English language)

At the other extreme, sharing of the device between a couple illustrates the intimate, even "fusional", nature of the social link.

In my case it was more a question of sharing the device itself. As for the earpieces, I would certainly share them, but before I used to borrow the iPod from my girlfriend, but now because this device also has a radio function and other stuff, it was now her who wanted to borrow my device to listen to her programs. (Student in Communication)

As these excerpts demonstrate, this diminutive device is infused with meaning. Its usages largely exceed utilitarian functions to reach symbolic significance that concern identity and relations. Given this *symbolic loading*, the codification of gestures remains subtle and sophisticated because it takes into account the different Others that may be on scene.

It's true but it really depends with whom I am talking and if it's just a semi-conversation with someone that I don't know too well or if it is with a really good friend, I will take them both off and I will talk with him properly. And that's the way it is... if I see my music is going to be disruptive I will shut it off. But I am also able to talk with music in the background and it's really nice. (Student in Pharmacy)

Yeah but it depends who it is... if it's with a friend, she knows I can listen to her while listening to my music but if it's with a Prof. I will take both earpieces off. (Student in Italian language)

It's like hierarchical, it looks like, I am walking along and I have my earpieces on and someone comes to talk to me, if I know I won't have a long conversation I will only take one off: 'Hi! Chit-chat.' If I know it will take longer, I will take both off, press on pause and stop my music. (Student in Communication)

You go to class; you don't feel great. You don't want to talk with the people around: you put your earpieces on, even if you don't put any sound on, and you wait for class to start... (Student in Communication)

The positioning of the earpieces is a social practice that simultaneously reflects and constructs the type of imminent interaction and the type of relationship. Thus, when an individual is with someone he/she does not know well or is having a conversation that is expected to be short, removing only one earpiece can indicate partial availability that more closely resembles a social obligation than a demonstration of authentic interest. However, when the other person is a close friend, the same gesture is not necessarily perceived as a lack of courtesy or availability: it certifies the friendly nature of the interaction and the informal nature of the relationship. Similarly, removing both earpieces can either indicate an authentic interest in someone who is close or simply formal respect. The codification of gestures therefore allows for a strategic and strictly communicational usage: by removing one, two or no earpieces, you express not only who the Other is but also the relationship between the user and this Other and in what type of interaction the user is ready to engage.

CODIFICATION OF GESTURES IN RULES OF COURTESY

Like any behavior charged with social significance, the use of a mobile learning device prompts questioning on the conditions of polite usage. One can therefore envision the usages that reflect in their practical unfolding the orientation of the user toward the Other.

The entry of a new technology in the everyday lives of students has an effect on their immediate environment and especially their social relations. The majority of students said that a certain code of conduct had to be adopted in the new social environment.

It might bug someone nearby because you're listening louder than you normally should, for example. And then the people around you are bothered or... me, that bugs me... You know, I'd like it a little quieter. Of course in class it's even more irritating when you're trying to follow or... At the library it happens sometimes. (Student in Pharmacy)

In their analyses of good manners for mobile learning device users, young people also go beyond the most elementary levels such as lowering the volume so as not to disturb the audience. In fact, they reflect on and deliberately choose to display non-verbal cues that convey the image that the Other is valued, or which, conversely, let the Other know he/she is a negligible player on the social scene.

One good example is when I work at the pharmacy and someone shows up with both earpieces on that really cuts off our communication. (Student in Pharmacy)

When you are talking to someone and he only has one earpiece on and listening to something and he's only listening to you with one ear ...I really feel like telling him listen to what I have to tell you and afterwards you can listen to your music! (Student in Pharmacy)

I take them both off [the earpieces]. I think that out of courtesy, when you meet someone, I don't know, you should be 100% with that person. And I don't know, I take them both off and I'm a little like Mathew, if the person I'm facing only takes one off, well in the end... it bugs me a little somehow. (Student in Pharmacy)

Well anyplace where there is a discussion, at a family supper I would not put my earpieces on.....I think it is not appropriate in a discussion, during a conference or in class. (Student in Pharmacy)

You enter class, you don't feel right, you're not tempted to talk with people and you put your earphones waiting for your course to start, and then... (Student in Communication)

I say that: when I'm in the subway and that I have my earphones, I want to be quiet. I am in my bubble. (Student in Design)

I train myself a lot at the gym, (...) and I don't want people to talk to me because I'm in my bubble. It [the iPod] allows that. (Student in English language)

Sometimes I put on my Zen and I don't even have music but they see my earphones and they don't talk to me. (Student in Pharmacy)

A *social filter* par excellence, the mobile learning device can put in perspective the importance of the Other to the extent of erasing their social presence. Consequently, the use of the mobile learning device in public space requires standards whose respect minimizes the occurrence of the social offence, ranging up to completely negating the threat. Thus, display of non-verbal cues certifies the presence of the Other by underlining his/her importance and by conferring value on them.

It's the same for me for example when in the subway I almost always meet one of my friends that is going the same route as I am. When we meet systematically it's funny we both take off our earpieces, put away our mobile devices it's always the same ritual. Then we talk till we get home. It's systematic every time... (Student in Italian language)

If I meet someone I know I automatically take both off I don't know if it's even polite to keep one on. Sometimes you go to a party; there is someone who is sitting alone on a sofa with an IPod on even though there is plenty of music around. Why? You go to a party to socialize then why do you bring something that is going to isolate you. (Student in Communication)

Removing both earpieces and even stowing the mobile learning device thus becomes a polite gesture of a subject oriented to protect the "face of the Other" and sensitive to the know-how that governs communal life. However, in this case as in many others, compliance with norms of politeness is not only a question of preventing the Other from loosing face. It is also or mainly a question of protecting the face of the actor. The lack of politeness or know-how threatens the use of technologies because it reveals their own incompetence at "being in society/socializing."

ETHICAL AWARENESS

Given their mobile nature, cell phones and iPods exemplify the profoundly social dimension of public spaces. Young people appear to have understood this quite well. Their use of these technologies is constantly evaluated in a subtle critical sense that takes into account a variety of nuances, of constraints and obligations that typify a social scene. This perception of the social scene where technologies are used by young people should not surprise us given it is so closely associated with the development of their identity and social persona from the perspective of the Other. As critical evaluators of their social appearances among others, young people see their use of technologies as a social performance where the stakes are profoundly symbolic. Young people must therefore find the difficult but right balance between the personal reasons and the collective reasons of their codes. They perceive and explore the inter-subjective dimension of social life. The taking into account of this inter-subjective perspective is the basis to all ethical reason. It is fascinating to discover that beyond the new practices proposed by mobile communication and information technologies in young peoples' life, it also appears to trigger a "social thought" oriented by a new conscious of the "Other".

Consequently, our thinking on the ethical dimension of the use of technologies in public spaces brings us to reflecting on codes that must regulate social performance. This link between moral preoccupations and courtesy codes is not banal. In fact, the codes of courtesy or politeness become a pragmatic answer to most everyday micro-ethical issues. Whether it is for simple good manners or for codified protocols, courtesy codes reveal themselves to be "shock absorbers": they prevent and absorb a great many offences that can deter an individual's action. As stated by Goffman, it is in some ways practicing a situational ethic (Goffman 1981: 132). When a new technology appears in everyday life, it raises, among other things, issues of politeness for which the habitus of a given community can only provide partial solutions. With the multiplication of actors that mobile devices bring about simultaneously and the confusion of the private/public spheres, traditional politeness codes can only be modestly appropriate. The habitus itself needs updating. In this case, as in many others, it is praxis that creates not only the norms, but also and mostly the social significance of new behaviours. Young people participate actively to the elaboration of these new social codes where each act, performed or not, acquire a social significance that leads to an evaluation.

CONCLUSION

The growing adoption of mobile information technologies by educational institutions as well as the increasing number of programs on distance education and e-learning makes in depth investigation of the consequences of these phenomena paramount. Contemporary research on mobile learning often focuses mainly on issues like the acquisition of knowledge, the development of cognitive skills, students' motivation to participate in teaching/learning processes, and the efficiency of these tools with respect to the achievement of particular educational goals.

Our research has shown that the consequences of the adoption, for educational purposes, of a technology within a learning context cannot be reduced solely to the cognitive dimension implied in its use, nor should it be measured only in terms of goal achievement. Even if conceived as purely educational tools, technologies are complex social objects that redefine the sense of the context, the activity and even the individual and social identity of the actors engaged in their use.

How does this process affect formal learning? The answer to this question strictly depends on two frames of reference:

- The definition of learning per se;
- How we consider the relationship between formal education and the larger developmental process of becoming a competent member of one's cultural community.

To grasp the richness and the complexity of the learning involved in using mobile information devices, we require a larger and holistic definition of learning that goes beyond acquiring knowledge on some particular topic or processing information for formal educational purposes. Learning is a complex process that involves an individual's psychological, cultural, social and cognitive development. Knowledge acquisition and skills development are part of this larger socialization process. In keeping with this holistic definition of the learning process, we may argue that activities mediated by mobile devices *are learning activities* because they affect the development of students' identities, sense of belonging (or not belonging) to a community of peers, and the construction of social awareness and social skills with respect to the use of mobile information in collective spaces.

Even if this chapter did not specifically focus on the relationship between the socio-cultural dimension and the instructional dimension of mobile learning, it nevertheless sheds light on how these two dimensions are strongly interwoven.[3]

Whatever the cognitive activity involved in the use of mobile learning device (i.e. acquiring knowledge, processing information condensed in podcasts, achieving particular educational goals) it does not take place in a socio-cultural vacuum. Rather, it depends on and gives rise to some images of the self (as for instance being more or less "cool" or "belonging to the community of iPod users"), and social interactions with peers, experts or novices that nurture the use of mobile devices for formal learning purposes.

Our findings concur with Traxler's (2008) observations that: "mobile learning should perhaps be defined most generally, durably and powerfully as learning aligned to these societies and cultures in motion (rather than, for example, being defined as learning delivered by these devices or to learners whilst in motion)" (2008, 236) This alignment however still needs to be mutually constructed before becoming another part of young people's "mobile culture".

When an educational institution adopts mobile information technologies it introduces a form of life, which is a repertoire of possible uses, actions, meanings, status and ways of participation, and even identities the users may adopt. In this sense, mobile learning is a highly cultural innovation whose stakes go far beyond simply learning curricular topics.

REFERENCES

Bachfischer, A., Lawrence, E., & Litchfield, A. Dyson, L. E., & Rabran, R. (2008). Student perspective about using mobile devices in their studies. *2008 IADIS International Conference on Mobile Communication Proceedings* (pp. 287-295), Algarve, Portugal: ACM.

Caron, A. H., & Caronia, L. (2007). *Moving cultures. Mobile communication in everyday life*. Montreal: McGill-Queens University Press.

Caron, A. H., & Caronia, L. (2008). The cultural dimensions of "IPODS" in higher education *IADIS International Journal on WWW/Internet, 6*(2), 73-89.

Caronia, L. (2002). *La socializzazione ai media. Contesti, interazioni e pratiche educative*. Milano, Italy: Guerini.

Caronia, L., & Caron, A. H. (2004). Constructing a specific culture: Young people's use of the mobile phone as a social performance. *Convergence (London), 10*(2), 28–61. doi:10.1177/135485650401000204

Caronia, L., & Caron, A. H. (2009). Mobile learning in the digital age: a clash of cultures. In Kleinman, S. (Ed.), *The culture of efficiency: technology in everyday life* (pp. 191–211). New York: Peter Lang.

Cicourel, A. (1982). Interviews, surveys, and the problem of ecological validity. *The American Sociologist, 17*, 11–20.

Colombo, F., & Scifo, B. (2005). Social shaping of the new mobile devices. Representations and uses among Italian Youth. In Haddon, L., Mante, E., Sapio, B., Kommonen, K.-H., Fortunati, L., & Kant, A. (Eds.), *Everyday innovators. Researching the role of users in shaping ICTs* (pp. 86–103). London: Springer.

Cook, T. D., & Campbell, D. T. (1979). *Quasi-experimental design and analysis issues for field settings*. Boston: Houghton Mifflin Co.

Deal, A. (2007). A teaching with technology white paper, *Educause Quaterly*. Retrieved September 6, 2007, from: http://connect.educause.edu/files/CMU_Podcasting_Jun07.pdf

Goffman, E. (1967). On facework: an analysis of ritual elements in social interaction. In Jaworski, A., & Coupland, N. (Eds.), *The Discourse Reader* (pp. 306–321). London: Routledge.

Goffman, E. (1981). *Forms of talk*. Philadelphia: University of Pennsylvania.

Gumpert, G., & Drucker, J. (2007). Mobile Communication in the twenty-first century or "Everybody, everywhere, at any time". In Kleinman, S. (Ed.), *Displacing place. Mobile communication in the twenty-first century* (pp. 7–37). New York: Peter Lang.

Huntsberger, M., & Statvistky, A. (2007). The New "podagogy": Incorporating podcasting into journalism education. *Journalism & Mass Communication Educator, 64*(4), 397–410.

Kasesniemi, E., & Rautiainen, P. (2002). Mobile culture of children and teenagers in Finland. In Katz, J. E., & Aakhus, M. A. (Eds.), *Perpetual contact. Mobile communication, private talk, public performance* (pp. 170–192). Cambridge, UK: Cambridge University Press. doi:10.1017/CBO9780511489471.014

Kasesniemi, E.-L. (2003). *Mobile messages: Young people and a new communication culture*. Massachusetts: Tampere University Press.

Katz, J. E., Lever, K. M., & Chen, Y. (2008). Mobile music as environmental control and prosocial entertainment. In Katz, J. E. (Ed.), *Handbook of Mobile Communication Studies* (pp. 367–376). MIT Press.

Kennedy, G., Krause, K.-L., Judd, T., Churward, A., & Gray, K. (2006). First year students' experience with technology: Are they really digital natives? *Australian Journal of Educational Technology, 24*(1), 108–122.

Lane, C. (2008). *UW Podcasting: Evaluation of year one*. Washington, DC: Catalyst.

Ling, R. (1999). « *C'est bien d'être joignable!* »: L'usage du téléphone fixe et mobile chez les jeunes norvégiens. *Reseaux*, (92-93): 261–291.

Lobet-Maris, C. (2003). Mobile phone tribes. Youth and social identity. In Fortunati, L., Katz, J., & Riccini, R. (Eds.), *Mediating the Human Body: Technology, Communication and Fashion* (pp. 87–92). Mahwah: Lawrence Erlbaum Associates.

Malan, D. (2007). *Podcasting computer science E-1. 38th SIGCSE Technical Symphosium on Computer Science Education* (pp. 389–393). Cambridge, Massachusetts: Harvard University.

Naismith, L., Lonsdale, P., Vavoula, G., & Sharples, M. (2004). *Literature Review in Mobile Technologies and Learning*. Birmingham UK: Futurelab & University of Birmingham. Retrieve May 1, 2009, from www.futurelab.org.uk/research/lit_reviews.htm

Pontecorvo, C. (1991). *Introduzione. La scuola come contesto sociale naturale. Discutendo si impara. Interazione sociale e conoscenza a scuola*. Berlin, Germany: NIS.

Traxler, J. (2008). Mobile learning: Moving through philosophical space. [Algarve] [Portugal] [ACM.]. *Proceedings of IADIS International Conference Mobile Learning, 2008*, 235–239.

Wetherell, M., Taylor, S., & Yates, S. J. (2001). *Discourse theory and practice*. A Reader London: SAGE.

ENDNOTES

1 "IPod" is the commercial label of a specific digital audio visual player created by Apple. Given the market share of this technology, its commercial name has become the common reference for this device. Most participants in our research used "iPod" as a general label to designate their digital audio visual players although when asked to comment on the brands they expressed their views on specific brands whether it was an Apple model or Zen Creative. For the purpose of this chapter we however generally use the label "iPod" to refer to all digital audiovisual players.

2 A complete discussion of the quantitative data collected in this study may be found in another of our publications (Caron, Caronia 2008).

3 A full discussion of the "clash of cultures" implied in the use of ipods for academic uses may be found in Caronia & Caron 2009.

Chapter 5
Activity in a Mobile Learning Environment:
Ubiquitous Personalized Learning Using Context and Social Presence Awareness

Ray M. Kekwaletswe
Tshwane University of Technology, South Africa

ABSTRACT

The practical contribution of the chapter is the understanding of activity in mobile learning environments and how learners use awareness to model their actions for the provision of personalized learning support. The chapter is about the advancement of the human-centric approach to personalized learning through enhanced learner-to-learner interaction – where context and social presence awareness is of vital significance to how learners decide and act on a learning task. It is an expedition towards understanding the phenomenon of mobile learning, where personalized learning and support is a result of social awareness activities of learners as they traverse varied learning contexts. Mobile learning, in this chapter, is signified by mobility of learners regardless of mobile technologies. Activity Theory, which draws attention to mediated activity within a social context, is used to explore how mobile learners use context and social presence awareness to facilitate their ubiquitous social interactions.

INTRODUCTION

The chapter posits that learning and personalized support decisions and actions are influenced by awareness, as mobile learners traverse varied learning contexts. Thus, the chapter shows that learners' awareness of a social presence and context facilitates ubiquitous personalized support and on-demand sharing of knowledge. Knowledge is not a fixed commodity, but a function of our interactions with external resources including tools, media, and other humans (Ryder & Wilson, 1997). In this chapter, 'other humans' is inferred to mean learners or peers who share what they know. This suggests that human knowledge transforms as people socially interact with others and the surrounding environment. Consequently, the chapter is premised on the notion that knowledge is created and transferred through the dynamic interactions among individuals and between individuals and

DOI: 10.4018/978-1-61692-849-0.ch005

their environments (Nonaka, 1994). Thus, mobile learning and sharing of knowledge are social and sensitive to context.

It is inferred from the notion of knowledge creation, sharing or transfer that learning and knowledge can be perceived to transform when context and social presence awareness interact. In this regard, context and presence awareness influence the interaction and the learning problems that could be solved and how they are solved. Although a great number of studies (e.g., Shariq, 1999; Polanyi, 1966; 1958) have shown that knowledge creation and transfer is essentially a human-to-human process or an outcome of social interaction (Nonaka, 1994), the relationships or roles of context and social presence awareness as catalysts for learning and knowledge sharing in a mobile learning environment has not been explored. This chapter contributes to that effect.

The chapter contributes by exploring and understanding how a mobile learner uses awareness to model his action and leverage personalized learning. It reveals the actual nature of ubiquitous mobile learning through social interaction where awareness of context and social presence is argued to be the underlying process of the activity. Essentially, this chapter is on how personalized learning support and knowledge sharing in a mobile learning environment is an outcome of social interaction coordinated by social awareness. Social awareness is synonymous with context and social presence awareness.

Context Awareness

Nardi (1997) defines context as what takes place in an activity system composed of object, actions, and operation. "Context is constituted through the enactment of an activity involving people and artefacts. Context is internal to people – involving specific objects and goals; and external to people– involving artefacts, other people, and specific settings" (op. cit., p38). To this point, context in this chapter is understood as the situation in which a learner or a group of learners find themselves. Subsequently, context is seen as that which can be used to characterize the situation of a learner. The chapter looks at context awareness where context is defined to include: learner's location and environment, learner as the participant (including socio-cultural background), plus social interaction as well as activities. The "environment" category contains context factors that are outside of the control of a learner, such as the weather. The "participant" category includes the status (mental and physical) of the learner or other learners. The "activities" category covers learner, participant, and environmental activities. "Interactions" deal with those characteristics that pertain to relationships between individual learners, their activities, and the environment.

Social Presence

Short, Williams and Christie (1976) asserted that different communication media express varying degrees of social presence based on their ability to transmit nonverbal and vocal information. Thus, they initially introduced the Social Presence Theory as "technical social presence," defining it as the capacity of the medium itself to present the "salience of the other person in interpersonal interaction" (p65). The claim by Short et al. (1976) that the quality of the communication media determines its social presence or richness was disputed by Ngwenyama and Lee (1997) who showed that the communication richness of a media is dependent on who uses the media and how they use it. In this chapter, social presence is re-defined and understood to be the mediated presence or the face-to-face presence of another learner who could provide personalized on-demand social support for a learning problem as the learner traverses varied learning contexts.

Social Awareness

Awareness is an understanding of the activities of others, which provides a context for your own activity (Dourish and Bly, 1992). Christiansen and Maglaughlin (2003) identified and grouped four types of awareness as follows: workplace awareness – which is knowledge of tasks within the virtual environment; availability awareness– which relates to the availability of people and objects; group awareness – which promotes the feeling of belonging to a group; and contextual awareness – which includes the physical, social, and mental contexts. Awareness, by the foregoing definitions, is both a perception of the users of a system, and an aspect of a system that facilitates that perception (Rettie, 2003). Numerous different definitions of social awareness have been put forward in the literature (e.g., Prinz, 1999; Tollmar et al., 1996; Gutwin et al., 1996). In general, however, social awareness is defined as awareness about the social situation of other people. That is, what they are doing, whether they are engaged in an activity and could be disturbed, who is around and what is going on. In a mobile learning environment, social awareness accommodates awareness of particular social situations at varied and specific learning moments. Dourish and Bellotti (1992) define social awareness as an understanding of the activities of others, which provides a context for your own activity.

Mobile learners use awareness of context and social presence as a means to access ubiquitous learning support, interpret and adjust their knowledge – sharing what they know with others through social interaction. The chapter, thus, focuses on the peer-to-peer interaction and the learning environment. The social interaction whose outcome is learning and provision of support is location and time independent. And since sharing learning experiences is a ubiquitous phenomenon, learners continuously use awareness of the environment and awareness of available social resources they can draw upon to facilitate knowledge consultation (Kekwaletswe, 2007; 2009).

Understanding how learner action in a mobile learning environment is a consequence of an awareness of context and social presence is significant. This awareness of context and social presence happens in the mind of the individual learner. The research argument and the subsequent objective of the chapter is motivated by the following three challenges for a mobile learning environment:

1. Learning activity that happens when a learner is mobile is considerably more situated than learning activity that happens in formal contexts.
2. Learning activity in informal settings renders familiar formal contextual awareness less practical. Context relies on environmental situations and availability of resources which in the case of mobile learning are not always consistent.
3. Unlike learning activity in a formal context, much of the learning that happens while mobile is often through support from socially present peers or class-mates.

With the three challenges facing a mobile learner – who needs ubiquitous personalized learning support – there is a need to understand how a learner uses social awareness for the purpose of social interaction whose outcome is learning. To this end, the objective of the chapter is to contribute by giving an informed insight of how learners act and transform knowledge in a mobile learning environment. The chapter, using the contextual inquiry research method and Activity theory, explores the possibilities awareness, in ways that are sensitive to the background and context of a learner, provides proximity of available social resources regardless of location of the learner and task at hand. This chapter tackles the ways that awareness of the environment and knowledgeable peers could be leveraged to enrich the learning and knowledge experience.

Background

Mobile learning research is still in its infancy, hence there are ongoing debates as to what really signifies mobile learning. To this point, mobile learning researchers often converge in conferences and workshops to grapple with mobile learning issues, for example, "Big Issues in Mobile Learning Workshop" (Sharples, 2006). The Workshop aimed to provide a forum for reflecting on the issues behind the growth in mobile learning and for identifying the big challenges facing mobile learning. One issue that became apparent is that mobile learning is not just about learning about using portable devices, but learning across contexts. Mobile learning is not something that people do; learning is what people do. With technology getting smaller, more personal, ubiquitous, and powerful, it is better able to support a mobile society (*op.cit.*).

In earlier research, the mobile learning concept was strongly linked to the device (e.g., Sharples et al., 2002) and the potential for enabling lifelong learning (e.g., Sharples, 2000). During the last few years, researchers were beginning to also view mobile learning as being about learners on the move rather than the device, with the focus therefore shifting to the mobility of the learner. This led to considering mobile learning from the learner's point of view, and to the definition that mobile learning is "any sort of learning that happens when the learner is not at a fixed, predetermined location or learning that happens when the learner takes advantage of learning opportunities offered by mobile technologies" (O'Malley et al., 2003).

At the Big Issues in Mobile Learning Workshop (Sharples, 2006) there was general agreement that a precise definition of mobile learning is unattainable. Instead, Winters (2006) and other workshop participants agreed on the following key characteristics of mobile learning: i) Enables knowledge building by learners in different contexts, ii) Enables learners to construct understandings, iii) Mobile technology often changes the pattern of learning/work activity, iv) The context of mobile learning is about more than time and space.

Again, discussions on the "Big Issues in Mobile Learning" (Sharples, 2006) allude to the on going debate on what really signifies mobile learning. There are generally three schools of thoughts on what signifies mobile learning: Learning mediated by mobile devices, Mobility of learners (regardless of their devices), and Mobility of content/resources in the sense that it can be accessed from anywhere (Taylor, 2006).

Mobile Learning Defined

Mobile learning is proving to be a multifaceted phenomenon; hence, there is still a debate as to what constitutes mobile learning and a clear definition of it is lacking. This chapter embraces the view point of a learner being mobile, thus, mobile learning is signified by mobility of learners regardless of their mobile devices. In the chapter, mobile learning is defined as any sort of learning and knowledge sharing that happens due to social awareness when the learner is not at a fixed, predetermined location but in varied learning contexts. The learner may or may not take advantage of learning opportunities offered by the mobile technologies.

Mobile learning conveys ubiquitous learning that is not confined to specific locations, and is time independent hence mobile learning still happens during face-to-face social interaction. Ubiquitous personalized support refers to the provision of context sensitive and anywhere, anytime support as mobile learners traverse varied locations. There are generally two types of universities: distance learning universities (where teaching does not happen in a classroom or no face to face contact) and contact universities (where teaching mostly happens in a classroom or laboratory - learners have face to face contact with the instructor). This chapter focuses on learning at a contact university. Learners in contact universities come from varied social backgrounds, with diverse languages and cultures. Hence, one of the prevailing educational challenges is that of providing personalized academic support to under-prepared learners (Jaffer et al, 2006). To this point, awareness of the social environment and social resources is fundamental to the provision of personalized academic support to a mobile learner. Learning is made easy when a learner has consistent awareness of context and presence of social resources (Kekwaletswe, 2009; 2007).

Context aware ubiquitous learning is meant to support learning by identifying a learner's surrounding contextual environment and social presence to provide a rounded and seamless learning experiences. In a contact university, there tend to be disproportionate access to available social resources between when learners are attending scheduled or formal classes and when they are away from scheduled classes (Kekwaletswe and Ng'ambi, 2006) especially as they move away from campus locations. It is therefore useful to distinguish between the learning contexts in which the learner's knowledge experience takes place. These are: formal contexts e.g. scheduled classes and laboratory sessions (where a learner's activity is modeled according to the university class timetable); semi-formal contexts e.g. libraries, walk-in laboratories, cafeteria and mingling areas (these contexts may represent informal spaces on campus used by learners while waiting for the next lecture to start); and lastly informal contexts (the characteristics of an informal learning context is not explicit, however, these contexts include working in informal environments during after-hours, week-ends or in university residences and private homes).

THEORETICAL UNDERPINNING

In this section, the main ideas and concepts of the theory of Activity are presented. Social presence theory and Context awareness, as pertinent to the inquiry of social awareness in a mobile learning environment, have been succinctly discussed in the previous section. The three theories were used as lenses for the empirical data collection and the analysis of the data. The section begins by discussing Activity Theory and its relevance to the Mobile Learning research.

Overview of Activity Theory

In Activity Theory (Leont'ev, 1974; 1978) the unit of inquiry is an activity. The activity is composed of subject, object, actions, and operations. A subject is a person (a learner, in this study) or a group engaged in an activity. An object (as in the objective) is held by the subject and motivates activity, giving it a specific direction. Actions are goal-oriented processes that must be undertaken to fulfill the object. That is, the object (objective) provides a need or a desire, to which the activity always answers. A goal is held in the learner's mind and different actions may be undertaken to meet the same goal.

In the process of an activity, the object itself can be transformed but not on a moment-to-moment basis, however. The way action is carried out has some operational aspect, since operations become routine and unconscious with practice. Activities may overlap in that different subjects working together in coordinated actions may have various and conflicting objects (Kuutti, 1991). Activity theory focuses on the dynamic relationship between consciousness and activity (Nardi, 1996). The constituents of activity are not fixed but can dynamically change with changing situations. A key concept in activity theory is the idea of mediation by artefacts. Artefacts – including instruments, signs, language, and machines – mediate activity and are created by people to control their own behaviour. Artefacts carry with them a particular culture and history (Kuutti, 1991).

Activity Theory and Mobile Learning Research

Activity Theory (AT) is an interpretive research approach that presents a framework within which to analyze the actions of people as they socially interact with each other in an attempt to accomplish a desired outcome. Activity Theory has been used in Human Computer Interaction (HCI) and Computer Supported Cooperative Work (CSCW) as a way to overcome problems of contextuality when designing and researching information systems (McMichael, 1999), for example, Korpela et al. (2002) and Kuutti (1996 and 1991). Activity theory has also been proposed and used as a research approach to study individual and social transformation in Information Systems development projects (e.g., Kuutti, 1999), coordination in networked organizations (Kuutti and Molin-Juustilla, 1998), cultural adoption and rejection of information systems (Gobbin, 1998) and as a framework to bridge the dichotomy between humans and technology (Hasan et al., 1998).

The preceding works suggest that ubiquitous mobile learning be studied using an activity based research focus to: 1) take into account the historical and ongoing contextual discovery of work practices and 2) acknowledge that mobile learning is a socially constructed tool subject to ongoing development by both users and designers. Activity theory provides a general integrated framework that can be used to reframe mobile learning by focusing on human activities and by highlighting the individual and collective contexts of learning.

The following discussion highlights the key principles of Activity Theory that are of particular significance to mobile learning research. These principles include activity as basic unit of analysis, artefacts and mediation, history and development, and the structure of an activity.

Activities as Basic Units of Analysis

Actions are always situated in a context and they are impossible to understand without that context (Suchman, 1987). The provision offered by Activity Theory is that a minimal meaningful context for individual actions, both physical and mental, must be included in the basic unit of analysis. The unit is called an activity. An activity is doing in order to transform something. In this chapter, the end result of an activity is learning. Since the context is included in the unit of analysis, the object of research is always essentially collective even if the main interest is in individual actions. An individual can and usually does participate in several activities simultaneously (Kuutti, 1996).

Artefacts and Mediation

A human activity always involves the use of various artefacts such as instruments, signs, procedures, machines, methods, laws, forms of work, etc. The fundamental feature of these artefacts is that they have a mediating role, since the relationships between elements of an activity are not direct but mediated. Artefacts have been created and transformed during the development of the activity itself and carry with them a particular culture – a historical residue of that development (Kuutti, 1996). An instrument, including human social processes, mediates between an actor and the object of doing. The object is seen and manipulated within the limitations set by the instrument (Engeström, 1999). The mediating tool makes it possible to mediate and change a supporting activity as subjects re-invent the context of their activities.

History and Development

Activities evolve over time within a culture; hence, an activity is a historically developed phenomenon (Jonassen and Rohrer-Murphy, 1999). Activities are not static but dynamic; they are under continuous change and development. This development is uneven and discontinuous so that each activity also has a history of its own. The historical analysis of the development is often needed in order to understand the current situation (Kuutti, 1996). A key task in historical analysis is periodization. Such an irreversible time structure is called an expansive cycle (Engeström, 1987).

Amid repetitive or expansive cycles, it is important to note that activity time is qualitatively different from action time. Action time is basically linear and anticipates a finite termination. Activity time is recurrent and cyclic. Action time corresponds to 'time's arrow' and activity time to 'time's cycle' (Gould, 1987). Activity theory recognizes the social and historical contexts as they evolve over time.

The Structure of an Activity

An activity is a form of doing directed to an object, and activities are differentiated from each other based on their objects. Transforming the object into an outcome motivates the existence of an activity. An object, e.g., knowledge, can be shared for manipulation and transformation by the participants in the activity. It is possible that the object and motive themselves will undergo changes during the process of an activity. The relationship between the subject and the object of activity is mediated by a tool. The tool is both enabling and limiting – it empowers the subject in the transformation process with the historically collected experience and skill, but it also restricts the interaction to be from the perspective of that particular *tool* or instrument only (Kuutti, 1996).

A *tool* is anything used in the transformation, including both the material tools and tools for thinking, e.g., wired and wireless devices, language, text, books, etc. In the study social awareness is, above all, seen as an over-arching tool that mediates transformation. *Rules* cover both explicit and implicit norms, practice, conventions, structures and social relations within a learning community. *Division of labour* refers to the implicit and explicit organization of a community as related to the transformation process of the object into the outcome (Kuutti, 1996). Since the actions of mobile learners are mostly situated and not necessarily delegated, the division of labour is seen, rather, as roles undertaken by peers in the provision of personalized learning during a social interaction.

Using Activity Theory as a theoretical lens requires that learner activities be studied in context – where the political, economical, social as well as cultural realities of a learner are taken into account. An activity is itself a context.

RESEARCH METHOD

In this chapter, Activity theory provides a framework for analyzing social awareness activities in a mobile learning environment – focusing on learning actions with or without mobile devices or technology. An Activity theory research approach requires that the researcher develops a complete understanding of all forces impacting the system and the changes in activity systems over time, by immersing himself in the system under observation for the entirety of the process (McMichael, 1999). For the researcher to see out all these steps or phases there may be a need to use a varied set of data collection techniques to elicit a complete picture of the activity system (*ibid*, 1999).

To the preceding point, the research employed four strategies for data collection, using Contextual Inquiry method – these strategies included interaction text, observation, interviews and qualitative questionnaire. Using multiple sources of information develops converging lines of inquiry; hence findings are likely to be more plausible and accurate. However, this technique also has limitations. For example, the use of multiple strategies within a single study has been known to nullify extraneous influences (Yin, 2002). Nevertheless, in this inquiry where learners move about different environments, such triangulation helps to address the problems of validity and reliability.

Observation allowed the gathering of rich data in real and natural settings. Rich data meant a better description and understanding of the phenomenon – socially-oriented learning contexts. In addition to observations, interaction text, and the questionnaire, contextual interviews were conducted with learners in authentic learning contexts. The interviews were situated, thus open-ended, and conducted in a conversational manner. That is, learners were asked to explain their decisions and actions as they went about an actual learning activity. The objective was to collect data – in natural settings – that would help to give a deeper understanding of how mobile learners use social awareness to facilitate actions and interaction with others for the purpose of personalized learning and support, in varied learning contexts. Contextual Inquiry was, therefore, employed as the main research method in the study.

Case Study Participants

The study was conducted at the University of Cape Town upper campus and student residences. Participants were drawn from full-time registered undergraduate learners. The following brief historical background of the university is meant to help the reader understand the participants' background. The University of Cape Town (UCT) is South Africa's oldest university, and is one of the leading teaching and research institutions in Africa. It was founded in 1829 as the South African College, a boys' school which also provided some tertiary education, and was formally established as a university in 1918. It moved to its present site in 1928. Apart from establishing itself as a leading research and teaching university in the decades that followed, the period 1960 to 1990 was marked by sustained opposition to apartheid. The university admitted its first small group of black students in the 1920s. The number of black students remained relatively low until the 1980s and 90s, when the institution – reading and welcoming the signs of change in the country – committed to a deliberate and planned process of internal transformation. From the 1980s to the early 1990s, the number of black students admitted to the university rose by 35 percent. By 2004, nearly half of the 20, 000 UCT students were black and just under half of the student body was female. Today UCT has one of the most diverse campuses in South Africa.

Since the UCT learners are mobile and engage in learning activities regardless of location and time, the collection of empirical data was done in semi-formal and informal learning contexts.

The case study began by giving out sixty qualitative questionnaires to learners on campus and at the off-campus university residences. Twenty of the learners who filled in the questionnaire were then given Wi-Fi enabled PDAs installed with *iMov* messenger, the Instant Messenger (IM) client. Contextual Interviews were done with four learners who were part of the group of twenty given the PDAs. Two of the four learners were also used as references in observing the situational learning actions and activities in authentic semi informal learning contexts.

Observations allowed me to learn as much as possible about the natural settings of a mobile learning environment and the participants so that I could then describe the social awareness and the environmental realities much more accurately. By staying in the background, I observed participants and their peers in order that I could have an informed understanding of how social awareness may be a catalyst for social interactions whose outcome is knowledge transformation, in authentic mobile learning environments.

A qualitative questionnaire was used to draw from a diverse UCT learner population. Learners, who either sat in a group or in isolation, were approached and asked if they would participate in the study by responding to the qualitative questionnaire. Contact with these learners was not prearranged. In other words, it was situational and the learners I approached could have been any other learners I came in contact with during that time. The participants completed the questionnaires on the spot. There was no participant selection criterion except that the qualitative questionnaire had to capture authentic experiences that represent a mobile UCT learner.

Mobile instant messaging (IM) was used to facilitate the text-based interaction amongst learners in authentic semi-formal and informal learning contexts through its social presence awareness mechanism. The learners, who kept the PDAs for the duration of the study, were encouraged to interact through mobile IM at any given opportunity. To make the mediated learning interactions more authentic, learners used their discretion and opportunity in using the mobile instant messaging.

The interviews were meant to allow participants to share their experiences, verbally, in authentic learning contexts. The interviews were based on the premise that the participants and I were equals. Thus the partnership provided a way for me to continually validate the assumptions made while observing learners as they interacted within a mobile learning environment. That is, I asked participants why they did things instead of interpreting these on my own. The contextual interviews happened at the UCT informal campus locations and the UCT residence locations. Firstly, I talked to learners as they were engaged and interacting with the environment, the learning material and during social interaction with peers outside formal classes, but within the UCT main campus. Secondly, I talked to learners in their residence rooms, where they do most of their studying in isolation.

Contextual Inquiry

Contextual inquiry is an interpretive field research framework that depends on conversations with users in the context of their work (Holtzblatt and Jones 1994). It is based on ethnography, where the researcher goes into the research participant's own environment. It is an explicit step for understanding who the user really is and how the work progresses from day to day (Beyer and Holtzblatt, 1998). Again, the contextual inquiry consisted of mobile technology mediated interactions, observation of learning environments, contextual interviews with learners engaged in authentic learning tasks, and a qualitative questionnaire. In contextual inquiry, the researcher stays in the background for most of the time, but also inquires about the events that are not obvious but may be significant regarding the focus of the research (Ruuska and Vaananen-VainioMattila, 1998).

The three key principles that further define Contextual Inquiry are context, partnership, and focus (Raven and Flanders, 1996); 1. Data gathering must take place in the context of the users' work, 2. The data gatherer and the user form a partnership to explore issues together. 3. The inquiry is based on a focus; that is, it is based on a clearly defined set of concerns, rather than on a list of specific questions (as in a survey). Data gathering took place in the context of the learners' actions and tasks. Context was the environment and the situation in which learners learn. Advocates of Contextual Inquiry realize that you cannot fully understand what people are doing or why they are doing it unless you can see – and perhaps experience – the interrelated situations in which they work.

Figure 1 shows different mobile learning actions in varied contexts. Participants were observed and interviewed while engaged in face-to-face and mobile device learning interactions. For these participants, context is the interrelated situations within which the personalized interaction occurs or exists. Observing and talking to participants in the context of performing learning tasks helped gather data that is different from the type of data one would obtain from questionnaires or surveys. The research was conducted at the University of Cape Town campus and residences, as noted previously. Participants, who were engaged with authentic learning tasks, were drawn from full-time registered undergraduate learners. The study was

Figure 1. Activity in a mobile learning environments: ubiquitous social presence and context aware personalized learning

not about how knowledge is created or applied in the minds of learners but rather about activities and interactions, influenced by social awareness, which result in the provision of personalized learning support.

AWARENESS OF CONTEXT AND SOCIAL PRESENCE: ACTIVITY IN A MOBILE LEARNING ENVIRONMENT

In this section, the activities and actions of learners, who are mobile, are discussed and interpreted. The actions were as observed and captured during the case study to see how the phenomena of social awareness and interaction decisions are closely related – in providing ubiquitous learning support. The empirical evidence were collected in the semi-formal and informal learning contexts, where learners needed to "help" or "were helped" through social interaction influenced by social awareness. In formal learning contexts, social interaction is mostly passive; hence no empirical evidence was sought in those contexts.

In semi and informal contexts, learners engaged in a learning activity are able to use implicit and explicit context awareness to increase or decrease the interaction whose outcome is learners "helping" or "supporting" each other. The notion of social interaction presupposes an existence of two or more learners speaking or acting. Embedded in this notion is the assumption that a learning community is socially present without which interaction – "help" or "support" – is impossible.

The following tables give examples of how learners use implicit and explicit context and social presence awareness to increase or decrease social interaction whose outcome is personalized learning. To help the reader understand how awareness manifests, the examples are structured as follows:

Table 1. How does a learner interact with others?

Researcher's intention	**Different Learners' Response / experiences**	**Researcher's Comments**
How does a learner usually communicate with peers, besides face-to-face meetings, when faced with a learning problem and needing help?	*.Through SMS,* *.I call or email,* *.I use my mobile phone,* *.We do online chat,* *.Mobile phone chat (via**miXit**).* NB: *MiXit* is a mobile phone service that allows friends to chat (send/receive text) instantly and in real-time.	Here the learner's intention is to consult about a learning problem, where learning is the object. He or she picks up an appropriate tool that best mediates the thought.
What is a learner's reason for communicating using a particular tool?	*.Depends on how desperate I am, if desperate I phone, if not I SMS.* *.Depends on what I want to say* *.It depends on the situation at the time* *.Depends on the seriousness of the situation* *.Availability of funds* *.Economic limitations and time available* *.It depends on where I am at the time that I need to contact them.*	The experiences show evidence of the learners' situation as awareness of context. Location, cost, need and reason to interact are taken as context awareness which influences how learners consult each other, e.g., calling will give immediate response.
How does a learner decide which method of interaction to use?	*.With email, it.s efficient because I can phrase the issues better and they can review and reply with specific and thorough answers.* *.It depends on the issues I need to discuss or message I need to get across.* *.How they will respond.* *.With SMS and calls, the response is faster and I can get help I need in time.* *.It is the fastest way to get feedback from them.* *.Whichever will reach the fastest in terms of being seen by the person, and the immediate need.* *.It.s quick and direct (Mobile phone calls).* *.They are always with them (peers always carry their mobile phones)* *.I know they will always have their cell* *.Depends on if they are available, and what will be quicker at that moment.* *.If they are available (via mobile phones)*	Learners use content of the interaction as context for which a mediating tool is chosen. Learners decide on the method of interaction with peers based on how quickly and fast they need the response. Where a short message is enough for the interaction, they text each other. Issues of time are noted as an influencing factor in picking a mediating tool, noting that most interactions are done on-demand to solve a learning problem instantly. Learners use social presence awareness as an influencing factor for choosing a specific mediating tool.

what the researcher wanted to know, the learner's experience and a brief discussion. The experiences are grouped in a table form per research theme and query. Showing the empirical evidence this way was to allow the reader to see where the low level research questions (elements of awareness) that support the main question are addressed.

Table 1 shows how learners are conscious or aware of the situation and the context in which they are interacting with others, e.g., how fast they would need the response, the content of interaction, etc. It also highlights how presence and availability help the decision regarding how to communicate. The different mediating tools offer different opportunities and accessibility.

Not all empirical evidence could be shared due to the length constraint of the chapter. The preceding tables showed just some of the examples of how awareness of context and social presence manifest in a mobile learning environment. In a mobile learning environment (MLE), the central activity is learning where peers provide personalized social support through social interaction. Learners acquire knowledge and experience through social interaction, drawing upon available social resources for the solution of practical learning problems. In discussing the empirical findings, the MLE was viewed as a learning activity where learners with goals and intensions engage in learning activities and their actions are driven by social awareness – looked at within the context in which it happens. The empirical evidence showed that social awareness of the social resources, the learning task, and the environment does influence the action a learner undertakes. Through Activity theory lens, which was applied differently to the different data, the analyses showed how social

Table 2. How does location as context awareness influence the social interaction?

Researcher's intention	**Different learners' response / experiences**	**Researcher's comments**
From which locations is a learner willing to contact his/her peers for a learning task or problem?	*.Home, library, hospital* *.UCT. basically anywhere on campus* *.Only at UCT* *.Everywhere* *.Malls* *.In the lab*	Learning tasks and problems are not confined to fixed locations, thus learners could consult about a problem from anywhere.
If a learner wanted to interact with a peer, how would s/he consider a peer's present location?	*.If they are in a lecture or the library and I have a small issue then I won't bother them. In essence, I would just take into consideration my need to contact them and their availability in terms of what they're doing and where they are.* *.If they were on campus or at the library or maybe at home, I.d feel free to contact them since they will probably be more flexible and relaxed to respond.*	Learners use awareness of location to determine if the peer's location favours a social interaction.
How does a learner's knowledge of where a peer is (their location) affect the social interaction?	*.Makes it easier because you would know who is closest to you for help* *.It would be great to know if they are nearby, but if they're not anywhere where I can reach them, then I'll have to call regardless of where they are.* *.The closer he or she is, the better, because it will be easier for me to go to her when push comes to shove.* *.If I knew they were in a club or something, I wouldn.t bother them with school work.*	The experiences show that awareness of a peer's location would determine if it is plausible or would make sense to consult a peer.

awareness and the action happen in varied mobile learning environments.

Figure 2 shows how awareness of context and social presence is used as tool that enables personalized learning. An aspect of social awareness includes the location factor and context, which manifests as valuable to how the interaction is influenced. Culture and social background manifest as playing a fundamental role in mobile learning contexts where learners often would rather seek help or support from peers who share a background. A common background allows better understanding and better communication and thus eases up learning interactions.

Awareness of context also manifests in how an interaction is influenced by a learner's emotional and physical state, including behaviours and actions of others. Environmental situations

Table 3. How does a peer's activity and emotional state as context awareness influence social interaction?

Researcher's intention	**Different learners' response / experiences**	**Researcher's comments**
How would knowledge of what a peer is doing (current activities) affect a learner.s interaction decisions?	*.Issues like a person sleeping or eating may make me wait a while.* *.I need to know if they got my full attention* *.If sleeping, I will leave them be, if studying will interact with them* *.Would respect that they are really busy and seek help elsewhere*	A peer's activity influences the decision about whether to consult with that learner or consult with someone else.
How do actions and behaviours of others affect a learner's decision to learn?	*.I am motivated by seeing others work and tend to slack when others aren.t working* *.When others are discussing work, it increases chances of grasping concepts.* *.It doesn.t affect me much, because we all have different learning capabilities.* *. I take seriously any feedback or advice from* *learners who have done courses I.m doing.* *.If am studying and people are noisy, I will not be able to concentrate. If they are serious, I too will be*	Presence of others, that is, seeing what they are engaged with influences the decision of a learner to get involved and do the same. For some, what others are doing does not influence their decision to work
To what extent do emotional states of a peer affect a learner's actions?	*.Very little, unless they're my friends and I am concerned, or if it's group work and they're slacking due to moods.* *.You know what you can ask and can't, their moods may limit your questions.* *.Lets you know how to approach them*	Learners could be sensitive to the mental state of a peer, thus altering the way they consult.
How does the awareness of a peer's current emotional situation (e.g., s/he is stressed, happy, sad, cheerful, etc.) help in interacting with them?	*.Very helpful, you don't want to interact and learn with a sad guy, so they ought to be happy, cheerful. If anything they can be stressed.* *.It's useful in the sense that I can know whether to bother them or not. It makes little* *difference in group work unless their situation is very serious.* *.It makes me understand their behaviour and the way in which they react to my questions or experience.* *.It makes me understand them better and assist in determining what things I can or cannot say.* *.Will try to approach them correctly, depending on what kind of mood they are in.*	The experiences of learners suggest that emotional states and activities of peers do influence a social interaction. The awareness determines how learners approach their peers for a learning purpose. In the Activity Theory perspective the emotional states could be regarded as signs and rules, thereby altering the context in which a learning activity happens. Although these are unwritten signs and rules, learners know not to disturb or pester a peer who is not in the best of emotion and spirit to help with a learning task.

Table 4. How do environmental situations as awareness of context influence learning actions?

Researcher's intention	Different learners' response / experiences	Researcher's comments
To what extent is a learner aware of his/her surroundings, during a learning activity, and how does that influence his/her decision to interact with peers?	*.Temperature, I don.t like studying alone when it.s too cold so I normally chat a lot when it.s cold.* *.The time of the day determines when I need to chat or where I need to work.* *.Noise, too much movements, then I can.t concentrate on my work.* *.When I talk I can tell who is an understanding person from a lost one.* *.When I am with people I am not so aware of my surroundings in the sense that I am when I am alone, however, it is when its silent and cold that I tend to SMS/call people with my problems.* *.If I see a lot of other students studying and discussing with their friends, they keep me motivated and I ask them questions.*	The environmental context and situations do influence how learners act and how a learning activity happens. The temperature during the day or ambience or noise levels may help a learner resort to a different learning activity. For example, a learner saying that she prefers to do more social interaction than studying in isolation when the temperatures drop. A learner consults with peers, taking advantage of their social presence and current activity.

Table 5. What is the role of Social Presence Awareness in the Mobile Learning Activity?

Researcher's intention	Different learners' response / experiences	Researcher's comments
How is it important for a learner to know a peer.s availability before he/she decides to interact with them?	*Very important* *For me, it is quite important*	Awareness of social presence or a peer's availability is essential for interaction decisions
How does awareness of presence influence a learner's learning action?	*.It helps me determine who I can get assistance from* *.Puts me at ease since I know my friends could help when I get stuck* *.I know I.m going to get help* *.If they are available and willing to help, I will contact them.* *.I have a choice to pick any depending on who is more knowledgeable on that particular topic.* *.I know when my academic questions arise I.ll be answered with certainty* *..Then I could reach them whenever I need them* *.It makes learning very easy, we could exchange ideas and information anytime and I will know that I am always able to ask for help*	Social presence of knowledgeable peers is important for interaction whose purpose is sharing knowledge experiences. Learners are consciously at ease when they know their peers are available to help with a learning task, at anytime. Knowing a peer is socially present (awareness of social presence) gives a learner a sense of having a personalized academic support that "follows" them regardless of a learning problem and their location.
How is a learner able to keep the sense of presence during a technology-mediated interaction, particularly when using text-based instant messaging.	*.Make the messages as short as possible* *.Instant or quick responses* *.I don.t want to wait for a response to my question, which is why I use IM.*	Short, quick and instant responses grab a learner's attention in a textbased social interaction.
How is a learner able to read emotions of peers during IM interaction and how can s/he sense if the peer is interested in talking to them or not?	*. Words. descriptive uninhibited words* *. I know when they have something else in their minds* *. One word answers would indicate that he isn't keen to talk.* *. The level of response and language used (enthusiasm)*	Even though they are not in the same location, learners are able to exhibit social awareness levels during a mediated social interaction

Figure 2. Summary of findings using the basic Activity theory diagram

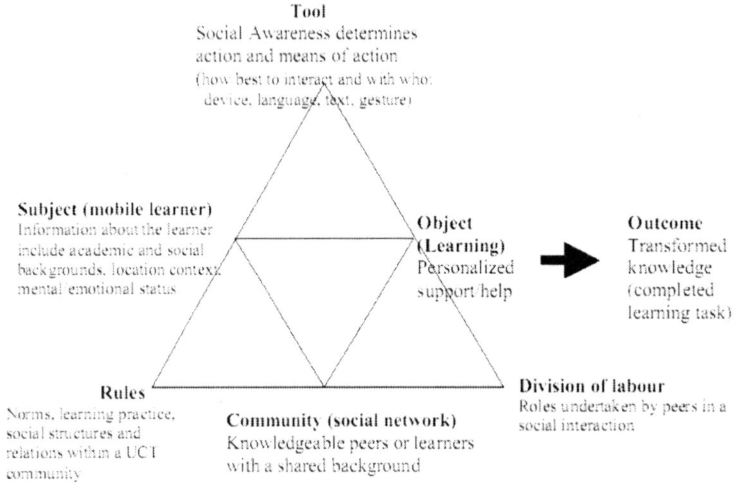

also manifest as playing a role in a learner's activity decisions. Social presence of peers manifest as fundamental with the role it plays in the learning activity decisions. That is, awareness of a social presence enables an opportunistic learning interaction and can also motivate a learner to engage with a learning task.

In Figure 3, awareness of context and social presence determines the appropriateness of a mediating tool. From the analyzed data, these mediating tools include varied languages, instant messaging text, wired and wireless devices. The mediating tools are used differently depending on the learner's context and needs. Social awareness is used as a means of coordinating and communicating a learning decision and action. Coordination and communication presupposes that a social network (knowledgeable peers, friends with shared background, tutors, etc.) is consistently available, without which a social interaction is impossible. The context in which a learner finds herself and the available social resources influence a learning action and the appropriate tool.

In order to understand the MLE as a learning activity there is a need to shed light on the motives and goals of learners as well as their expectations about the outcome. In doing so, the goal is to understand the context in which a learning activity happens, the motivation for the activity and the subsequent contradictions. With the understanding of the learners' motivations and situated actions, the MLE – as an activity system viewed from the learner's perspective – will be seen to enhance mobile learning by way of personalized academic social support (social interaction).

In the MLE learning activity, the learner has a network of peers with which she socially interacts about a learning problem. The relevance of the social network is that its social presence impact on the learner's work by providing personalized support with the intention that the learner accomplishes the goal of the activity. Therefore, to provide relevant social support, it is important that the social network also sees the learner's goal and motivation as significant. Thus, the nature of the social interaction is influenced by awareness of context and social presence. The social awareness is steered by prior knowledge, experience, beliefs and instincts.

The mobile learning environment has activity structure that consists of activity, action and operation. For any learner's intentions and motives, it is necessary to identify all of the actions and operations that support the learning activity. The

Figure 3. Elements of mobile learning activity model as a systemic entity.(Korpela et al., 2002; 2004; modified)

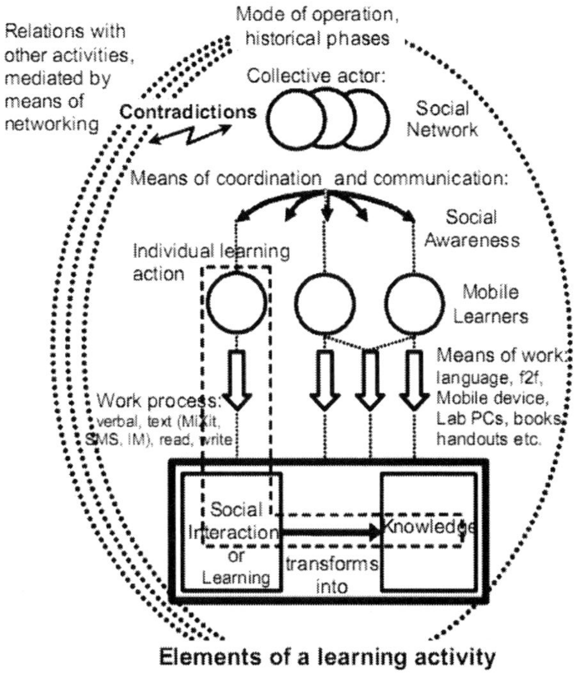

learning activity is what defines the mobile learning environment since the learning environment focuses on an activity. It is, thus, also necessary to understand the intentionality of the action for the mobile learner, where there has to be an understanding of why and how they engage in a situated learning action.

Figure 4 shows a simple mobile learning activity structure where a learning activity is oriented to motives that satisfy a need. An action is the learning process functionally subordinated to activity and directed at specific awareness goals. In mobile learning what motivates a learning activity and the goals to which the activity is immediately directed is of fundamental significance in that, for example, it changes the context for the learning activity and how it is carried out. Actions are realized through operations that are themselves determined by the actual awareness of context and social presence; that is, actual conditions of activity.

Figure 4. A mobile learning activity system structure

Activity	Action/task	Operation
Isolated learning	Reading, writing, term project, e.g., building a database	Learning from Handouts, books Website, PCs
Learning via social interaction	Face-to-face meeting, phone call, text message	SMS or MiXit on a mobile phone, instant message on PDA or PC

The finding from the inquiry showed that the elements of the mobile learning environment – the learner, the social network and the object – do not always interact or act on each other directly. In fact, their interactions are often mediated by signs, conditions and tools. Tools and mediators basically describe the kinds of approaches that constrain, enable or shape a mobile learning activity. In the mobile learning environment observed, the mediators consisted of varied languages (spoken and written), text including messages, handouts and books, as well as mobile devices that included mobile phones, laptop PCs and PDAs.

In the inquiry, the tools – language and mobile devices – mediated a learning activity by enabling social interaction between mobile learners. University of Cape Town learners have diverse social and language backgrounds. Whereas the English language served as a collective mediator for most learning activities and actions, for some mobile learners it hindered social interaction – prompting these learners to resort to seeking social support mostly from peers who can speak their mother tongue. The widespread mediating tool in the mobile learning environment was a mobile phone. The mobile phones unlike PDAs and laptop PCs are readily available to almost all learners. Thus, mobile phones mediated most of the textual and verbal interactions with others not in the same location. Although PDAs are not as common as mobile phones, they were made available to participants to mediate opportunistic and spontaneous textual social interaction facilitated by the instant messenger social presence indicator.

Interpreting Awareness for Ubiquitous Personalized Learning Support

Ubiquitous mobile learning is characterized by the provision of insightful ways for identifying the right social presence and the right learning experience. To this point, resourceful and effective learning and provision of personalized support relies a great deal on the learner's awareness of context and social presence – social awareness (Kekwaletswe, 2007). Social awareness stems from two sources: firstly, from the learner's experience of and interaction with the environment and the learner's social background; and secondly from the social resources or service delivery standpoint which incorporates access to ubiquitous learning services such as mobile or immobile interaction devices, network, university schedules or calendars as well as other personalized service delivery activities (*op. cit.*). Services delivery is synonymous with personalized academic social support. Personalized academic social support through social interaction among learners enables ubiquitous learning and subsequently, knowledge transformation.

A learning action is an interpretation of a situation and it reiterates the learner's place in it. Anything a mobile learner does and says is an action. An action is, therefore, a performance within a social awareness milieu that has called for or shaped that performance in some way. It is a course for dealing with mobile learning needs and events of the day by using the interpretive or mediating tools the social system provides. A learning action is a performance enacted within a particular situation or context that is constructed within the surroundings.

Observing the mobile learning environments revealed that learners are aware of their own and other's activities. Accordingly, actions reflect an individual learner's learning. These actions serve as cues to others regarding appropriate sharing of learning experiences. The suggestion, here, is that a mobile learner can use the actions of others as cues from which he or she learn explicitly. By watching what happens to others when they engage in different activity patterns, the learner comes to understand that a certain option leads to success while another leads to failure, without engaging in either option personally. In cases where neither the environmental cues nor cues from models are available, a mobile learner can still resort to proactive action to obtain information and can create cues by trial and error.

CONCLUSION

The main characteristic of a mobile learning environment activity system is that the learner has increased freedom of mobility. The increase in mobility creates situations where the learner's context such as location, presence or non-presence of others as well as things around him or her, are dynamically changing. Thus, in the mobile learning activity system, presence and context awareness proves to be a useful and fundamental concept to understanding how the activity structure shapes. In the inquiry, context awareness allowed mobile learners to adapt their behaviour and actions to the circumstances in which they found themselves.

The role of context awareness, and how it is experienced, is significant to ubiquitous social interaction whose outcome is personalized mobile learning. The context awareness in the MLE activity system manifests in three categories: (1) Mediating tools as context of the activity - learners use context awareness to decide on the appropriateness of the mediating tool. For example, if a learner is faced with a time-driven learning task she cannot resolve in isolation then s(he) uses a PDA or a mobile phone to call or text message a peer as opposed to writing and sending an email to a peer. In this situation the learner is aware of the need to get instantaneous response, aware of how best to get the instant support, aware of how best to get social presence, as well as aware of the best mediating tool to exploit. (2) location of a learner as context of the activity - learners need awareness of where they are themselves and where their peers are located, which is significant in determining the plausibility of a social interaction. (3) learner's situation –refers to context awareness dealing with culture and language, the learner's current activity as well as environmental states.

This chapter has shown that awareness of context and social presence (synonymous with social awareness) is a useful and significant characteristic for sharing learning experiences and the consequent learning support that occur independent of location and time. When learners socially interact face-to-face or through mediation, they are able to use inherent situational information, or context awareness, to increase or decrease the knowledge sharing experience. Thus, in a mobile learning environment, the learners' use of context and social presence awareness is important since it influences the mobile learning decisions and actions, as the learning contexts change.

REFERENCES

Beyer, H., & Holtzblatt, K. (1998). *Contextual Design, Defining Customer-centred systems*. San Francisco: Morgan Kaufmann Publishers, Inc.

Christiansen, N., & Maglaughlin, K. (2003). Crossing from physical workplace to virtual workplace: be AWARE! In *Proceedings of HCI International Conference on Human Computer Interaction*. Lawrence Erlbaum Associates, NJ, pp.1128 – 1132.

Dey, A. K., & Abowd, G. D. (1999). *Towards a better understanding of context and context-awareness*. GVU Technical Report GIT-GVU-99-22, College of Computing, Georgia Institute of Technology, Atlanta, Georgia.

Dourish, P., & Bellotti, V. (1992). Awareness and coordination in shared workspaces. In proceedings of *ACM CSCW Conference*, pp. 107-114.

Dourish, P., & Bly, S. (1992). Portholes: Supporting Awareness in a Distributed Work Group. Proceedings of the *CHI '92 Conference on Human Factors in Computing Systems*, pp. 541-547.

Engeström, Y. (1987). *Learning by expanding: An activity theoretical approach to developmental research*. Helsinki: Orienta-Konsultit.

Engeström, Y. (1999). Activity theory and individual and social transformation. In Engeström, Y., Miettinen, R., & Punamaki, R. (Eds.), *Perspectives on Activity Theory*. Cambridge, UK: Cambridge University Press.

Gobbin, R. (1998). Adoption or rejection: information systems and their cultural fitness. In Hasan, H., Gould, E., & Hyland, P. (Eds.), *Information systems and activity theory: tools in context, University of Wollongong Press* (pp. 109–124). Wollongong: N.S.W.

Gould, S. J. (1987). *Time's arrow; time's cycle: Myth and metaphor in the discovery of geographical time*. Cambridge: Harvard University Press.

Gutwin, C., Greenberg, S., & Roseman, M. (1996). *Workspace awareness in real-time distributed groupware: Framework, widgets and evaluation. In proceedings of* (pp. 281–298). HCI on People and Computers XI.

Hasan, H., Gould, E., & Hyland, P. (1998). *Information systems and activity theory: tools in context, University of Wollongong Press*. Wollongong: N.S.W.

Holtzblatt, K., & Jones, S. (1993). Contextual inquiry: A participatory design technique for system design. In Schuler, D., & Namioka, A. (Eds.), *Participatory design: Principles and practice*. Englewood Cliffs, NJ: Prentice-Hall.

Jaffer, S., Ng'ambi, D., & Czerniewicz, L. (2006). The Role of ICTs in Higher Education in South Africa: One Strategy for addressing teaching and learning challenges. *Proceedings of Emerge Online Conference: Learning Landscapes in Southern Africa*, 10-21 July. Retrieved from http://emerge2006.net

Jonassen, D. H., & Rohrer-Murphy, L. (1999). Activity theory as a framework for designing constructivist learning environments. *Educational Technology Research and Development*, *47*(1), 62–79. doi:10.1007/BF02299477

Kekwaletswe, R. M. (2007). *Knowledge transformation in a mobile learning environment: An inquiry of context and social presence awareness*. (Unpublished PhD thesis). University of Cape Town.

Kekwaletswe, R. M. (2009). Conceptualizing ubiquitous learning through context-aware wired and wireless web services. *In proceedings of the IADIS Mobile Learning Conference*. Barcelona, Spain. 25 – 28 February.

Kekwaletswe, R. M., & Ng'ambi, D. (2006). Ubiquitous Social Presence: Context-Awareness in a Mobile Learning Environment. In Proceedings of the *IEEE International Conference on Sensor Networks, Ubiquitous, and Trustworthy Computing*. Taichung, Taiwan. 5-7 June. pp. 90-95.

Korpela, M., Mursu, A., Soriyan, A., Eerola, A., Hikkinen, H., & Toivanen, M. (2004). IS research and development by activity analysis and development: Dead horse or the next wave? In Kaplan, B., Truex D.P. III, Wastell D, Wood-Harper A.T., and DeGross J.I (eds). *Information Systems Research: Relevant Theory and Informed Practice, IFIP TC8/WG8.2 Working Conference,* (pp. 453-470), Manchester, UK, July 15-17. Boston, MA: Kluwer Academic Publishers.

Korpela, M., Mursu, A., & Soriyan, H. (2002). Information systems development as an activity. *Computer Supported Cooperative Work, 11*, 111–128. doi:10.1023/A:1015252806306

Kuutti, K. (1991). Activity theory and its implications to information systems research and development. In Nissen, H. E., Klein, H. K., & Hirschheim, R. A. (Eds.), *Information systems research* (pp. 529–549). Amsterdam: Elsevier Science Publishers.

Kuutti, K. (1996). Activity theory as a potential framework for human-computer interaction research. In Nardi, B. A. (Ed.), *Context and consciousness: Activity theory and human computer interaction* (pp. 17–44). Cambridge, MA: MIT Press.

Kuutti, K., & Molin-Juustilla, T. (1998). Information system support for 'loose' coordination in a network organisation: an activity theory perspective. In Hasan, H., Gould, E., & Hyland, P. (Eds.), *Information systems and activity theory: tools in context* (pp. 73–92). Wollongong, NSW: University of Wollongong Press.

Leont'ev, A. (1974). The problem of activity in psychology. *Social Psychology, 13*(2), 4–33.

Leont'ev, A. (1978). *Activity, Consciousness, and Personality*. Englewood Cliffs, NJ: Prentice-Hall.

McMichael, H. (1999). An Activity based perspectives for Information Systems research, In Proceedings of the *10th Australian Conference on Information Systems*, Wellington, New Zealand, pp. 610-621.

Nardi, B. (1996). Studying context: A comparison of activity theory, situated action models, and distributed cognition. In Nardi, B. (Ed.), *Context and consciousness: Activity theory and human computer interaction*. Cambridge, MA: MIT Press.

Nardi, B., Whittaker, S., & Bradner, E. (2000). Interaction and outeraction: instant messaging in action. *Proceedings of CSCW '00*. Philadelphia, PA, pp. 79-88.

Ngwenyama, O. K., & Lee, A. (1997). Communication richness in electronic mail: critical social theory and the contextuality of meaning. *Management Information Systems Quarterly, 21*(2), 145–167. doi:10.2307/249417

Nonaka, I. (1994). A dynamic theory of organizational knowledge creation. *Organization Science, 5*(1), 14–37. doi:10.1287/orsc.5.1.14

O'Malley, C., Vavoula, G., Glew, J., Taylor, J., Sharples, M., & Lefrere, P. (2003). *MOBIlearn – Guidelines for learning, teaching, and tutoring in a mobile environment*. Retrieved November 5, 2007 from http://www.mobilearn.org/download/results/guidelines.pdf.

Polanyi, M. (1958). *Personal Knowledge,* Chicago: The University of Chicago press.

Polanyi, M. (1966). *The tacit dimension*. London: Routledge and Kegan Paul.

Prinz, W. (1999). NESSIE: An awareness environment for cooperative settings. In S. Bødker, M. Kyng, and K. Schmidt (eds.), *ECSCW"99: Proceedings of the Sixth European Conference on Computer-Supported Cooperative Work.* (pp.391–410). Copenhagen. Dordrecht: Kluwer Academic Publishers.

Raven, M. E., & Flanders, A. (1996). Using contextual inquiry to learn about your audiences. *ACM SIGDOC Asterisk Journal of Computer Documentation, 20*(1), 1–13. doi:10.1145/227614.227615

Rettie, R. 2003. Connectedness, Awareness and Social Presence, 6th Annual International Workshop on Presence 2003.

Ryder, M., & Wilson, B. (1997. From center to periphery: shifting agency in a complex technical learning environments. Paper presented at the *Meeting of the American Educational Research Association*, Chicago, IL. Ruuska, S., & K. Vaananen-VainioMattila, (1998). User needs for mobile communication devices: Requirements gathering and analysis through contextual inquiry. In C. Johnson (ed.*) Proceedings of the first Workshop on Human Computer Interaction with Mobile Devices.*

Shariq, Z. S. (1999). How does knowledge transform as it is transferred? Speculations on the possibility of a cognitive theory of knowledgescapes. *Journal of Knowledge Management, 3*(4), 243–251. doi:10.1108/13673279910303998

Sharples, M. (2000). The design of personal mobile technologies for lifelong learning. *Computers & Education, 34*, 177–193. doi:10.1016/S0360-1315(99)00044-5

Sharples, M. (2005). Learning as conversation: Transforming education in the mobile age. In *Proceedings of* (pp. 147–152). Seeing, Understanding, Learning in the Mobile Age.

Sharples, M. (Ed.). (2006). *Big Issues in Mobile Learning: Report of a workshop by the Kaleidoscope Network of Excellence Mobile Learning Initiative*. Retrieved November 29, 2009 from http://telearn.noekaleidoscope.org/ warehouse/ Sharples-2006.pdf.

Sharples, M., Corlett, D., & Westmancott, O. (2002). The design and implementation of a mobile learning resource. *Personal and Ubiquitous Computing, 6*, 220–234. doi:10.1007/s007790200021

Short, J., Williams, E., & Christie, B. (1976). *The Social Psychology of Telecommunications*. Toronto: Wiley.

Suchman, L. (1987). *Plans and situated actions: The problem of human–machine communication*. Cambridge, UK: Cambridge University Press.

Taylor, J. (2006). Evaluating Mobile Learning: What are appropriate methods for evaluating learning in mobile environments? In Sharples, M (ed) *Big Issues in Mobile Learning: Report of a workshop by the Kaleidoscope Network of Excellence Mobile Learning Initiative:* University of Nottingham. Retrieved November 29, 2009 from http://telearn.noe-kaleidoscope.org/ warehouse/ Sharples-2006.pdf.

Tollmar, K., Sandor, O., & Shömer, A. (1996). Supporting social awareness at Work, design and experience. In *Proceedings of CSCW '96* (pp. 298–307). New York: ACM Press.

Winter, N. (2006). What is Mobile Learning? In Sharples, M. (Ed.). *Big Issues in Mobile Learning: Report of a workshop by the Kaleidoscope Network of Excellence Mobile Learning Initiative*. Retrieved 29 Nobember from http://telearn.noekaleidoscope.org/warehouse/Sharples-2006.pdf.

Yin, R. K. (2002). *Case Inquiry Research, Design and Methods* (3rd ed.). Newbury Park, London: SAGE Publications.

Chapter 6
Insights into Students' Thinking with Handheld Computers

Wan Ng
University of New South Wales, Australia

Howard Nicholas
La Trobe University, Australia

ABSTRACT

The handheld computer as a pedagogical tool has the capacity to enable students to demonstrate understanding through different modes of representations, for example, verbal, text, tables and graph, drawings, writing or written formulas, concept mapping and animations through Flash or Pocket Slides PowerPoint. Its impact as a motivational learning tool has been described in numerous articles. The purpose of this chapter is to describe its use as a research tool for capturing students' thinking processes as they construct representations in science and mathematics, or solve problems in these learning areas on the handheld. By using an avi-screen capture software operating in the background to do this, the research is a non-intrusive method of capturing the verbal and screen-based (visual) elements of students' thinking as they use the handhelds to complete individual or collaborative tasks.

INTRODUCTION

Despite their basis in business practices, handheld computers (also known as PDAs, pocket PCs or Palms) have appealed to schools as educational tools because they have the potential to offer school children unique educational affordances (Kimber & Wyatt-Smith, 2006; Perry, 2003, 2006; Sharples, 2003, 2006; Waycott, Jones, & Scanlon, 2005) that include (i) personalised learning (ii) portability (iii) social interactivity (iv) context sensitivity (v) connectivity to networked learning communities and (vi) affordability.

Handheld computers have evolved dramatically over the past decade from electronic organisers to devices with a wide range of features and capabilities that enable them to be useful pedagogical tools in education. These features provide the capacity for students to learn individually and collaboratively in activities that include note taking, audio recording, writing short texts, researching on the Internet, constructing graphs

DOI: 10.4018/978-1-61692-849-0.ch006

and presentations, organising daily activities, taking images, video recording and working collaboratively using Bluetooth in class or emailing outside of class (Nicholas & Ng, 2009). These capacities, together with the small size and reduced price of handhelds compared to other technologies such as laptops or tablet computers offer students and teachers flexible access to the powers of computing.

Most research into handhelds to date has focused on the identification of pedagogical uses of the devices, resulting in the development of a vast array of software for teaching and learning (Becta, 2003; Finn & Vandenham, 2004; Ng & Nicholas, 2007; Palm, 2002; Perry, 2003; 2006; van 't Hooft, 2006). Only a few studies have reported on the impact of handheld computers on conceptual development in the school years. For example, Swan, van't Hooft, Kratcoski and Jason (2007) investigated the impact of ubiquitous computing on representations, conceptualisations and uses of knowledge. They found that the availability of technology, including a variety of handhelds, encouraged students to construct representations in multiple forms through which significant gains in conceptual understandings among the students were demonstrated.

This chapter describes the capturing of the construction of multimodal representations by school students learning science and mathematics through the exploitation of the multimedia capacity of the PDA. To capture the thinking and decision-making processes of students as they construct representations to demonstrate their understandings, the use of a video capturing software operating in the background of the PDAs to record verbal and screen-based actions will be described and discussed. The discussion will address the capacity of the handheld computer as a research tool in recording the dynamic (real-time) development of multiple and multimodal representations and recording of the cognitive processes of these students in solving problems in science and mathematics.

THEORETICAL FRAMEWORK UNDERPINNING RESEARCH INTO CAPTURING THINKING WITH HANDHELD COMPUTERS

Science and Mathematics Learning

Many school students find learning mathematics and science concepts difficult, lack confidence in their own abilities and are disengaged from learning the subjects (Dowker, 2004; Goodrum, Hackling & Renniel, 2001; Hart, 1981; Parliament of Victoria Education and Training Committee, 2006; Pierce & Stacey, 2006; Steward & Nardi, 2003). Contributing to the disengagement are factors such as textbook-based work, repetitive exercises, memorisation of facts, formulas and procedures and a lack of connection between classroom content and the real-world context. These factors leave gaps in the students' knowledge and understanding of mathematics and science. One of the means of overcoming the gaps is for teachers to have a greater insight into *how* their students learn and a greater capacity to engage with students' development of representations of those concepts. Ultimately, learning about science and mathematics is not only about test scores but students' educational and practical functioning in mathematics (Dowker, 2004) and science, so as to produce mathematically and scientifically literate future citizens who are able to apply mathematical and scientific knowledge and understanding to enable them to think critically about decisions that affect their lives (Goodrum, et. al, 2001; Hurd, 1998; Miller, 1998; Thomson, Cresswell & De Bortoli, 2004)

Modes of Representations

Nuthall (1999) proposes that in order to establish more permanent understandings, learners will require multiple exposures to the same concept. Russell and McGuigan (2001) assert that in order to learn, learners need to be provided with opportunities to construct a variety of representations of a concept through which they refine their representations in different modes as they make more explicit their conceptual understandings. Helping students to understand concepts requires both teachers and students to work with the students' representations of their understanding and to make what they understand explicit and to provide means of exploring their understanding. As different modes of representations have different strengths and weaknesses in demonstrating associative meaning, students need to be taught the use of different modes in accordance to their purposes in order to be able to effectively demonstrate their understandings of concepts learned (Prain & Waldrip, 2006). In science and mathematics education, learning involves the production and exploration of multiple representations of the concepts under study as well as the ability to create connections between the different modes of representations in order to demonstrate reasoning processes and understandings (Ainsworth, 1999, 2005; Cai & Wang, 2006; Bardini, Pierce and Stacey, 2004; Prain and Waldrip, 2006; Stylianou and Pitta-Pantazi, 2002; Waldrip, Prain & Carolan, 2006).

According to Bruner (1960), thought processes can be divided into three distinct modes of representation (Figure 1): enactive representation (action), iconic representation (images) and symbolic representation (language – for example, words, numbers, graphs). While he viewed each mode as being dominant at different times during a child's development, learning is essentially represented by a combination of these modes, reinforcing the multimodal options for knowledge representation. In science and mathematics learning, enactive representations can be manifested through role-plays and manipulating objects through experimentations and modelling; iconic representations can be demonstrated using pictures, videos, simulations, illustrations and diagrams while symbolic representations can be shown through text, verbalisations, numbers, graphs, tables and charts. Acknowledging multimodal communication reinforces how thinking involves transforming these representations and understanding the connections between the different knowledge representations.

Multimodal Representations

In this chapter, multimodal refers to the simultaneous use of two or more spoken or written texts (symbolic), drawn or photographed images (iconic) or movies of role plays (enactive) to represent understanding and reasoning (Prain and Waldrip, 2006). Handhelds permit students to use a number of different modes of representations and to combine these different representations. For example, in the diagram of the cell by an eleven year-old student in Figure 2a, both image and text are displayed, and in Figure 2b, a six-year old child made use of the voice recording function of the handheld to record an explanation of the butterfly

Figure 1. Modes of representation of knowledge

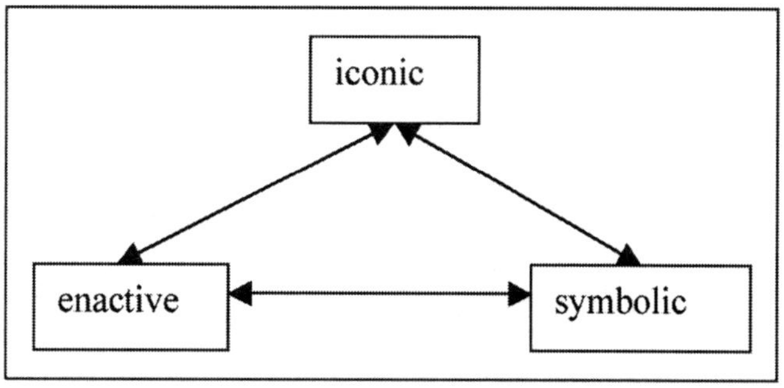

Figure 2. Examples of multimodal representations (a) image and text and (b) image and verbal (voice recorded)

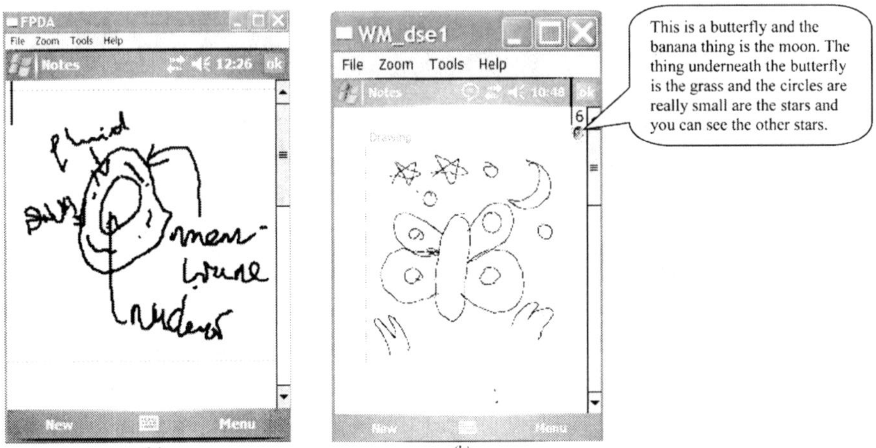

(a) *(b)*

and the environment it is in. In science, a practical report containing diagrams of equipment, images, videos or graphs of results and text to describe and discuss results is representing understanding in a multimodal manner.

Multiple Representations

The general term 'multiple representation' in this chapter refers to the re-representing of the same concept in different ways, including through different modalities. For example in studying about the 'particle theory' and changes of states when learning about solids, liquids and gases, the multiple representations could include:

(i) ***role play (enactive representation).*** Students themselves experience 'being' water particles in this example of a role play to represent the three states of ice, water and water vapour. In the solid (ice) state, by linking each other's arms at the elbows, students will automatically fall into a neat orderly arrangement because the 'bonding' (arm-linking) between the students is strong and prevents further movement. The only room for movement is the vibration of each particle while fixed in its position. In the liquid (water) state, the arm-linking between the students is less strong and this is represented by holding hands. In this representation, a greater degree of freedom for students to move in front of or behind each other can be demonstrated. The transformation to the gaseous state is represented by the students releasing their hand-holding state to break completely free from each other and spread themselves around the room, (including some possibly going out the classroom door if it is opened). The disappearance of the students in the gaseous state represents the dimension of understanding that the forces of attraction between gas particles are almost non-existent and that the particles spread themselves to take up all the available space

(ii) ***creative writing (symbolic representation).*** The student could write a creative story about the adventures of a block of ice and the transformation it undergoes in different environments. The student will need to demonstrate the concept of energy as well as how this affects the strengths of the bonds holding water molecules in the different states.

Figure 3. Text (symbolic) and images (iconic) representations

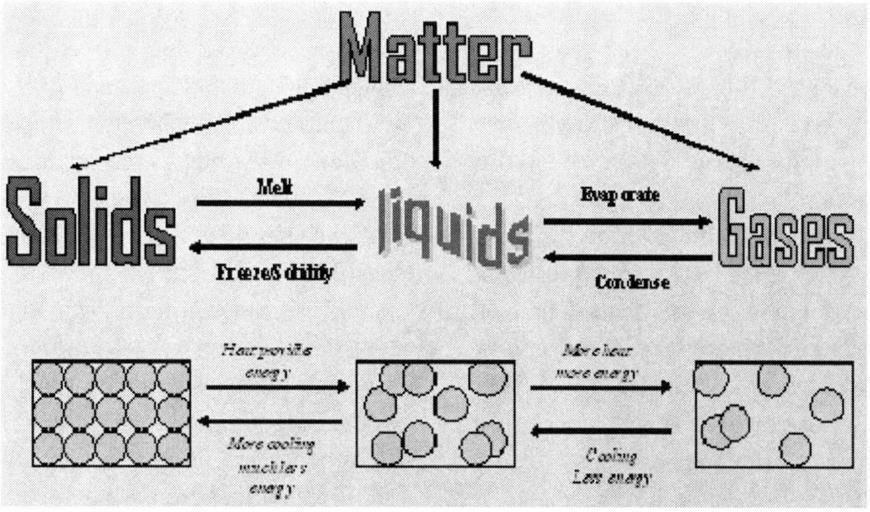

(iii) ***text (symbolic) and images (iconic).*** A summary of the changes of states and their relationships with energy and arrangement of particles is shown in Figure 3. The modes of representations in this figure are text (symbolic) and images (iconic).

(iv) ***experimenting (enactive representation) and explaining (either through text or verbally)***. An example includes conducting the activity of 'making clouds in a bottle'. This would require students to carry out the activity as well as observe and explain the phenomenon that took place in the plastic bottle.

CAPTURING KNOWLEDGE REPRESENTATIONS WITH THE HANDHELD COMPUTERS

The handheld computer is both a pedagogical (Becta, 2003; Finn & Vandenham, 2004; Ng & Nicholas, 2007; Nicholas & Ng, 2009; Palm, 2002; Perry, 2003; 2006; van 't Hooft, 2006) and a research tool (Ng & Nicholas, 2008). As a pedagogical device, it offers multiple ways of representing content. Its open-ended capacities enable students to both choose from a range of modes of representation and to incorporate multiple modalities of representation in the one 'package'. Some of the modalities of representation afforded by handheld computer software are verbal (through voice recording), text (through Word Mobile), tables and graphics (through Mobile Excel and other graphing software), drawings or written formulas (through Notes), concept mapping (using pocket Inspiration), and animations (through Flash or Pocket Slides PowerPoint editing software) (Ng & Nicholas, 2009). An additional important dimension of the handheld as a pedagogical tool is its ability to access the Internet for independent research purposes and also its communicative capacity via Bluetooth for collaborative learning.

Within the context of this chapter, the handheld computer is also being used as a research tool to capture students' representations and the thinking processes they engage with in creating them. We will describe three small research studies that recorded the strategies and processes used by the students:

i. in solving decimal-based mathematical problems that the student undertook as a game with a learning object called *Wishball*

ii. in an integrated task involving science and mathematics that required the students to find the averages of individual and class pulse rates. Other aspects of this task included answering questions on the circulatory system but this aspect will not be discussed in this chapter and
iii. working on science and mathematics tasks that required students to construct individually a diagram of a cell on the handheld computer and to solve a percentage problem, as well as to work collaboratively to demonstrate understanding of the concept of 'gravity'.

The capability of the handheld computer in capturing how students thought or used strategies as they solved problems or demonstrated understanding of specific concepts will be evaluated and discussed.

Video Capture Software as a Research Tool

The handheld 'avi' screen and voice capture application that was used in these studies is called *CoolCapture*. It was purchased over the Internet from Ateksoft. The software operates in the background recording all of the transitional and ultimate representations of students' thinking (through dialogue and actions on the screen) as they work individually or with others. Transitional representations are those that students 'try' and 'change/erase' or 'toy with' before deciding not to use them. These transitional representations show areas of uncertainty and exploration in students' thinking and indicate, for example, if the arrival at a solution is via systematic processes, or through trial and error and/or as a result of just being lucky. We refer to these recordings as 'videosnags' Videosnags captured in this manner are able to distinguish in real time between the input of one learner and of others without intruding into the learning or thinking process. It offers a non-intrusive way of getting inside students' cognitive and decision-making processes without drowning the students in recording technology. Since the handheld computer is also the learning tool, the research technology supports rather than disrupts the learning process as in the case of using traditional recording devices, where their obvious presence has the potential to disrupt learning and teaching. In addition, analyses of data are cumbersome, requiring multiple data sources and time consuming transcriptions. Furthermore, it has not been easy to track students' transitional thinking where they 'try' and 'change/erase' or 'toy with' ideas before deciding to use or not to use them. These transitional representations show areas of uncertainties/exploration in students' thinking and indicate, for example, if the arrival at a solution is via systematic processes, or through trial and error and/or just being lucky.

By reviewing videosnags and playing back the sequences of actions (thinking) and dialogue captured, (a) researchers can gain insight into the evolution of a student representing a new concept and identify patterns of learning of different ability students (b) students are able to 'see' and 'hear' what they are thinking, enabling them to engage in metacognitive thinking to improve their own learning via inspection of any of the modalities of representation that they have used, (c) teachers will have access to better information to offer individualised support for their students' meta-reflections. Since viewing students' videosnags could be a time consuming process for teachers, viewing them with more able peers for support should be encouraged (see Figure 4).

RESEARCH METHOD

The participants in the research studies were year 6 and 7 students from two primary and one secondary government schools in Victoria, Australia. At the time the research was conducted, the students were familiar with the use of the handhelds. While

Figure 4. Captured learning by handheld computer for improved learning

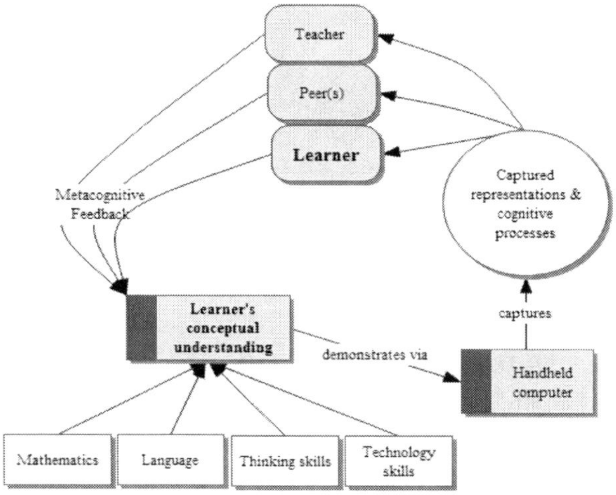

the secondary school students purchased their own handhelds (PDAs), the primary schools had only class sets of them. The handheld computers the schools used were either Dell Axim 51 or a HP iPAQ hx2490b type of handheld computer (Figure 5). Both brands of computers ran on the Windows platform and contained similar software, for example, Word Mobile, Mobile Excel, Windows Media, Notes for drawing or writing and audio recording, Powerpoint presentation capability, Internet Explorer, an organiser and a calculator. These versions of Dell and HP handhelds did not have cameras or video recording capabilities.

In conducting the research, volunteer students were provided with handhelds that had the *Cool-Capture* software installed in them. Prior to commencing the tasks, the video capturing software were turned on in their handhelds and the students were informed that everything they would say and do on the handheld screen would be recorded. For each study, between 7 and 10 handheld computers with the recording software were distributed to students who were willing to participate in the research. As the avi-files were quite large, the videosnag recordings were automatically programmed to be saved on secure digital (SD) memory cards. Due to a limitation of the *Cool-Capture* software in its inability to automatically save when exiting the program, the students were instructed not to turn off the handhelds at the end of each session but to leave them on the table for the researchers to save their work. Inevitably, some students turned off the handhelds either intentionally or unintentionally, hence full recordings on all the handhelds distributed were never achieved. For two of the research sessions, five recordings were captured while seven were captured for the third research session.

Observations of how students used the handhelds and how they collaborated with each other were made as field notes and triangulated with the videosnag data. The videosnags were viewed 2-3 times and analysed by interpreting screen-based

Figure 5. Dell Axim 51 (left) and HP iPAQ (right)

actions and associated verbal conversations. Analyses were made of a combination or all of the items below:

- the images on the screen
- the sequence in which varied applications were accessed
- the written text produced on the handhelds by the students, the relationship between the texts, images and uses of applications with any spoken comments recorded - the relationship between the different modes and modalities of representation
- any changes in representation or strategies to problem-solving tasks over time

Similarities and differences in how students represented their understanding or solved problems will be described in each of the three research studies reported below.

RESEARCH STUDY 1

This research study captured how year 6 students from the first primary school solved decimal-based mathematical problems with a learning object called *Wishball* that is produced by the Australian Government Learning Federation initiative. *Wishball* problems were presented to them in a game format. A snapshot of *Wishball* is shown in Figure 6. A student playing this game is required to reach their target number in 20 turns or less. In Figure 6, the target number is 8.058. The student begins the game with a randomly generated number which has a thousandth place value. At each turn, the student spins the ball on the left and a number is generated. The student will be required to add to or subtract this number from the starting number by clicking on the + or – button. (S)he will keep doing this with each spin of the ball and with a newly generated number until the target number is achieved. The ball with 'W' on it is the wishball. Spinning this ball will provide the student with a final number to use and the game will end after this. The game requires the student to use the numbers generated strategically to arrive at the target number in as few turns as possible.

Five of the students' videosnags were viewed at least twice and brief descriptions of the processes the students undertook were made. Com-

Figure 6. Wishball mathematics game

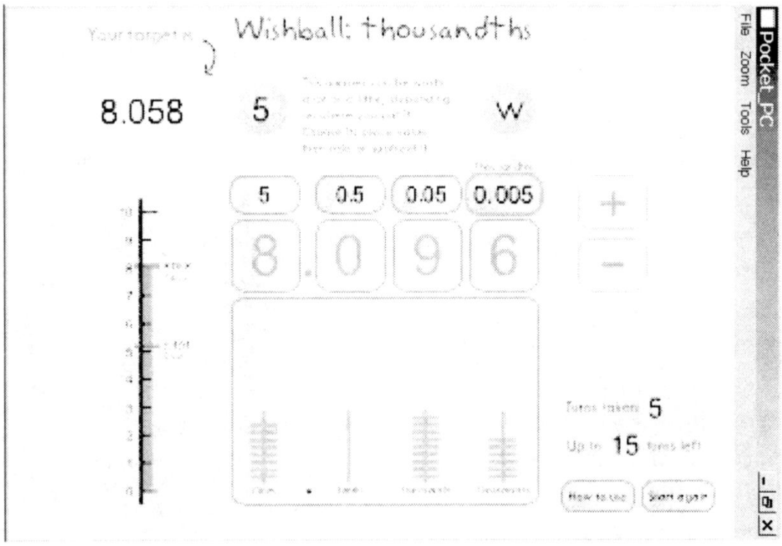

mon themes that were deduced from the videosnags are described below.

The game required the students to first understand the rules of the game and how the game was to be played. Not every student could grasp them the first time and the audio-recording function of the software on every handheld captured collaborative discussions and explanations of how the game was to be played. Spontaneous conversations that promoted collaboration between peers when using handheld computers were a common feature that was captured by the videosnags and reinforced by the researchers' observations.

A common theme that emerged is that students had the tendency to give up or restart the game when they have not achieved any of the target numbers after 5-7 turns. Four of the students demonstrated this trait although two of them successfully reached their target numbers in 5 and 8 turns in subsequent trials. The student who was successful with 8 turns used the wishball in the eighth turn to complete the game. These data appear to suggest that students need to feel success quite early in the piece to want to continue with the game. This need is further supported by the frustration captured in one of the videosnags in which the student said several time 'I just need one number'. This student however persisted until he completed the game at the tenth turn.

For three of the students who successfully arrived at the target number in 5-10 turns, apart from obvious addition or subtraction to get a target number, the general strategy was to systematically select the place value to add if their number was smaller than the target number or subtract if it was larger. For students who did not successfully arrive at their target number, it appeared that the selection of the place value and the addition/subtraction were random. Evidence of this was found when there were obvious additions or subtractions to arrive at one of the targeted numbers, but these options were bypassed for placement selections and modes of operation that were less obvious.

For two of the students who were confident with solving the problems in the *Wishball* games, obvious additions or subtractions did not deter them from achieving targets in a more subtle and less obvious manner. An example is shown in Figure 7. In Figure 7a, the number generated was 9 at this fourth turn. This number was placed over the different decimal places. The obvious selection would be the thousandth place which was 0.009 and subtraction of the '1' underneath it would have yielded the '8' in the thousandth place of the target number. However, the student selected the hundredth position and added the 0.09 to the number '5' below it, allowing him to achieve the number '4' in the tenth place of the target number.

Using mobile technology to engage students with learning about mathematics in this manner was motivating for the students. As the Learning

Figure 7 (a) – (c). Sequence of actions to obtain the tenth place value of '4' in the target number of 5.498

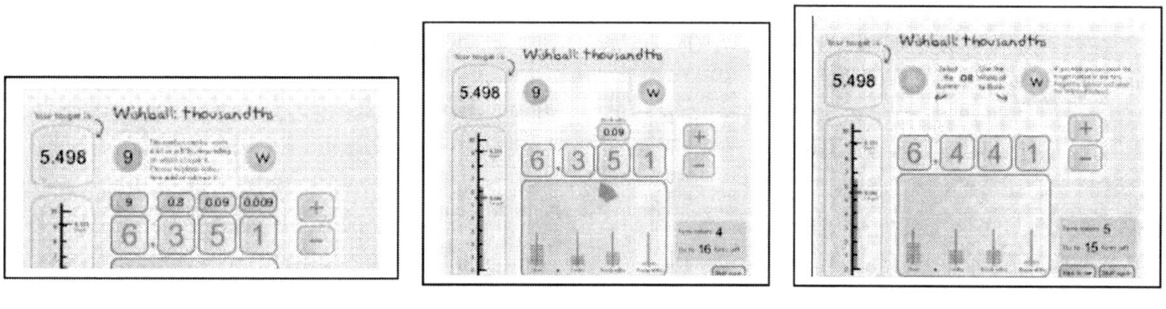

(a) *(b)* *(c)*

Federation learning objects had not been constructed for mobile devices, the small screen size of the handhelds meant that some of the letters could not be read as clearly in 'normal' full screen mode without using the 'zoom' function. However, being unable to read clearly did not appear to be an issue for the students as they continued with playing the game. When asked why they persisted even though some of the words were too small to read clearly, their replies were 'because it is fun' and 'this is a pretty fun way to learn'. Evidence from the videosnags to confirm this eagerness to play came from a couple of students' recording which showed that they did not wait for the *Wishball* game to finish congratulating them but started on a new game as soon as they had successfully achieved their targets.

RESEARCH STUDY 2

This research study was conducted in a school with a pocket PC program. The eighteen year 7 students in the study owned personal pocket PCs, which were HP iPAQ hx2490b handheld computers. The eight students participating in the research were provided with Dell Axim 51 handhelds with the *CoolCapture* software while the rest of the class worked on the task provided with their own personal handhelds. This did not seem to be a problem since both brands of handheld computers ran on the Microsoft platform. The software was turned on before the handhelds were distributed to the participants. However, issues with incorrect saving and other technical problems (software freezing in two of them) meant that the videosnags of only five of the students were captured. The videosnags captured showed how these students worked on a science and mathematics integrated task on the topic of the circulatory system. The task that the class of year 7 students was required to undertake consisted of several parts as indicated below:

1. determine and compare the averages of their own individual pulse rates before ('resting') and after exercises
2. obtain the 'resting' and 'after exercise' pulse rates from every other member of the class
3. determine and compare the average class pulse rates for 'resting' and 'after exercise' with their own individual pulse rates and
4. construct graphs or histograms for the number of students with pulse rates in each of these categories: 71-80; 81-90; 91-100; 101-110; 111-120; 121-130

In the next sections we will describe what the videosnags were able to inform us of in terms of how students made use of the tools on the handhelds and the strategies they used to complete the task.

Pulse Rate Data Collection and Recording

Individual pulse rate data was recorded in either Word or Excel. The two students who used Word either scribed or used the keyboard to enter the data. Three of the students entered their individual and class members' pulse rate data directly into Excel.

The videosnags indicated some attempts by the year 7 students to Bluetooth their individual average pulse rates to other class members. Based on observations, this stopped after a little while as the students indicated that the process was taking too long. The mobility of the handhelds allowed the students to move around the class to obtain their peers' pulse rates and to enter them directly into their handhelds.

Finding the Averages of Individual and Whole Class Pulse Rates

The videosnags from the handhelds showed that students used three ways of obtaining averages of individual and whole class pulse rates. Averages of two individual pulse rates were obtained using the calculator in the handhelds. Where there was no evidence of using the handheld calculator in the videosnags, it can be assumed that the students calculated that mentally, manually on paper or with a separate calculator. The latter two options were observed with several class members in the class.

The videosnags captured three ways in which the students calculated the averages of the class members' pulse rates. For student who entered their data in Word, the calculator was used for calculating the class averages. For students who entered their data in Excel, the two ways of calculating the class averages were:

i. using the formula bar where the formula of the sum of the class pulse rates divided by the number of students was entered
ii. using the 'average' function in the Excel spreadsheet as shown in Figure 8. Figure 8 (a) shows the students highlighting (selecting) the 'average' function. Figure 8(b) shows the average of the 18 students pulse rates as 70.833.

The videosnags provided insights into both the mathematical and technical skills of the students in finding the class averages. In addition, the understanding of one of the students was elicited through a conversation that was recorded in the videosnags. The student had asked (her partner) 'how can you have one average for the whole class?'. This student appeared to know the mechanics of finding averages but lacked the understanding of what 'average' meant.

Comparing Pulse Rates

The year 7 students were asked to compare their own average 'resting' and 'after exercise' pulse rates with that of the class. This was done in text modes in Word or Excel. For example in Figure 8(b), the display was

(iii) me = 86, (class) average = 70.833

While the students were not asked to compare their class members' individual 'resting' and 'after exercise' pulse rates, two of the students constructed graphs to do this. Figure 9 shows the saved files for the two graphs. The two methods of graphing demonstrated in the Figure are: in 9(a) each class member's 'resting' and 'after exercise' pulse rates have been plotted next to each other and in 9(b) all the 'resting' pulse rates for each member

Figure 8 (a) – (b). Using the 'average' function in Excel to find the class average pulse rate

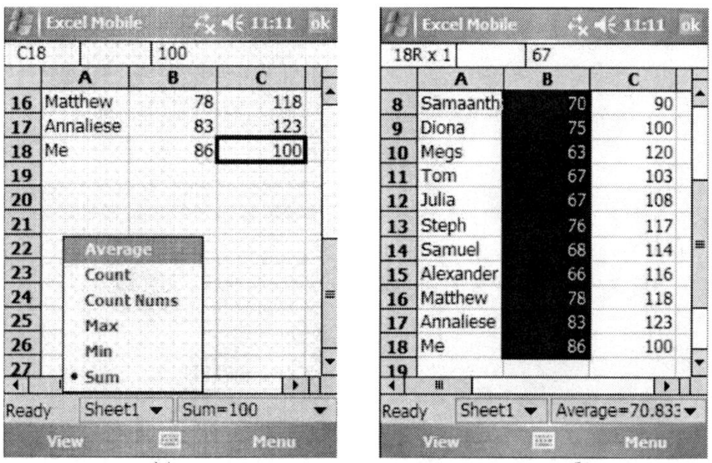

(a) (b)

of the class have been plotted together on the left in the figure and all the 'after exercise' pulse rates on the right. The students also carefully chose the colour scheme for representing the bar graphs. For these students, the videosnags showed their competent technical skills in constructing these graphs. However, the manner of presentation of the two sets of data for 'resting' and 'after exercise' is different and indicates different interpretations of the task. Visually, the representation in 9 (a) makes it easier for viewers to compare individual before and after exercise pulse rates. While these representations were saved in Notes and do not require video-capture software to view them, the videosnags have the potential to identify, through transitional representations, whether a students' inability to produce graphs on Excel is due to a lack of technical skills or a lack of understanding of the task undertaken leading to the inability to choose the appropriate modes of representation in mathematics.

For part (iv) of the task, students were required to categorise the pulse rates of the class member into categories 71-80; 81-90; 91-100...etc. The videosnag showed that one student used the 'tally' method of doing this. The tally strokes in each category were subsequently erased and replaced by the number. Another student verbally counted the pulse rates from the spreadsheet and entered the total in each category. The rest of the videosnags did not indicate that the students had started this task when the period ended.

RESEARCH STUDY 3

Seven year 6 students from the second primary school participated in this research study. The class had just finished studying about biotechnology and other science concepts, as part of a federally-funded science project. The modes of representations that the students were exposed to in learning about science and biotechnology included:

- **Verbal:** Powerpoint presentations, discussions
- **Written:** Worksheets, assignments, glossary
- **Visual:** Labelled diagrams, videos
- **3D modes:** Models (DNA double helix); Experiments (DNA extractions, yeast-based experiments)

Figure 9. Two different ways of graphing to compare individual class member's before and after exercise pulse rates

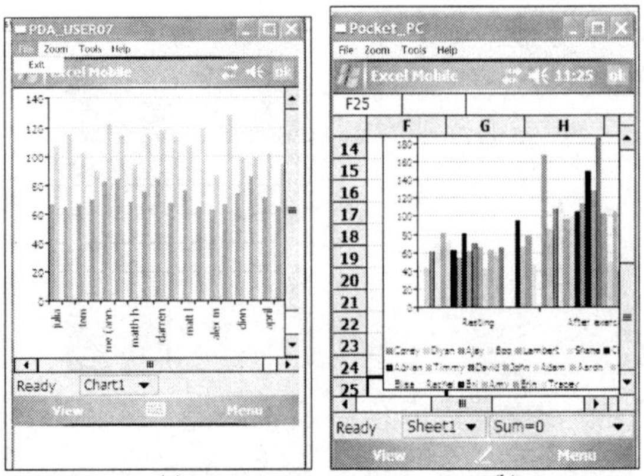

(a) (b)

- Numerical: Mathematics (measurements, estimations in experiments)

Some of the tasks that the students were required to undertake with their handhelds involved working individually to

(i) solve a mathematical problem: a manufacturer claims that a new biogenetically engineered motor oil saves 10% of the petrol used by a car. If you drive 24,000 km a year and get 16 km a litre of petrol, how many litres of petrol would you save in one year?
(ii) construct a diagram of a cell and label at least four features that they have learnt about it.

The students were also required to work collaboratively with their handhelds to show the relationship between gravity, weight and mass.

The sections below will describe selected cases to demonstrate the capacity of the video capture software to provide insights into how students represent their thinking in the various tasks.

Individual Task: Solving the Mathematics Problem

In solving the problem in question 1, one of the students, John (names of students are not real), made two attempts. The first (Figure 10a) was discarded and a second attempt was made, dividing 24,000 km (the total distance travelled) by 16 (the kilometres per litre of petrol). As no indication of the use of the handheld calculator was captured, the answer of 1,377 (Figure 10b) is indicative of incorrect (possibly mental) division. John proceeded to find 10% of 1,377, which as shown in Figure 10c is the correct process, indicating that he is familiar with percentages, fractions and

Figure 10. Transitional representations during the process of solving question 1

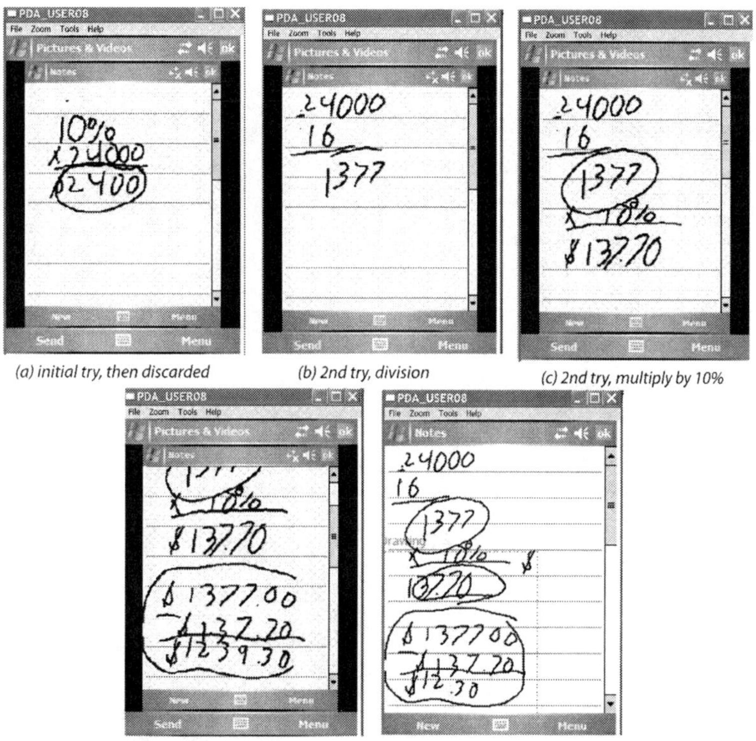

(a) initial try, then discarded
(b) 2nd try, division
(c) 2nd try, multiply by 10%
(d) final answer captured in
(e) answer saved in Notes

decimals. The subtraction for the final step (10d) is mathematically correct but the reasoning is incorrect since 137.70 is the 10% of the (incorrect) 1,377 litres of petrol instead of dollars as there is no reference to money in the problem.

The captured recording revealed several weaknesses demonstrated by the student in regard to his mathematical skills and reasoning and in the strategies he employed to solve this problem:

(i) While in the second attempt, he understood the need to divide 24,000 by 16 followed by finding 10% of the answer obtained, the last step of substraction demonstrated that he either did not carefully read the question, which asked for the amount of petrol saved or he made the general assumption that 'savings' have to do with money and that a subtraction needs to be done (as would have been the case with many problems that deal with gains and losses in buying and selling).

(ii) John was confused by the different quantities, for example, interchanging between litres and dollars - see Figure 9c where 10% of 1,377 litres became $137.70 and in Figure 9d, subtracting $137.70 from 1,377 litres which was changed to dollars by the addition of the $ symbol. He does appear to have some understanding of 1,377 being required to be in dollars (same quantity) in order to do the last subtraction.

(iii) He did not make use of the calculator in the handheld although he knew that it was there as shown in one of the images captured later, where he did use the calculator for question 5.

Only student Liz correctly solved this problem. She combined the use of the handheld calculator and Notes to represent her solution to the problem. She did the division of 24,000 by 16 with the calculator and used Notes to manually work through the percentage calculations showing the cancelling down working (see Figure 11) representing her thinking by using both numerical and linguistic symbols. She then re-checked the manual calculations on a separate Notes notepad.

For another student, Kim, the *CoolCapture* software could capture and explain the discrepancy between the answer recorded in the Word file named 15000 in her handheld and the answer to Q1 in the document itself, which was written as 150. Kim's initial answer was 15000. Approximately nineteen minutes into the task, she discussed the answers with her partner, Jane. The discussion was recorded on the handheld:

Kim: have you done the first question...this one? I just guessed

Jane: you can't just guess. Divide 24000 by 16 then take 10% off....I got 150.

Kim: I did different, I divide by 16 then times 10.

Figure 11. Using Notes to show calculations manually

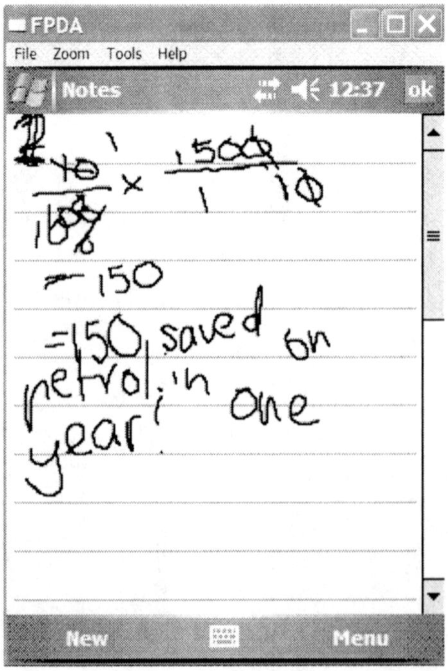

At approximately 21 minutes into the task, Kim opened up the Word file 15000, and changed the answer from 15000 to 150. Being able to capture the students' conversation and understand that Kim did not actually understand the problem, even though the answer appears correct, is important for teachers seeking to help students to improve. The application permitted this insight to be gained without additional intrusion or personnel.

Individual Task: Constructing the Cell Diagram

All students used the Notes function in their handheld computers to hand draw the cell diagrams. Six of the students drew cell diagrams similar to the one shown in Figure 12, varying only in the number of circle-like organelles inside the cell and either correctly or incorrectly naming them. Most of these students took about two minutes to construct the diagrams at first go, with one student erasing the first diagram and re-drawing a similar diagram later. All students labelled the cell by scribing even though typing in labels using the keypad is possible in Notes.

Figure 12. Diagram of cell with labels membrane, nucleus, DNA and fluid

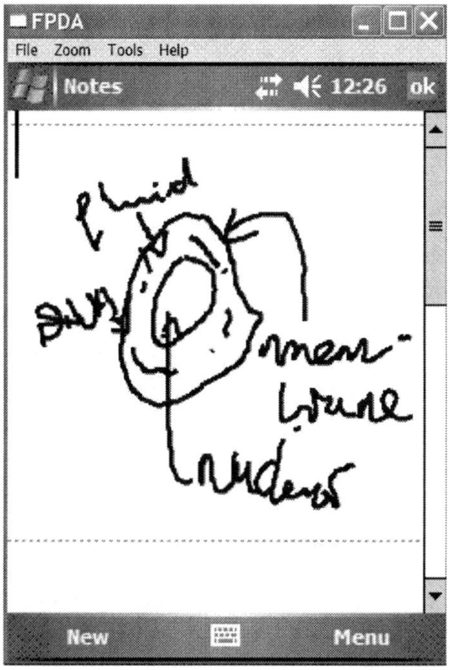

The process of constructing the cell diagram differed with one student. She erased the first diagram and started again on a second, which was not a lot different from the first. Some stages from her diagram are captured in the images below (see Figures 13a-h). In each of the images shown in Figure 13, the non-full screen is shown. In full screen, the video could be viewed in the correct orientation. In Figure 13a, the ID of the device is PDA_USER01, shown in what would be the 'top left' corner of the device. The number 40:19 at the top right of the device is the total time recorded while the time of 12:56 immediately to its left is the time at which this part of the videosnag was captured. Hence the first image was recorded at 12 minutes and 56 seconds into the total recording time of 40 minutes and 19 seconds.

The sequence of the images in Figure 13 should be read according to the rows. (a) and (b) are two initial versions of the student's diagram, with (b) building on (a). Before the (b) image was deleted, the biotechnology module consulted (c) then (d). The diagram was started again (e) and (f), and the module consulted for labels (g) and the diagram annotated (h). This student started with a diagram of two circles (one inside the other), then deleted the file and started over again with a diagram of three (almost concentric) circles. She referred to the biotechnology module in her handheld several times to assist with some labels. She took about five and a half minutes (time 12:56 – 18:30) to complete this task. The videosnag captured both the interim attempts by this student to develop a representation, the resources that she consulted and the relationship between her ultimate representation and the model representations available to her – including some of the differences in formats of representation (lack of colour, non-standard spellings).

Figure 13. Composite diagrams showing sequence of actions in drawing and labelling a cell

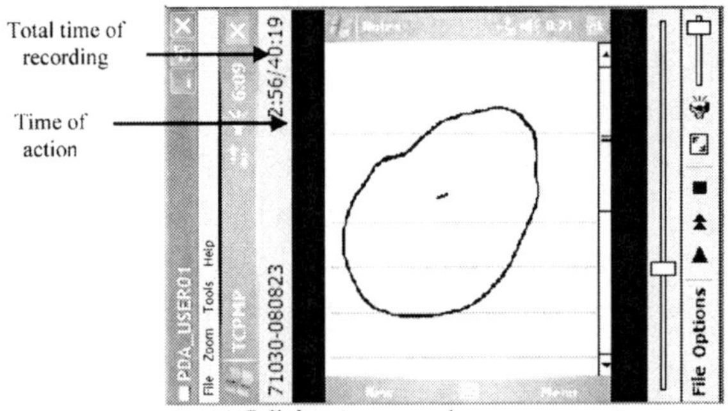

(a) Cell drawing: stage 1

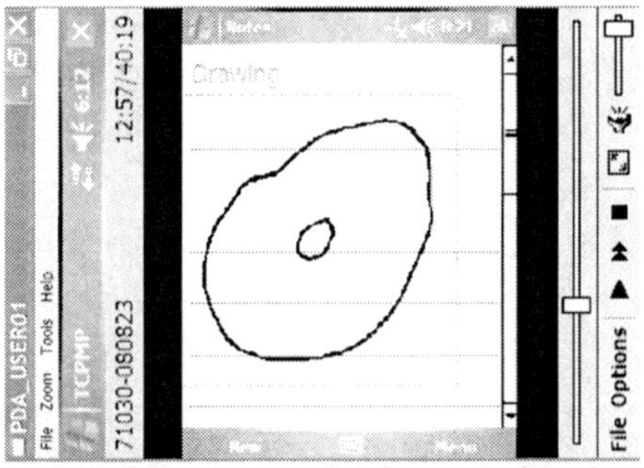

(b) Cell drawing: stage 2 (subsequently deleted)

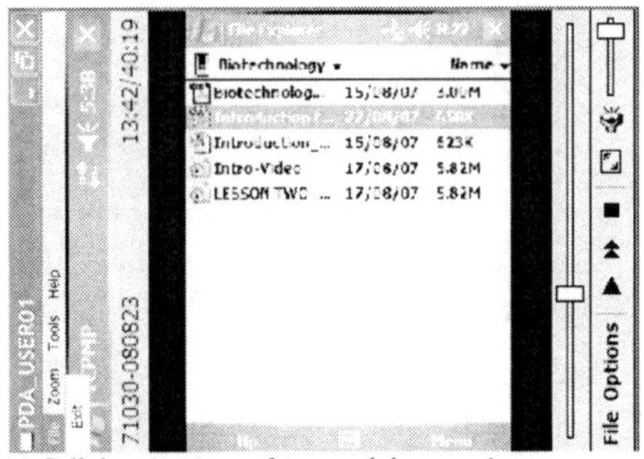

(c) Cell drawing: consulting module stage 1

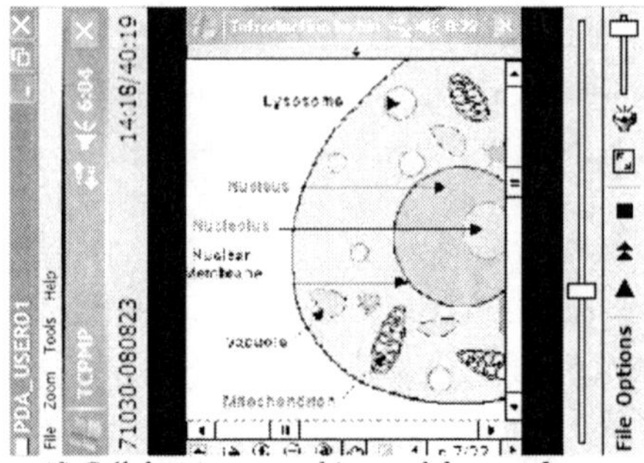

(d) Cell drawing: consulting module stage 2

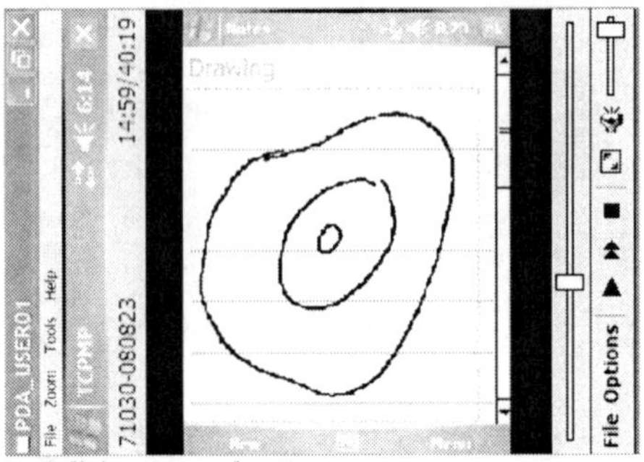

(e) Cell drawing: redrawing

(f) Cell drawing: labelling

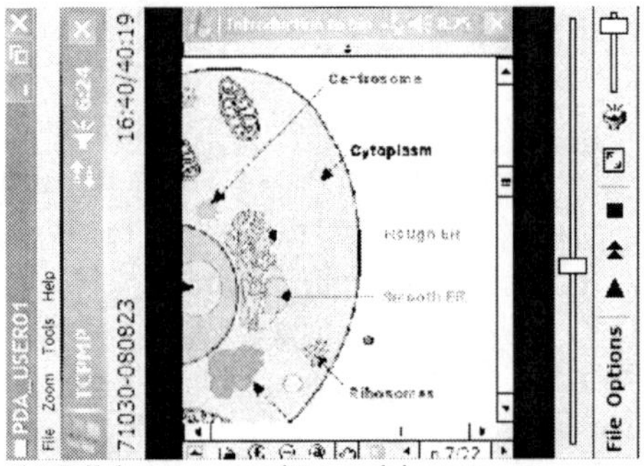

(g) Cell drawing: consulting module again

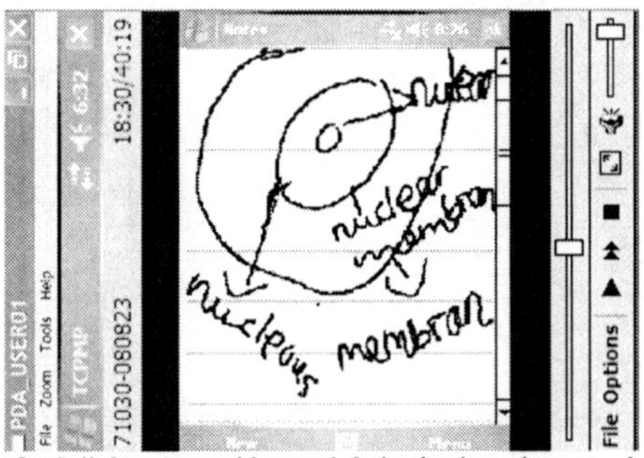

(h) Cell drawing: add more labels (final product saved)

Collaborative Task: Representing Conceptual Understanding of 'Gravity', 'Weight' and 'Mass'

The collaborative task stipulated that the students had to work in small teams and to use Bluetooth to show the relationship between gravity, weight and mass. The students formed two groups of 3 boys and 4 girls for this task. The girls started on the task by discussing how to define 'gravity' but had not come to a consensus when the period ended. An attempt to seek assistance from the other group was shown in the Bluetooth messages that one of the girls sent. The message 'figured out group task yet?' was sent but she subsequently deleted the file with the message from her handheld device. What was captured in this task undertaken by the group of girls included verbal representations and a strategy (seeking assistance) to solve the problem.

In contrast, the strategy that the three boys adopted for this task was for one student to construct a sentence about gravity. The first student constructed the sentence 'Gravity pulls stuff down'. This was saved as a Word file and sent by Bluetooth to the other two students. This first student also drew a diagram of earth and a person standing on it with a downward arrow to show gravity – indicating combined modes of representation (text and visual).

Using the file received, the second student undertook to show the relationship between the two words 'gravity' and 'weight' in a phrase 'no gravity no weight'. This was typed below the first sentence, saved and the file transmitted to the other two students. The third student constructed the last sentence 'mass times gravity = weight' to show the relationship between mass, weight and gravity. This was typed underneath the second sentence. Hence in the same file, the three sentences constructed by three different students was shown (see Figure 14). This process of collaboration that was captured by the *CoolCapture* software showed that the boys had adopted a systematic way of collaborating in completing this task.

Figure 14. Collaboration to show the relationship between gravity, weight and mass

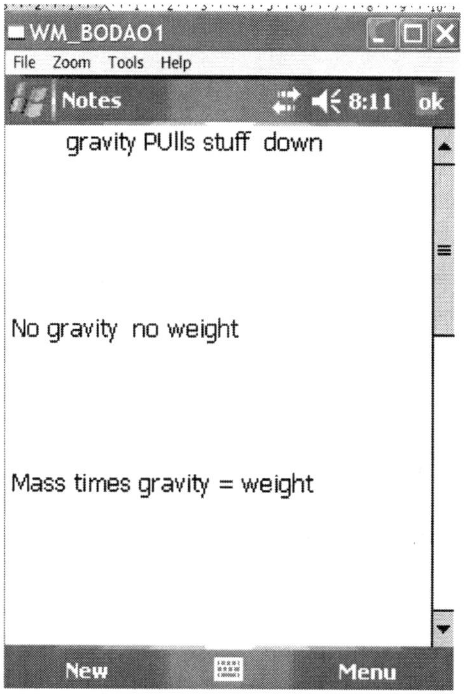

CONCLUSION, LIMITATIONS AND IMPLICATIONS FOR FUTURE RESEARCH

This chapter has reported on the capacities of handheld computers and screen/audio-capture software to create a record of students' thinking and learning in science and mathematics. The videosnags captured how students worked individually and collaboratively to solve mathematics and science problems or construct representations of mathematics and science understandings. Videosnags could potentially serve several purposes: (i) provide researchers with insights into some of these students' thinking and decision making processes (ii) assist researchers/teachers to identify the weaknesses demonstrated in the processes and (iii) allow for playing back to students to help them review their own thinking and reflect on their own understanding as a way to actively re-construct thinking and understanding with or without assistance from peers or teachers.

In summary, videosnags captured using the *CoolCapture* software were able to provide quite a rich source of data that enabled us to: (i) track the sequence of actions students took in creating a science representation or solving a mathematical problem. This allowed us as researchers to get into some aspects of the minds of students to follow their processes of construction, strategies used and choices made in constructing representations to demonstrate understanding of the concepts under study (ii) capture the student's choices of applications and tools, for example Excel, Word, Note, audio-recorder or calculator (iii) track the amount of time spent on each problem and the order in which students undertook a set of problems. For example, in some videosnags, students spent considerable amounts of time colouring in fonts and formatting their presentations (iv) capture the transitional representations that students 'try' and 'change/erase' or 'toy with' before deciding not to use them. This transitional thinking, which is often not evident in pen and paper based representations, could help with identification of areas of strengths and weaknesses in students' thinking (v) monitor how collaboration between students works in creating understanding through representations and (vi) gain insights into the IT skills and proficiencies of students using the features of handheld computers.

A limitation of the software is that the amount of illumination of some of the keys, especially the operation keys (+, -, x or /) are brief and often almost unnoticeable. Another limitation of this type of usage is that for teachers, it would be too time-consuming to watch all the students' videosnags, even as a means of assessment. The best use of these videosnags would be to assist weaker students to identify their areas of weakness and be engaged in metacognitive processes to improve their learning performance. One way to do this is to recruit the assistance of more capable peers to help the student to reflect on and work through his/her thinking and the strategies employed in presenting that thinking and hence understanding.

In terms of the actual potential of the technology, we acknowledge that this method does not capture *completely* a student's thinking if it is based entirely on screen-captured actions. However, structuring tasks that require students to represent understanding in multimodal and multiple modes, for example, verbally recording an explanation of their actions as well would assist in obtaining a more complete view of the students' thinking. Alternatively and as a research strategy, the 'videosnags' could be replayed to students in stimulated recall interviews to elicit the 'why' of persistent or inaccurate actions. Observations on how students work with the tools in the handhelds would further enrich the data sources.

One of the implications of using the handheld computer as a pedagogy and research tool is that teachers will need to be professionally developed to prepare their students to use the applications and software on their handhelds, including multimodal representations so that lack of technical skill does not mask data on students' learning difficulties

While there are similar avi-screen capture software for desktops and laptops, the need to book computer labs or sets of laptops, the lack of Bluetooth/beaming functions and just the sheer size of the computers and laptops are disadvantages when compared with handhelds. While there are limitations with handhelds, this chapter has shown that they are sufficiently sophisticated for students to demonstrate representations in written and drawn forms to convey scientific and mathematical understandings. Future research will encourage the use of the handheld's audio recording, PowerPoint and image displays, video-playing capacities as well as other software such as concept mapping and Excel (for graphing display) as means of representing understanding that could be tracked through videosnags. The research could potentially lead to identification of patterns of learning that demonstrate strengths and weaknesses in the learning and thinking processes. The research will also assist with improving pedagogy in the classroom.

ACKNOWLEDGMENT

This research was supported by a grant from the Faculty of Education, La Trobe University.

REFERENCES

Ainsworth, S. (1999). The functions of multiple representations. *Computers & Education*, *33*, 131–152. doi:10.1016/S0360-1315(99)00029-9

Ainsworth, S., & Iacovides, I. (2005, August). *Learning by constructing self-explanation diagrams.* Paper presented at the 11th Biennial Conference of European Association for Research on Learning and Instruction, Nicosia, Cyprus.

Anderson, J. R. (1983). *The architecture of cognition*. Cambridge, MA: Harvard University Press.

Bardini, C., Pierce, R., & Stacey, K. (2004). Teaching linear functions in context with graphics calculators: students' responses and the impact of the approach on their use of algebraic symbols. *International Journal of Science and Mathematics Education*, *2*, 353–376. doi:10.1007/s10763-004-8075-3

Becta, (British Educational and Technology Agency) (2003). *Becta ICT Research Online*. Resource document. Retrieved May 27, 2008 from http://www.becta.org.uk.

Bruner, J. S. (1960). *The process of education*. Boston: Harvard University Press.

Derbentseva, N., Safayeni, F., & Canas, A. J. (2007). Concept maps: experiments on dynamic thinking. *Journal of Research in Science Teaching*, *44*(3), 448–465. doi:10.1002/tea.20153

Finn, M., & Vandenham, N. (2004). The handheld classroom: Educational implications of mobile Computing. *Australian Journal of Emerging Technologies and Society*, *2*(1), 1–15.

Ng, W., & Nicholas, H. (2007). Ubiquitous learning with handhelds in schools. In Norman, A. & Pearce, J. (Eds) *Proceedings of the 6th International Conference on Mobile Learning*, Melbourne, Australia. 16-19 October, 2007, p 186-193.

Ng, W., & Nicholas, H. (2008). Capturing individual thinking and through collaboration with handheld computers. *British Educational Research Association*, Edinburgh 3-6 September, 2008

Ng, W., & Nicholas, H. (2009). Introduction of pocket PC in schools: attitudes and beliefs in the first year. *Computers & Education*, *52*(2), 470–480. doi:10.1016/j.compedu.2008.10.001

Nicholas, H., & Ng, W. (2009). Ubiquitous learning and handhelds: An overview of theory and pedagogy. In Rogers, P., Berg, G., Boettcher, J., Howard, C., Justice, L., & Schenk, K. (Eds.), *Encyclopedia for online and distance learning* (*Vol. IV*, pp. 2171–2176). Hershey, USA: IGI Global Publishers.

Nuthall, G. (1999). The way students learn. *The Elementary School Journal*, *99*, 303. doi:10.1086/461928

Palm Inc. (2002). *Palm Education Pioneers Program Final Evaluation Report*. Resource document. Retrieved 21 October 2008 from http://www.palmgrants.sri.com/PEP_Final_Report.pdf.

Perry, D. (2003). *Handheld computers (PDAs) in schools*. Coventry: Becta (British Educational Communications and Technology Agency). Retrieved May 12, 2008 from http://partners.becta.org.uk/uploaddir/downloads/page_documents/research/handhelds.pdf

Perry, D. (2006). *Learning2Go Wolverhampton Mobile Learning: Report – end of Phase II.*

Prain, V., & Waldrip, B. (2006). An exploratory study of teachers' and students' use of multi-modal representation of concepts in primary science. *International Journal of Science Education, 28*(15), 1843–1866. doi:10.1080/09500690600718294

Resource document. Retrived May 12, 2008 from http://www.learning2go.org/media/download_gallery/dpa%20L2G%20Ph2%20FINAL%20Report.pdf.

Russell, T., & McGuigan, L. (2001). Promoting understanding through representational redescription: an illustration referring to young pupils' ideas about gravity. In D. Psillos, P. Kariotoglou, V. Tselfes, G. Bisdikian, G. Fassoulopoulos, E. Hatzikraniotis, & E. Kallery (Eds.), *Science education research in the knowledge-based society. Proceedings of the Third International Conference of the ESERA* (pp. 600–602). Thessaloniki, Greece: Aristotle University of Thessaloniki.

Stylianou, D., & Pitta-Pantazi, D. (2002). Visualization and high achievement in mathematics: A critical look at successful visualization strategies. In Hitt, F. (Ed.), *Representations and mathematics visualization* (pp. 31–46). Mexico: Cinvestav-IPN.

Swan, K., van't Hooft, M., Kratcoski, A., & Schenker, J. (2007). Ubiquitous computing and changing pedagogical possibilities: Representations, conceptualisations and uses of knowledge. *Journal of Educational Computing Research, 36*(4), 481–515. doi:10.2190/B577-7162-2X11-17N5

van 't Hooft, M. (2006). Tapping into digital literacy: Handheld computers in the K-12 classroom. In Tan, L. W. H., & Subramaniam, R. (Eds.), *Handbook of Research on Literacy in Technology at the K 12 Level* (pp. 287–307). Hershey, USA: Idea Group Publishing.

Waldrip, B., Prain, V., & Carolan, J. (2006). Learning junior secondary science through multi-modal representations. *Electronic Journal of Science Education, 11*(1), 87–107.

Chapter 7
Using Mobile Technologies as Research Tools:
Pragmatics, Possibilities and Problems

Ria Hanewald
The University of Melbourne, Australia

ABSTRACT

This chapter describes the use of mobile technologies as research tools while discussing both the benefits and limitations of this mode. It maps practical approaches to the management of a mobile, digital data collection process and its accompanying challenges, namely the potential and pitfalls in gathering, coding and archiving the research data. The use of mobile technology devices and the amalgamation of hardware, operating systems, networking and software to facilitate and support this process are discussed. The lessons learned from the technical aspects underpinning fieldwork for research purposes will assist others in maximizing the potential of mobile technology to support their research projects.

INTRODUCTION

The Use of Mobile Technology to Support Research

This chapter discusses the arrival of mobile technologies, types of digital research methods and issues in exploiting the capabilities of the technology. Practical suggestions for technical strategies and processes are made so that qualitative researchers may employ those for and in their research contexts. It is argued that mobile technologies need to be adopted and that their integration will require a rethinking and reshaping of social research and its practice.

New developments in Information and Communication Technology (ICT) and their pervasiveness in contemporary culture constantly change the way people communicate and interact. This widespread availability and accessibility of ICT has implications for investigating and understanding a technology-saturated society. Without a doubt, mobile technologies transform the way we do things including the way we do research.

Mobile technologies provide an eclectic assortment of sources (text, audio, still images and

DOI: 10.4018/978-1-61692-849-0.ch007

film clips, which enable new ways of creating, collecting, processing and analyzing material. Although this diverse combination adds richness to the research materials, it also highlights the need to integrate and blend the mixture properly. In consequence, in dealing with digital forms of data, researchers will need different skills, expertise, resources and tools. Given that they are the growth platform of the future, it is imperative for evolving research practice to incorporate these new techniques into the existing research repertoire. Furthermore, the implications of multimedia for qualitative research in general and the impact of digital technologies on social research in particular will not only transform the way in which research is done but also the way in which research narratives are constructed.

BACKGROUND

Definition of Key Terms

Prior to tackling the methodological, theoretical, technical and ethical aspects of dealing with digital technology as research tools, it seems appropriate to briefly clarify the terminology. New technologies, emerging technologies, digital technologies, mobile technology and portable technology are terms that are often used interchangeably to describe advanced technologies in recording and manipulating information and communication. In this chapter, Information and Communication Technology (ICT) will be used as an umbrella term to refer to the use of computer-based information systems (i.e. PCs, servers, mainframes, networked storage) to convert, store, protect, process, transmit and retrieve information. Digital technology has the same functionalities but the associated devices (i.e. tape recorder, still cameras, camcorder) are not computer-bound. Their main advantage is the freedom of mobility, hence mobile technology is exactly that: technology that is portable. Thus, they are ideally suited for conducting fieldwork, yielding audio-recordings, photographs, video footage, graphics and so on. These digitally-generated research materials can be integrated through the computer - together with traditional written interpretation – into hypermedia environments. Hypermedia meshes audio, video, text, graphics and hyperlinks to create an interactive, non-linear stream of information (for example the world wide web).

Qualitative and Quantitative Instruments

Technology can be used for and in research projects. Both quantitative and qualitative research can be conducted with and through new technologies. The major research traditions (e.g. grounded theory practice, narratology, storytelling, ethnography or shadowing) and their associated techniques (e.g. surveys, questionnaires and field notes, observations, interviews, focus group discussions, reflexive journals, drawn responses) all hold potential for the use of mobile technologies to do research more time and cost efficiently.

The Table 1 gives an overview of traditional instruments and their contemporary versions for each data collection technique.

Mobile devices such as digital voice recorders, PDAs (Personal Digital Assistants), still cameras, camcorders, hand-held scanners and digital pens open new ways for research practices due to their affordability, lightweight and user-friendly qualities. These hardware components are particularly helpful in the collection of data, capturing rich materials that can be easily processed, stored and presented.

"The single most appealing advantage of the electronic method of data collection is the elimination of the tedious data entry process." (Benfield and Szlemko, 2006:12)

ICT offer the capacity and potential for building new qualitative methods in the field of social research. Computers and specific software are basic requirements for digital data storage or analysis,

Table 1. Overview of research instruments with traditional and contemporary tools

Instrument	Traditional	Contemporary
Survey	Pen & Paper	online, ie. survey monkey, ahaSurvey, Zoomerang
Questionnaire	Pen & Paper	online, ie. Question Pro
Interview	Tape recorder	digital voice recorder, camcorder
Observation	Pen & Paper	camcorder, mobile phone, digital pen
Field notes	Pen & Paper	camcorder, mobile phone, digital voice recorder, digital pen
Focus Group Discussion	Tape recorder	camcorder, digital voice recorder
Reflexive journal	Pen & Paper	digital voice recorder, PDA, digital pen, hand-held scanner
Drawn response	Pen & Paper	Interactive Whiteboard (IWB), hand-held scanner, digital pen
Photo elicitation	Paper/ analogue photos Pen & Paper and/ or tape recorder	camcorder, digital voice recorder, digital photos/ camera, computer, portable DVD player

while peripherals such as printers and scanners afford means of data presentation.

The Technology Adoption Challenge

Qualitative researchers are now increasingly including new technologies for data gathering and analysis into their work. The incorporation of these new tools and techniques alongside the gradual evolution of digital, multimodal and hypermedia methods are reflected in recent publications (Hine, 2000; Dicks, Mason, Coffey & Atkinson, 2005; Kozinets 2006; Garcia, Standlee, Bechkoff & Cui, 2009).

These developments raise major methodological and technical issues, but also the need to address theoretical and ethical concerns. Methodological issues refer to the exploration of technology-rich societies themselves by studying hypermedia environments such as the Internet (research into ICT). The methodological issues also encompass the capacity to add multi-media data to traditionally collected research narratives. Integrating a mixture of sources adds additional dimensions to the research and yields a richer representation (research with ICT). An audio-visual data collection enriches text-based explorations. For example, non-verbal images in the form of digital photos and videos add to verbal investigations such as interviews. "We propose using the digital and wireless communication revolutions as platforms for rethinking ethnographic principles, methodologies, and analysis." (Masten & Plowman, 2003:75).

Figure 1. Research into, with and through new technologies

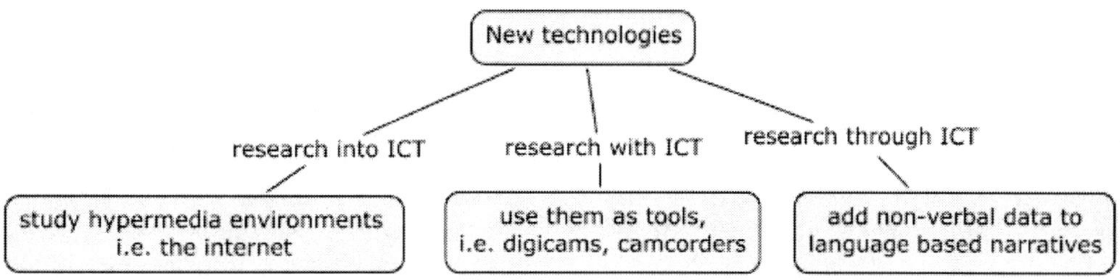

Apart from the need to re-conceptualize theoretical frameworks and methodological issues, researchers need to conquer the technical issues (i.e. handling equipment, physical security of devices) of using digital tools to generate and manage data (research through ICT). Obviously, visual data collection is a growth platform for the future. Using imaging technology in addition to traditional qualitative sources (i.e. interviews, field notes, surveys) adds another dimension to capture, explore and represent phenomena and experiences, especially in the social sciences.

Visual Methods

The next section outlines information about how to do visual research and to apply visual methods. Here, the key concepts of visual research are explained, specific examples of projects using visual methods and mobile technologies are given later on in the chapter. For the sake of brevity and relevance to this chapter's topic, readers interested in visual methods are referred to Bateson and Mead (1947), Collier and Collier (1986), Harper (1998) and Wagner (2006) as key referential texts.

Visual methods are often coined as 'beyond text', offering alternatives to expand and support research and produce knowledge. Weber (2008:47) outlines a number of ways in which visual images can be useful

- Images can be produced by participants
- Found or existing images can be used as data for theorizing
- Images and objects are useful to elicit or provoke other data
- Images can be used for feedback and documentation on the research process
- Images are useful as a mode of interpretation and/or representation

Images can be either *researcher-generated* or *respondent-generated*, which lends itself particularly to ethnographic studies. For *visual ethnography*, researchers can use mobile devices such as a digital camera, mobile phone, PDA, smart phone video camera and the like to record data as part of participant observation or classroom studies. The video camera functions like a tape recorder but with pictures and the visual record collects and preserves this richer data for further detailed study. Respondents using mobile technologies to record phenomena or experiences can take ownership of the research process. They are able to show their own perspectives and use the visual tool to display their own knowledge. At times, researchers will give guidance for the taking of images - either formally via written instructions or verbally and informally - to participants, suggesting particular topics, settings, sequences, length of contributions and so on. For participants with a limited vocabulary range or less sophisticated language skills (i.e. children, people using a second or third language) visuals add another layer to the narrative, thus giving a richer picture of their experience. Respondents who are intellectually immature or have cognitive limitations (i.e. children, intellectually disabled people, some of the elderly) may find it easier to create digital images and videos as a means of reflection on and exploration of their feelings and behaviors rather than just producing a narrative account (Prosser, 2008)

Photo-Ethnography in particular meets the needs of participants who prefer visual expressions, are uncomfortable and thus uncommunicative in new situations or have relatively short-attention spans and need stimulation to keep them focused. Participants can learn to be self-observant, using photos and storytelling as a self-observation tool while capturing their own lives and behaviors. Basically, researchers put a still camera into the hands of participants and ask them to use photographs to address specific questions directly related to the research issue. Participants explore the questions and take photos over several weeks, observing their own lives without the bias of an outside observer. Participants then debrief individually or in small groups, which often yields insights that are not achievable with traditional focus groups or classical ethnography.

Photo-elicitation is the approach used to enable and engage participants in conversation through visual stimuli. A photograph or film is an integral part of the interview, where participants are asked to discuss the meaning of the photo, video or film. Harper (2002) draws on three uses of photo elicitation: as an inventory of objects, people and artifacts; as part of collective events or institutional paths (i.e. historical photographs of proceedings or buildings to evoke a response); as intimate dimensions of the social (i.e. the family, one's own body). Photo-elicitation moves beyond the conventional language-based interview protocol as it uses photos of people, places and events as a focus for the conversation. Before the actual photo elicitation interview, the researcher has to decide who will take the photos. Photos can be either archival material, purposefully taken by a professional photographer or by the researcher, who will present the images to the participants (Hurworth, 2003). "This option allows the researcher to frame, select, develop, organize, and present the images to the interviewees based on their own research questions" (Stanczak, 2007:171). The disadvantage of *researcher-produced photographs* is possible omission of important aspects of the research. The images are chosen based on the researchers interest and not based on the interest of the participants. Alternatively, *auto-driven photo elicitation* uses images during the interview that were taken prior to it by the participants. Stanczak (2007:173) believes that this auto-driven photo elicitation is "…an ideal methodology to engage young people" as it conceptualizes children as active agents. Photo-elicitation accesses the child's perspective about specific issues and uncovers their views of experiences.

In addition auto-driven photo elicitation overcomes the awkwardness of interviews, lessens the power dynamics of an adult interviewing a child, caters to the linguistic and cognitive level of children thus stimulating verbal responses, empowers the interviewee, produces a richer interview experience and richer data (Stanczak, 2007).

Photo Novella (meaning picture stories) or *photovoice* is another form of photo-interviewing that encourages participants to talk about daily routines and events. It is well suited to participatory action research, allowing participants to create and speak about photos to facilitate personal and community change. In the production of knowledge, photovoice has the goals of enabling people to record and reflect their community's strengths and concerns, to promote critical dialogue about important issues through discussion of photographs and to reach policymakers (Wang and Burris, 1997).

The focus on the use of imagery such as photos and video footage created a need to think about arranging and disseminating ethnographic work and visual ethnography. The use of digital tools and skills has stimulated the emerging field of hypermedia ethnography. *Hypermedia ethnography* moves beyond linear forms and is well suited to illustrate diverging viewpoints, multiple perspectives and alternative voices.

Traditional and Contemporary Research Tools

The implications of hypermedia for the production, presentation and dissemination of social science research data moves beyond the limits of a solely written approach by offering an integrated representation. In doing so, Dicks, Soyinka and Coffey (2006) point out that amalgamation of several media rather than keeping and viewing these media separately can construct different kinds of meaning. These fusions into multimodal forms have the capacity to convey changed understandings but they require the development of ways to deal with this new potential.

Research Tools in the Digital Age

Mobile technology devices are a combination of hardware, operating systems, networking and software. Hardware includes PDAs (i.e. Palm Pilot or Handspring), mobile phones, laptops, iPods, digital cameras, camcorders, video game players, global positioning system (GPS) devices, digital pens, hand-held scanners and the like. The operating system (i.e. Mac OS X, Windows NT/XP/Vista7/Mobile, Linux, SunOS) acts as a host for applications (i.e. Web 2.0 based) that run on the device. Applications are programs such as 'calendar', 'phone book', word processors, spread sheets, media players and so on. Networks are the infrastructure that supports the transfer of information (i.e. WiFi, Bluetooth, 3G, SM, dial-up service or virtual private networks). Software is the general term used to describe programs in a computer system.

To exemplify the use of mobile technologies, the following hypothetical scenario has been created: Professor X is studying music making in remote indigenous communities and gets a reminder from her 'to-do-list' on the PDA (i.e. Blackberry, Apple iPhone, Nokia N-Series) that the next field visit is due. She accesses her contact list, makes a quick phone call to confirm the appointment before grabbing a portable GPS (loaded with software such as TomTom, NavMan, Garmin, iGO or ViaMichelin) and attaching it with the suction cap to the windscreen of her car. While driving to the destination she follows the instructions from the GPS and listens on her iPod to the recording of the latest faculty meeting. Minutes are no longer taken but audio files are usually podcasted on the university intranet for staff unable to attend meetings in person. Professor X arrives early and uses the waiting time to review the video recording (from her camcorder) of her last field trip on the portable DVD player (i.e. Panasonic, Telefunken, Soniq, Winta, Lenoxx). Initially, she was going to use a collection of didgeridoos from various aboriginal mobs for the discussion with the participants. These artifacts allow three—dimensional exploration by using the senses to feel the surfaces of the didgeridoo, smell the aroma of the wood, and handling it to feel the carvings on it. However, she decided in favor of digital photos of didgeridoos to have a standardized set and focus on their painted decorations only. She realizes that she had forgotten to bring them and so accesses her university's secured server to download the images onto her PDA. During the photo-elicitation, respondents talk about the significance and meaning of each didgeridoo's decoration while Professor X uses her digital voice recorder to capture the interview and takes field notes with her digital pen. Later, she will connect it to her PC or Smartphone to digitize her handwritten notes for downloading, this basically transfers her handwriting to print in a word document. Before she leaves, Professor X makes another appointment on the spot via her PDA's calendar function (i.e. in outlook) while checking other commitments. Then, she links to her university network to upload all the research data gathered during her field trip. Finally, she emails her research assistant to put the voice recording of the interview through print software (i.e. Dragon Naturally Speaking or IBM ViaVoice 10), which generates a transcript. By the time Professor X arrives back on campus, her research assistant has entered and coded the interview data (audio file and transcript from the voice recorder and print-based field notes from the digital pen) into a data analysis program (i.e. NVivo) and Professor X is ready to write up her findings for publication.

Astute readers will notice that actually a smaller number of mobile devices are necessary to accomplish the tasks, as multiple functions are available on one unit. Nowadays PDAs can have a GPS incorporated, thus eliminating the need for a separate device. Also, a digital camcorder can be used to freeze frame visuals and produce still images (i.e. photos) from a film clip while the capacity to separate the audio track for transcribing makes a separate camera and voice recorder superfluous. However, to highlight the usefulness of mobile technologies in research a variety of mobile devices were used in the above scenario.

Advantages

The main reward of mobile technology for researchers is increased flexibility in the way that they work. Portability of equipment facilitates working from home and while traveling, for example to conferences, with 'hot spots' providing connection to base universities or while gathering research data during field trip. Shrum, Duque and Brown (2005:5) point out one of the major advantages of using digital materials

> Unlike analogue tape, which degrades slightly with each copy, digital recordings may be copied without loss of audio and video quality...media can be captured from a camera onto the hard drive of a computer, then edited and copied indefinitely without any degradation of the original images or audio tracks... it can be viewed, copied, published on the web or attached to an email for delivery anywhere in the world, any number of times. This characteristic of recording in the digital era means that researchers do not need to fear the destruction of their original work. But it also raises a host of ethical issues regarding distribution and transformation.

Disadvantages

The drawbacks include costs involved in setting up equipment and staff training to manage devices properly. Effective use of mobile devices reduces actual damage and potential loss of data, while training on security issues can help to overcome confidentiality and privacy concerns related to the research data. Nevertheless, these concerns are particularly relevant for wireless connections or when connecting to the network from outside locations. WLAN signals can travel significant distance and a wireless intruder could obtain critical information if sufficient security is not implemented.

Risk Management

Secure technologies such as virtual private networks (VPNs) and security measures such as VPN authentication and encryption are necessary as are strict access privileges for mobile users to protect sensitive information. Usually, "acceptable use" policies are already in place to delineate the boundaries within which academic staff is allowed to operate Internet and email, although these may need revisiting and reinforcing to ensure awareness and compliance. An equipment register needs to be established to capture all mobile devices, their serial number, a short description of all accessories as well as the date of issue and signature of the staff member. This confirms that they have taken possession of a particular device, understood policy and privacy issues to prevent any misunderstandings. Physical security for mobile devices consists of cable chains or Kensington locks for laptops or notebooks, serial numbers and engravings for identification purposes. Explicit instructions on theft prevention, reporting mechanism in case of loss or theft, mandated password protection (with minimum length, complexity and update frequency) and inactivity timeouts ensure that data is secure. Insurance policies may cover the cost of replacing a mobile device but only regular backing up enables recovery from a 'worst case' scenario. After all, lost data is irreplaceable.

USING MOBILE TECHNOLOGIES AS RESEARCH TOOLS

1) Data Collection Using Mobile Technology

ICT are constantly developing and evolving, thus changing the way we do research. Building research capacity through and with digital technology has been made easier through cheap, user-friendly and compact mobile devices. Mobile technology devices are generally available "off the shelf" to create a suite of equipment and software, including digital still cameras, video camcorders, voice recorders, scanners with OCR (Optical Character recognition) software to put various forms of non-digital information (newspapers, posters, participants drawings) into the computer and Corel Photo Paint for image editing. Computers will need to be equipped with a sound card, CD burner, digital video capture card (for video capture and editing) and software such as StorySpace (for hypertext authoring), Adobe Premiere (or video editing) and Macromedia Authorware 4 (for hypermedia authoring). Using these tools to collect audio, still imagery, video footage, transcripts and scan documents, a rich ethnography could be created on CD-ROM. However, each of these various data collection techniques and devices have their specific uses and limitations and only a skillful combination to suit each particular project accompanied by systematic data collection provide appropriate insight into the issue being studied (people, objects, phenomena).

Visual Data Variety, Pervasiveness and Sourcing

"Images are 'everywhere'. They permeate our academic work, everyday lives, conversations and dreams." (Pink, 2001:17). Accessible, affordable and pervasive photographic and video technology enables images to be used both as sources of data and as instruments for data collection. This requires re-thinking the way in which researchers manage these types of data and how they share and publish it.

Visual technologies can be used for observing human behavior or to record interviews or focus group discussions. Although observation is a much-used data collection technique it is also time-consuming and difficult to carry out. Hence, video cameras may allow a more flexible and cheaper method of observing. Research participants who may not be interested in linear textual forms can be engaged with visual devices such as such as digital cameras and video camcorders as they are fun and exciting. In particular, young research participants have grown up with technology and may have developed a preference for it, and may be enticed into research participation through the prospect of having access to digital technology. Visuals are a non-linguistic representation, able to support marginalized participants' communication (i.e. children, people using an unfamiliar language, the mentally challenged, some of the elderly) and able to add another dimension to a verbal narrative.

Drawn responses, concept or mind maps, artwork or children's drawings are other categories of visual data. While artifacts, artworks, children's drawings and drawn responses may be photographed for research purposes (after clearing copyright issues), scanning provides an attractive alternative. Working with a scanner avoids distortion, reflections, shadows, low contrast, low resolution and shaky images often experienced by working with a digital camera. Scanning converts hand drawings, handwriting (i.e. diaries, reflexive journals), printed text (i.e. in books, documents, newspapers) or an object (i.e. artifacts) into a digital image. The most commonly used devices are desktop or flatbed scanner. Hand-held scanners come in two forms, as 3D scanners and document (text scanning 'wands') although the latter has waned in popularity over the last decade due to the difficulties in getting an even scan and the proprietary interface. Nowadays, 3D scanners are combined with digital cameras to produce multi-colored, three-dimensional models of object although shiny, mirroring or transparent surfaces are still difficult to capture.

Mapping (either through mind maps or concept maps) is a valuable technique for visually displaying relationships and resources 'at a glance'. Mind maps use a central word, idea or other item and arrange others intuitively around it to branch into trees. Concept maps are used for brainstorming activities to generate ideas, communicating complex ideas, presenting findings and demonstrating progressive understandings. Maps can be constructed individually or in a group to represent key concepts in text, images, sound, multimedia format or a combination of those. Hand drawn maps by research participants could be scanned or photographed for electronic progressing. Alternatively, mapping software such as *Inspiration*, *Kidspiration*, *Gliffy*, *Conceptshare*, *Cmap and Thinkature* could be used, all of which provide means to communicate and share thinking and understanding visually with research participants or other researchers on the team. Maps could also be easily displayed on overhead or data projectors for peer-review or sharing at conference presentations (Ng & Hanewald, 2009).

Practical Issues in Collecting Visual Data

The inventorying, insuring and calibration of cameras are considerations, which can be avoided by using disposable cameras. Nevertheless, purchase, dissemination and retrieval of any cameras has to be coordinated.

Still images such as pictures (e.g. photo-elicitation) or moving images (video clips) can be taken either by the researcher or by the participant. Basic free hand shooting will often yield shaky movies and in the case of still images, probably limit photos to objects other than the participants. A tripod allows the research participants to take photos of themselves and reduces rapid motion in videos by providing a static position. The obvious disadvantages are weight, cumbersome set-up and additional cost.

Cameras or other recording devices may make it harder for the researcher to gain ethics approval, obtain permission from institutions or to recruit participants. Audio-visual material such as photos and video clips constitute special consideration in terms of ethics so that special measures need to be included to protect the participants' anonymity and privacy, particularly when they are used for publications or general dissemination of research data and findings.

The use of digital equipment in research may be an obstacle to participation as participants may see it as an additional layer of complexity and a task that may be beyond their skills level or interest. On the other hand, participants may be attracted by the prospect of owning a camera and the novelty of taking photos or making films.

The potential inappropriate use of (particularly, digital) cameras (i.e. pranks) is of concern, especially in research with children or teenagers. Stanczak (2007) cites the example of one young participant who was taking photos of his naked sister, which prompted the parent to destroy the camera. Issues of confidentiality and ethics have to be considered and addressed with the equipments' users before the deployment of any devices.

Analysis of Visual Data

There is a substantial body of published work on the details of analysis for written material (transcripts) yet - in comparison - there is very little available on the principles of analysis for visual material (i.e. images collected during fieldwork), with some studies suggesting the same approach to coding for both video footage and written transcripts (Nastasi, 1999). Banks (2001) extends the purpose and the content of the image as more than a record of natural action and stresses the relationships between the viewer and the image as well as the context in which the image was taken. Pink (2001) concurs while pointing out that the analysis of visuals may be interpreted rather differently, depending on their role and context. Pink (2001) cautions against using an approach to analysis where visual evidence is translated into verbal knowledge "…but to explore the relationship between visual and other knowledge" (p.96). This is particularly tempting with video footage as conventional analysis dissects segments and records them as verbal explanations. Pink (2001) cautions that ethnographic representations ought to show images in relation to words or other images.

2) Data Management of Digital Materials

The development of new technologies – including mobile technologies - and their availability, accessibility and pervasiveness within society and for research purposes enables new forms of data and the subsequent challenges for the appropriate handling of those materials. Researchers need to draw upon technical strategies and processes to handle technology-supported data collection. In doing so, a number of issues arise at each stage (i.e. collection, processing, analyzing and presentation). As indicated earlier, the necessary tools are generally available "off the shelf", easily enabling the creation of a suite of equipment and software.

Technical Issues

The most common technical issues in managing electronically-generated data are the compatibility of devices and operating systems; the transferability of data from mobile devices to a computer, server and/or back-up (to prevent corruption, accidental deletion or loss of data); limited or insufficient memory space or storage capacity of individual devices and security (protection of confidentiality and privacy of research data). In planning research of this kind, additional money and time for technical training and support with trouble shooting needs to be calculated.

Transfer of Digital Material to a Computer

Usually, the necessary software is supplied with a digital camera, camcorder or digital voice recorder and only needs to be installed on a computer to transfer and view images or listen to sound recordings. Often, software also allows a variety of other tasks such as simple editing tasks, printing pictures, viewing slide shows, copying pictures to CD or DVD, sending pictures by email. However, operating systems need to have particular features to meet the needs of the mobile device, and these need to be checked and confirmed before purchase.

The actual process of downloading the photos, film clips or audio recording for safe storage needs to be clearly demonstrated to the person operating the equipment, and possibly followed up with additional training to ensure full understanding and some ability to troubleshoot. A quick start guide with step-by-step instructions might be sufficient for more technologically proficient users.

Editing Software

A number of editing software packages are available, depending on computer configuration, budget, editing goals, technical expertise in using software and so on. Popular options include Windows Movie Maker, iMovie (for Macs), Adobe Premiere Elements, Ulead Video Studio or Final Cut Pro.

"Editing software makes it possible to manipulate audio and visual data as if it were text, that is, to edit, cut, paste and modify data for analysis, comparison and presentation." (Shrum, Duque and Brown, 2005:5) The three-dimensional representation (image, sound and text) of digital data can be maximized through editing work by organizing tracks to create a lucid and lively portrayal of research data or findings.

Skills Sets of Users

Technical skills and familiarity with mobile devices affect researchers' and participants' ability to collect data. A more advanced skill set can be expected to produce higher quality data in terms of clearer voice recordings and sharper images in the visual data collection. Within one research team, there might be a range of skills from basic through intermediate to advanced levels and tutorials may need to be offered to train individual researchers and perhaps even participants. Before offering training, the prior knowledge, technical experience and proficiency of the researchers and the research participants (if they are asked to use cameras, camcorders and the like) need to be assessed. Based on their existing skills set, a simple demonstration of the equipment or more substantial training in using particular devices will be necessary.

Figure 2. Digital voice recorder with microphone (top) and ear style microphone (left)

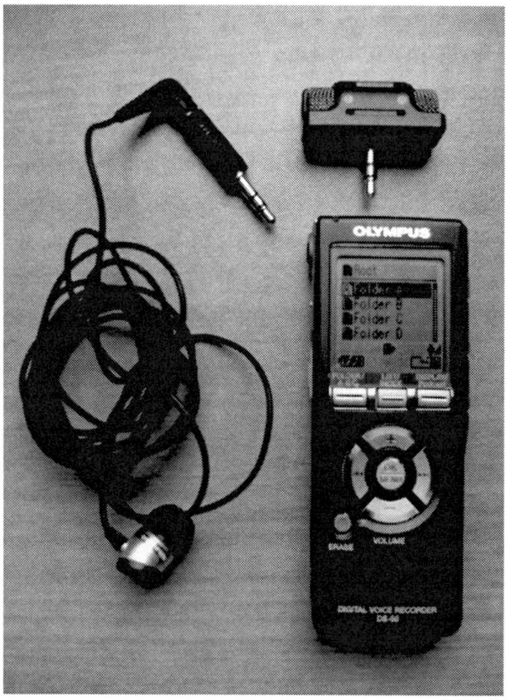

Demonstrating the Use of the Digital Device

This can be as simple as showing the researcher and the participant how to turn a device on and off and to press the record button and save files. This may sound basic to readers of this chapter but may be the most important aspect in getting the data in the first place.

Some initial shoots to test the zoom and various settings (landscape, portrait, night/day) on a camera and a simple sound test on the digital voice recorder will often suffice. If recordings of telephone interviews are planned (i.e. to save time and money in travel to remote areas), accessories need to be ordered and their usage explored and explained. Very minor matters can cause significant disruptions to data collection. In one instance, the incorrect use (i.e. the earphone style microphone in one ear and the phone's receiver on the other ear) caused a failure to record the participant's responses and thus a worthless recording. To successfully record a telephone interview, the microphone and receiver would have to be placed on the same ear.

Getting the Mobile Device Ready

Setting up the device prior to the first use by the researcher or the participant includes charging the device, (including the button sized lithium battery which is often overlooked). If this is not done, the camcorder will work for a short time only. Also important is inserting the memory card and entering the date and time on the digital camera or camcorder to provide an accurate record of the footage.

Figure 3. Digital camcorder with lithium battery, battery pack, two tapes and charger cable

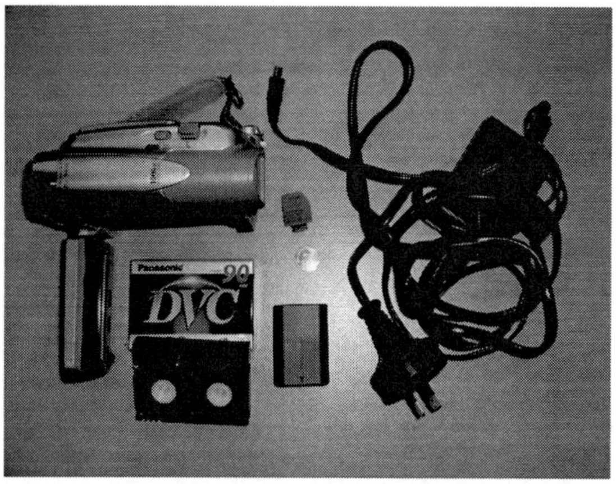

Some memory cards must be formatted before use; again these issues need to be dealt with in advance. Most still cameras nowadays will store pictures in the camera's internal memory. If desired, a removable Secure Digital (SD) memory card can be used for additional storage. Disposable cameras require less set up, which might be an attractive alternative. In choosing a suitable device, weight, ease of use and size might also be considerations, especially if participants are children and people with a chronic illness or disability.

Simple Maintenance

If participants are 'co-researchers' in the study and have little experience with mobile technology, they need to be trained to perform simple maintenance tasks. These include basic jobs such as changing the batteries, which varies on each camera and often involves opening the battery-chamber cover. This may need to be demonstrated to avoid breakage caused through inappropriate handling or use of excessive force. Inserting the batteries correctly (positive and negative poles inserted the right way) may again sound too simplistic to mention but may make the difference in getting usable data. Pictures will help participants remember the tasks necessary to keep the mobile device functioning. For example, refugees in a study from a third world country may be asked to take photos of their life to investigate community integration. They may be adults but illiterate as they may never have had the opportunity to attend school because they grew up in a war-torn country. It could well be the case that they have had no experience in handling mobile devices and are not able to read manuals. Hence, when the mobile device stops working (due to a lack of power) participants would not know how to overcome this issue unless they had some prior hands-on instructions.

Figure 4. Weight and size comparison of digital camcorder (left) and Flip Mino Video (right)

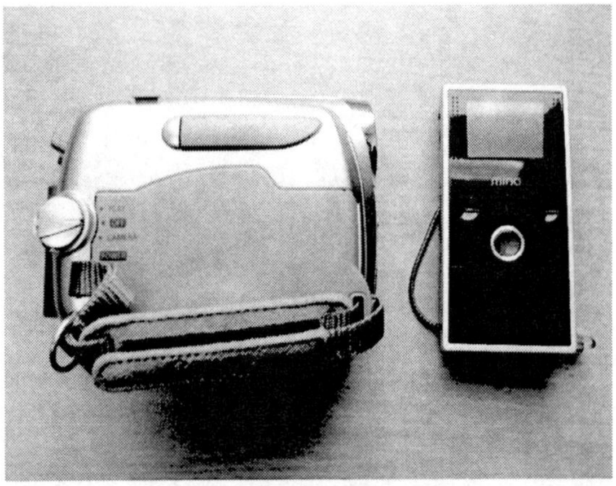

Additional Supplies

Sufficient additional supplies should be included with each device, for example an extra memory card or additional recording tape for camcorders, depending on configuration. This prevents loss of opportunity, for example a crucial interview being prematurely terminated (especially annoying if time and effort went into arranging it and meeting at a distant location) because the battery went flat and there were no replacement batteries within reach.

Equipment Documentation and Instructions

Often, products have inadequate documentation or instructions and a manual that is too lengthy and difficult to understand. These may need to be modified to assist the researcher or participant in using the equipment. The quality and adequacy of the documentation and instruction will make the difference between a positive experience and successful research outcome or a frustrating, unsuccessful one.

Easy to understand instructions and tip sheets (with handy hints for using equipment) should accompany the digital devices as a quick reference and reminder of the main points in using the equipment correctly.

Technical Support and Trouble-Shooting

A single page for handing out with basic instructions for error diagnosis and simple trouble shooting suggestions (i.e. what to do when…) would help in handling technical issues encountered by users in the first place. While a key contact person to provide technical assistance for any equipment needs, training and process (i.e. security advice, setting passwords, encryption) involving digital and mobile devices is advantageous, over reliance on one person can also lead to a support gap if that helper is temporarily absent due to illness, recreational leave or even decides to terminate employment or engagement in the project all together. For a large team, research packets handed out to each team member (containing a simplified version of the manual, simple trouble shooting instructions, additional supplies like batteries, tapes, backing up options) would ease over reliance on any particular individual. The research packet could also include phone numbers and contact details of people to ring in case of technical problems, theft of devices or other emergencies (i.e. virus-induced loss of data, recovery after accidental deletion or interrupted power supply).

Quality Issues

Collecting visual data such as photographs or video footage either by the researcher of by the participants is an action as much as a process. While random snapshots or filming ad hoc are spontaneous ways of capturing situations and respond to the design issue of authenticity, they may need some mechanical guidance (i.e. avoid sweeping movements because of instability and blurred images, avoid extreme close-ups because of too much distortion, use the flash when lightening is poor). When there is nothing staged, impromptu images are accurate presentations of events or situations captured as they occur. Subject to ethical issues, filming as unobtrusively as possible makes the researcher 'a fly on the wall'. However, handheld video cameras can be shaky and limit the perspective of the participant to observer behind the camera rather than actor in front of the camera. Tripods on the other hand offer stable shots and easy handling but the drawback is a limitation on how the camera is positioned and moved. Nowadays, most cameras have ten-second timers for self-portraits. These timers allow participants to take photos of themselves while resting the camera on a flat surface in front of them. This process of composing a photograph or a video clip involves pre-planning. The subject matter (a person or object), the location (indoors/outdoors) and its environmental constraints (weather conditions, natural or artificial light, background noise) have to be taken into account. Digital voice recorders need decisions such as the positioning of the device, which influences the sound quality and the participants' perceptions. A microphone on the table is more visible and obtrusive than a lapel one.

A single sheet with instructions on the care of the equipment (i.e. keep in a secure, locked up place when not in use, keep it in a dry, cool place/out of the sun, when taking photos, put your hand through the strap to avoid accidentally dropping the camera, charge the camera after each use) ought to accompany any device.

Directional Issues

To guide participants, researchers may issue tip sheets to focus the recording of images (i.e. what things the researcher wants to be captured or is of interest). It may also remind participants of the time commitment (how long to finish each particular task, the total length of involvement and the topics (e.g. take as many pictures as you like of your pets and your toys). If researchers have particular expectations, they need to be specific rather than generic. For example, in a study on health eating participants need to be asked to take photos for the duration of a week of their daily breakfast, lunch, dinner and each snack rather than photos of "what you eat". In more open-ended projects, if a camera is given to a participant, the participant may be left relatively free to construct their own interpretations of events relating to the research, context, subject matter, length of film clips or number of photos and so on. However, this might be counterproductive as one instance in a large research project with young people showed. Participants were given still cameras to take photos of their lives, which were discussed with the researcher to generate response beyond the language-based conventional interview protocol. In eagerness to contribute, one of the teenagers took 1030 photos, which was the maximum amount possible on the camera. However, the photos were of very little interest and use as too many were blurry, indistinct or 'taken on the run' and thus showed only partial or shaky objects and people. A briefing and written notes for future reference might contain tips for the participants such as guidance on how long and what to film, the placement of the camera (i.e. hand held or on a tripod, facing the participant or facing away from the participant) choice of frame angle (wide shot, close-up) and so on.

Ethical Issues

Visual images can be randomly shot or be the outcome of selections. Participants need to be given the option to remove unwanted pictures and need to be shown how to delete pictures currently displayed in the monitor. Participants should look at their photos and obliterate unwanted images before the meet up with the researcher(s). In one particular case, a teenager had taken photos of their friends without asking their permission first. The friends objected to their images being used and the photos had to be removed.

Both audio and video data can be stopped and rewound for re-examination and editing out of unwanted elements, particularly at the end of the recording. This might have happened easily if people continued their conversation after an interview, assuming that the digital recorder was turned off when it fact it had been left running because the researcher forgot to press the stop button. Likewise, accidental recording of private conversations will be avoided if the researcher indicates that the start button of the mobile device is about to be pressed and that the interview will formally commence. Researchers and participants (for the purpose of voice recorded reflexive journals) should adopt a practice of starting each recording with the date, time, place and relevant details of the participants. This avoids confusion, especially if there are multiple recordings of the same participant or on the same day.

Legal Issues

Researcher need check out the policy of particular sites and organization in terms of their restrictions for taking photographs or recording film footage. A variety of venues (i.e. museums, theatres, art exhibitions, indigenous events or sacred sites) do not allow recordings of any kind for copyright protection, religious or cultural reasons.

In Australia for instance, there is legal protection for literary, dramatic, artistic and musical works, sound recordings and performances, which is covered under *The Copyright Act 1968.* Other restrictions apply according to the locations used. For example, a researcher trying to investigate gambling addiction by filming inside a casino will need to get special permission from the owner(s) of the premises in addition to the participants' permission. Likewise, many schools object to recording behavior inside their grounds and during school events (e.g. swimming carnivals, school excursions, camps) due to privacy concerns for students and their families. One Australian secondary school for example prohibited taking and transmitting any audio or visual recordings (i.e. transmitting images, taking photographs with a phone, taking photographs with a camera, using a mobile phone on record setting, using an MP3 player on record setting, video recording). The school policy states that "Mobile phones, MP3 players and other technological devices can cause significant disruption to the learning and teaching rights of others." (Gladstone State High School, 2009:3). The identified consequences of violation of the rule include suspension for repeated offences. So, if a researcher would ask student participants to take recordings in their school settings, special permission would have to be obtained from the educational authority, which in many cases is the principal of the respective school. Separate consent would have to be sought for sensitive uses of student images (i.e. for publication on a website).

In Australia, the State & Territory Privacy Laws manifested in the various Information Privacy Acts (i.e. Victoria 2000; Queensland 2009) cover the collection and use of personal information. This relates to information that identifies a person, which can be a photograph, video or digital image of that person. Infringement of Information Privacy Acts can result in legal recourse against individual researchers. Hence, it is crucial for researchers to understand the institutional restrictions and environmental constraints associated with visual recordings and account for these in their initial research design.

3) Data Analysis of Digital Materials

Digital technology now enables new ways of analyzing data through computer assisted qualitative data analysis via a range of software. Increasing demand has extended the range of available software, their features and functions.

Accompanied by computer assisted qualitative data analysis, these tools and techniques enable new ways of creating, collecting, processing, archiving and analyzing data. However, one of the drawbacks of electronic data according to Benfield and Slemko (2006) is the need for it to be transformed, merged or even reformatted before use.

Data Coding for Archiving

In any research project, data accumulates quickly and has to be coded for archiving and easy retrieval. The larger the project gets, the more important an accurate system will be to keep an overview of the data. It is crucial that the archiving system is set up beforehand as later changes or amendments will be confusing, time intensive and frustrating. A hierarchical system based on numbers only or a combined letter and number system can be used, whichever is easier to follow.

A project with various participant groups and researchers might have the following hierarchical coding list and structure to differentiate their data subsets:

```
--- 1  - Meta data file for coding,
data overview and running record
      |
      |-- 1 - Case Studies (meta
data, overview and running record)
      |     |-- 1 - Researcher (ex-
ample Jane Doe)
                  1 - Participant
                                 1 -
Interview date
                          2 - Inter-
view data
3 - Interview data
4 - Interview date (for example
2009/08/27)
                  2 - Participant
                  3 - Participant
                  4 - Participant
      |     |-- 2  - Researcher
      |     |-- 3 - Researcher
      |
      |-- 1 - Focus Groups (meta
data, overview and running record)
      |     |-- 1 - Focus Group: Sci-
ence Teachers
                  1 - First Focus
Group - date
```

```
                    2 - Second Focus
Group - date
                    3 - Third Focus
Group - date
      |     |-- 2 - Focus Group: Maths
Teachers
                    1 - First Focus
Group - date
                    2 - Second Focus
Group - date
      |     |-- 3 - Focus Group: His-
tory Teachers
                    1 - First Focus
Group - date
                    2 - Second Focus
Group - date
                    3 - Third Focus
Group - date
```

So, collecting data on case studies (1) researcher Jane Doe, who is the first member in a research team of three researchers interviewing first case study participant (1) for the fourth time (4) on 27 August 2009 will code the audio file as

```
1-1-1-4-2009-08-27.wav
```

It might be worth pointing out that writing the date in the sequence of year-month-day allows for consistent and automatic sequencing in file, with first recordings always being first.

However, in large research teams, the individual researcher might want their data easily identified. Hence, each separate piece of data would start with the initials of the researcher. Assuming researcher Joe Doe (JD) has 12 participants, each participant would have to be allocated a running identification number to protect their identity (01 to 12). If they are interviewed multiple times, say over a period of 4 visits (V1 to V4) the running number of the interview and date would be incorporated into the coding. The data type is indicated by the respective extensions, for example interview transcript (word.doc), recording of conversations (audiofile.wav; audiofile.WMA) and possible photos (picture.jpg) or film clips (movie.mov).

Let's assume Joe Doe's (JD) contact with Participant No. 3 (03) on the second visit (V2) happened on 27th August 2009 (2009_08_27) and resulted in a recording of their interview, where they discussed a photo, which the participant had taken. The code for the recorded interview sound file (wav) would be:

```
JD_03_V2_2009_08_27.wav
```

The photo used during that interview/ discussion would be coded as:

```
JD_03_V2_2009_08_27.jpg
```

Following that, the interview is transcribed and coded as:

```
JD_03_V2_2009_08_27.doc
```

The consistent coding and extensions indicate clearly the type of collected data (sounds, text, image). It also shows the connection between them and their place within the overall data collection (Joe Doe and his third participant at visit 2 on 27th August 2009) without having to open up the file and searching for that information.

Regardless of the particulars, if a coding system is systematic and hierarchical it will afford easier retrieval of raw data, especially for analysis preparation.

Equipment Standardization

There is now a diversity of products and platforms available (web-based, software, open source/ PC/ MAC), which generates a strong need for standardization to facilitate exchange and cross analysis within a research team. For example, if everybody on a large research team works on a PC, it would not be advisable to change for one researcher to change to an Apple Macintosh during an equipment up-grade or to switch to Linux OS when the remainder of the team uses Windows.

Data Analysis Software

Computer Aided Qualitative Data Analysis Software (CAQDAS) includes ATLAS/ti, Code-a-text, Hyper RESEARCH, MAXqda, and NVivo. One of the most widely used is NVivo, now in its 8th version. NVivo is a software tool to manage, code, analyze and report on text, visual and auditory data.

For quantitative research, written questionnaires and surveys can now be administered online, through tools such as Survey Monkey and answers can be fed into a statistical package. There are a range of software packages available such as SPSS, SAS, SYSTAT and Minitab. SPSS is a full-featured, customizable package that performs and organizes in-depth statistical analyses (e.g. data mining, database analysis and surveys). SAS combines the tasks of data access, data management, data analysis and data presentation into one system. SYSTAT is a comprehensive statistics, graphics and data management program. Minitab is an easy-to-use statistical software package with basic and advanced data analysis capabilities; providing a system for organizing and analyzing data and reporting results of statistical analysis.

Issues in Digital Data Analysis

Regardless of which software package is used, the hidden cost of the software package has to be calculated into the research project budget. The initial license fee may increase because of multiple users (in the case of a large research team and several researchers working on the same data) or changed configuration. Software upgrades may also require additional fees, as may a consultant or the services of a contractor to implement the software, train or give tutorials.

In a longitudinal project, data from an old system may need to be converted to a new system as technologies emerge and vanish. Sometimes, projects have to deal with legacy systems such as a particular system being used for a pilot project, with the technology being adequate but not ideal but budget constraints forcing the team to continue with the inherited / legacy system.

4) Presentation of Digital Data

The employment of editing software packages enables the unlimited re-use of digital data. It allows researcher re-purposing to produce multiple versions for diverse audiences (i.e. their clients, other stakeholders, the participants themselves, policy makers, funding bodies, other academics) or purposes (i.e. training of professionals, educational tool for the general public, presentation at a conference). "When digital media are edited and distributed, they are an alternative means of presenting scholarly results for an audience…as a point of entry for discussing a research project, comparing notes with colleagues working on similar topics and addressing queries about strategies and lines of arguments" (Shrum, Duque and Brown, 2005:6).

Security Issues

Physical security of data is a major issue in any research project, but with digital devices physical security is more than just a locked filing cabinet in a secured room. Mobile devices such as digital voice recorders, still cameras, camcorders and the like are tempting targets for theft in the field. Therefore, data on those devices should be transferred to a computer hard-drive and external back up as soon as possible and then erased from the collection device. Large amounts of electronic data can now be stored on USB drives, which are less than half the size of a matchbox and thus can be easily lost. If transporting data on USB or CDs, it has to be ensured that the data is edited and coded to remove any identifying names or details so that the potential thread to the anonymity of the participants is minimized. In emailing data to other team members, care has to be taken that the correct email addressed is used, especially if there are other close possibilities (i.e. Joe Doe@ hotmail.com or Jo Doh@ hotmail.com or Jodo@ hotmail.com).

If a research team wants to share data (i.e. for cross case analysis), storing and accessing it from a server – protected by a firewall - is preferred to carrying it on loose storage units (USB, CD).

Ethical Issues

Research ethics committees in universities and research organizations check that the rights and privileges of participants are being upheld, usually through plain language statements informing participants about the purpose of the research, the potential risks and benefits as well as the rights of participants. Standard written consent forms may only cover the verbal information that has been gathered. Visual data and their use need to be made explicit by explaining to participants where, how, why and when data will be used. This could be covered in a differentiated authorization document on the usage of photographs and / or video footage, which might allow the use of images containing peoples' faces (1) for academic publications & conferences only, or (2) for other purposes (e.g. on websites, in research reports) if specific permission is sought for each case and (3) if the actual face of each person is blurred out.

Mechanical Issues/ Preparation for Presentation

Respecting a person's identity and anonymity in any visual reference includes also attending to identifying items of clothing (i.e. a teenager wearing a school jumper with a schools emblem and VCE badge on it) or an employee wearing a staff badge displaying their full name, their professional position and the organization they are working for. Editing software (i.e. Photoshop) enables image manipulation to blur faces or cover identifiers on solid background. In addition, cropping of images and corrections (i.e. overexposure or red eyes) can be undertaken.

CONCLUSION

Mobile technologies have been widely available for more than two decades and dedicated data analysis programs are able to assist in dealing with digital data. Anecdotal evidence and personal observation indicate that their actual use for research purposes remains relatively small, compared to their full potential. This might be based on researchers' lacking interest or expertise in dealing with the challenges of integrating new technologies into their work. It appears that many researchers have not yet recognized - much less responded - to researching with mobile technologies. Consequently, a comparatively small number of publications is available that outline experiences of research with, for and through mobile technologies so that others may benefit. Nevertheless, researchers need to start accepting that considerable cultural change is taking place around them, which is affecting fundamental understandings of research methodologies.

Shrum, Duque and Brown (2005:17) summarise the implications of digital technology as twofold:

first, researchers must learn a new set of technological and cognitive skills (planning, shooting, and editing...); second, they must negotiate more complex research protocols that include human subject considerations, copyright... and the institutional inertia of accepting text as the major form of scientific communication.

The goal should be to set in motion a development cycle of technology-rich research projects. This would raise awareness and visibility of this emerging field. Tools to experience new media and training to use them would empower more academics to engage with these technologies. The lessons learned from handling the technical aspects, the procedural skills and the methodological impact could be shared through ongoing discourse. Academics need to be in a position to both recognize and take advantage of the new opportunities for research afforded by mobile technologies.

REFERENCES

Australian Government. (2009). *State & Territory Privacy Laws*. Office of the Privacy Commissioner. Retrieved August 19, 2009, from http://203.210.106.3/privacy_rights/laws/

Banks, M. (2001). *Visual methods in social research*. London: Sage.

Bateson, G., & Mead, M. (1947). *Balinese character: A Photographic Analysis*. New York: New York Academy of Science.

Benfield, J. A., & Szlemko, W.J. (2006). Internet-Based Data Collection: Promises and Realities. *Journal of Research Practice*., 2 (2), Article D1.

Collier, J., & Collier, M. (1986). *Visual Anthropology: Photography as a Research Method*. Albuquerque: University of New Mexico Press.

Dicks, B., Mason, B., Coffey, A., & Atkinson, P. (2005). *Hypermedia Ethnography*. London: Sage.

Dicks, B., Soyinka, B., & Coffey, A. (2006). Multimodal ethnography. *Qualitative Research Journal*, *6*(1), 77–96. doi:10.1177/1468794106058876

Garcia, A. C., Standlee, A. I., Bechkoff, J., & Cui, Y. (2009). Ethnographic approaches to the Internet and computer-mediated communication. *Journal of Contemporary Ethnography*, *38*, 52–84. doi:10.1177/0891241607310839

Gladstone State High School. (2009). *Recording Device Policy.* Retrieved August 19, 2009, from www.gladstonshs.eq.edu.au/.../School%20 Information/Recording%20Device%20Policy.pdf

Harper, D. (1998). An Argument for Visual Sociology. In Prosser, J. (Ed.), *Image-based Research: A Sourcebook for Qualitative Researchers* (pp. 24–41). London: Falmer Press.

Harper, D. (2002). Talking about pictures: a case of photo elicitation. *Visual Studies Journal*, *17*(1), 13–26. doi:10.1080/14725860220137345

Hine, C. (2000). *Virtual Ethnography*. London: Sage.

Hurworth, R. (2003). Photo-Interviewing for Research. *Social Research Update*, 40, Spring. University of Surrey. Retrieved October 12, 2009, from http://sru.soc.surrey.ac.uk/SRU40.html

Kozinets, R. V. (2006), Netnography 2.0, in Russell W. Belk (Ed) *Handbook of Qualitative Research Methods in Marketing* (pp. 129-142), Cheltenham, UN and Northampton, MA: Edward Elgar Publishing.

Masten, D., & Plowman, T. (2003). Digital Ethnography: The next wave in understanding the consumer experience. *Design Management Journal*, *14*(2), 74–81.

Ng, W., & Hanewald, R. (2009). Concept maps as a tool for promoting online collaborative learning in virtual Mapping. In Marriott, R., & Torres, P. (Eds.), *Handbook of Research on Collaborative Learning Using Concept Mapping* (pp. 81–99). Hershey, PA: IGI Global Publisher.

Pink, S. (2001). *Doing Ethnography: Images, Media and Representation in Research*. London: Sage.

Prosser, J. (2008). *Introducing Visual Methods.* ESRC National Centre for Research Methods Review Paper. Retrieved 12 October 2009 from http://www.reallifemethods.ac.uk/publications/ articles/

Schrum, W., Duque, R., & Brown, T. (2005). Digital Video as Research Practice: Methodology for the Millennium. *Journal of Research Practice. 1*(1), Article M4.

Stanczak, G. C. (2007). *Visual research methods: Image, society, and representation*. Thousand Oaks, CA: Sage Publications Inc.

State of Victoria. (2009) *Information Privacy Act 2000*. Victorian Consolidated Legislation Australasian Legal Information Institute. Retrieved August 19, 2009, from http://www.austlii.edu.au/au/legis/vic/consol_act/ipa2000231/

The Parliament of Queensland. (2009) Information *Privacy Act 2009*. Act No. 14 of 2009. Retrieved August 19, 2009, from www.legislation.qld.gov.au/LEGISLTN/ACTS/2009/09AC014.pdf

Wagner, J. (2006). Visible materials, visual theory and images of social research. *Visual Studies, 21*(1), 56–61. doi:10.1080/14725860600613238

Wang, C., & Burris, M. A. (1997). Photovoice: Concept, Methodology, and Use for Participatory Needs Assessment. *Health Education & Behavior, 24*(3), 369–387. doi:10.1177/109019819702400309

Weber, S. (2008). Visual Images in Research. In Knowles, J. G., & Cole, A. L. (Eds.), *Handbook of the Arts in Qualitative Research* (pp. 44–45). London: Sage.

Section 2
Design and Implementation

Chapter 8
Mobile Phones, Developing Countries and Learning

Elba del Carmen Valderrama Bahamóndez
University of Duisburg-Essen, Germany

Albrecht Schmidt
University of Duisburg-Essen, Germany

ABSTRACT

The Internet and computers are accessible to only half of the population in the world. For the other half, computers and the Internet are almost alien concepts. This half has no medium for gathering information, and they are computer illiterate. In addition, it is well-known, that the use of computers and the Internet, directly and indirectly, enhance the learning process. Therefore, students from under privileged areas of developing regions of the world are, clearly, at a disadvantage compared to their peers in developed countries. However, mobile phones could change this situation. In developing countries, mobile phones are far more accessible than computers or Internet access. This high accessibility together with the multiple functionalities of mobile phones, allow for the potential to build feasible educational applications that enhance the learning experiences of students in developing countries. Such opportunities enable the students' experiences to be made proportionate to the other half of the world, with a real mechanism for gathering information.

INTRODUCTION

Even though, there has been an increasing usage and access to Information Communication Technology (ICT) worldwide during the last few decades, the digital divide between developed and developing regions still remains (ITU, 2009).

DOI: 10.4018/978-1-61692-849-0.ch008

In developed countries a well established ICT infrastructure exists in industrial, business, governmental and home environments. The infrastructure is available in big cities as well as in small villages. In contrast to developed countries, with a few exceptions, developing countries have little access to ICT infrastructure. The access and availability of ICT among developing countries is unequal, even within the same country. While some developing regions face a lack of global ICT infrastructure in their territories; other developing countries have a relatively satisfactory ICT infrastructure in urban areas (and/or for the wealthier populations), and a poor or non-existent ICT infrastructure in rural areas (and/or for the poorer populations). Sub-Saharan Africa and Latin America are representative examples of the former and latter situation respectively.

Although the earlier personal computer dates back to the end of the 80s and the Internet has been publicly available since the 90s, access to computers and the Internet in developing regions remains behind compared to developed regions. According to the data from ITU (2009) and the UN (2008) the proportion of household with computers in developed economies is over 75% in comparison to roughly 24% in developing countries. The proportion of households with Internet access in developed countries average 63% and only 16% in developing regions. Therefore, the majority of the population in developing regions has no access to computers or to the Internet, and it is very probably that many of them have never used or even seen a computer in their lives. Table 1 shows a summary of ICT access in households in developed and developing regions.

However, there has been a substantial advance in access to computer technology with significant relevance to developing countries through mobile phones. In contrast to developed regions, for many citizens of developing countries (specially the rural and poor areas) mobile phones represent the only way of remote communication, the only way of gathering information and the first and sole means of interaction with computer technology and digital information (ITU, 2009; Valderrama, Döring & Schmidt 2008). According to ITU (2009) *"the penetration of mobile phones worldwide in 2007 was around 40%. Moreover, around 64% of the mobile phone subscriptions worldwide during 2007 came from non-developed countries"* (pp. 3-4).

Table 1. Access to Information and Communication Technology (ICT) in households in Developed and Developing Regions (ITU, 2009).

ICT Access	Developed Countries	Developing Countries
Proportion of households with computers	75,50%	23,65%
Proportion of households with Internet access	63,00%	15,83%
Proportion of households with mobile phones	84,37%	45,72%
Proportion of households with dedicated phone lines	88,67%	36,25%

As shown in Table 1, in developing countries the proportion of households with mobile phones is almost twice that with computers, and around three times that with Internet access. Even though Table 1 does not include all the countries in the world, hence some data is incomplete and it is not representative of all the developing countries (ITU, 2009), it still gives a representative overview of ICT access in developing regions of the world. The lack of infrastructure for dedicated phone lines in rural villages in developing countries has given rise to the growing use of mobile phones where the signal coverage of cell phones covers vast areas. So mobile phones are making the difference for telecommunication in developing countries.

Additionally, in developing countries the rate of increase of the penetration of mobile phones is high. For example, from 2006 to 2007, Africa presented the highest mobile growth rate of 32% (ITU, 2009). The current high use of mobile phones in developing regions and their trend for increased use of mobile phones could represent a realistic alternative for introducing computer technology to people in underprivileged areas in developing countries. Mobile phones have the potential to overcome the shortage of access to computer technology there and hence the access to information and knowledge.

As well as the high penetration rate of mobile phones, mobile phones also introduce three main advantages over conventional computer technologies: complete portability, availability of a range of network connection and extensive multimedia capabilities. These advantages coupled with the high penetration rate of mobile phones – surpassing computer availability – gives them great potential to be used for educational proposes as well as for business use. Mobile phones can be easily carried everywhere where students are. They can be used in indoor as well as outdoor environments and interaction with them does not require a specific work setup (e.g. desk) or a specific infrastructure (e.g. power supply). In particular, the transportation systems within many cities and megacities in developing countries are chaotic and many children have long travel times to and from school. In rural areas they often have to travel long distances to and from school in the countryside. Using mobile devices children could take better advantage of this travelling time to learn and revise, e.g. learning English or to review learning content on mobile phones. As well as this, outdoor learning activities such as excursions to Museums, Natural parks or the Zoo can be supported by mobile phones as for example in MyArtSpace (Sharples et al., 2007) or GidderProject (Pierroux, 2008). Children can use mobile phones during these outdoor visits to retrieve further information (about a specific paint, sculpture, animal, plant, etc.), or record their visit for future use in the classroom (using the photo/ video camera, sound recorder) for example.

In this chapter we will first discuss some educational approaches which have explored ways in overcoming the lack of access to technology in developing countries. Then we critically analyze the main challenges that the education systems in developing countries face. After that, an overview of current m-learning projects for children using mobile phones in developing countries will be described. Finally, some guidelines about mobile learning for developing countries will be discussed.

The Learning Context in Developing Countries

In contrast to developed countries, the education system in developing countries faces many challenges, most of which are not directly concerned with educational issues but which have a direct effect on the quality of the learning process. The majority of these problems are caused by the social context of the country.

Poor infrastructure of schools. The infrastructure of school buildings as well as the equipment within classrooms are generally in precarious conditions, especially within those schools located in areas which are difficult to access. Electricity supply and dedicated phone lines are often unavailable in the remotest areas. Moreover in the poorest communities, there may not be any school buildings at all. Instead there are either makeshift huts or no closed infrastructure at all.

Long distance travelling to schools. Most children in remote areas of developing countries have to travel long distances to attend classes. The travelling can take two or more hours and is usually done on foot and/or public transportation. Even in cities, travel time to school may be long due to poor transport infrastructure.

Multigrade schools. In remote and rural areas of developing countries there are not enough students per grade and/or there are not enough teachers, therefore schools are usual organized as multigrade. This means that children with different levels of knowledge and different learning objectives share the same classroom. Teachers have to distribute time and tasks for every student or group of students with similar learning targets.

Multiple local languages and dialects. Another feature of developing countries – also valid for some developed countries, is that besides the official language there are many different languages or dialects spoken by different natives, tribes or communities. In many cases, in these communities only the local language is spoken and therefore children have serious problems learning the official language of their country. The performance of the children in schools is affected by this. Not being able to speak the official language obviously leaves them at a serious disadvantage later for integrating into society in comparison to those who speak the official language. Limitations in speaking the official language of the country are a common occurrence, especially for the native population, who keep their language, but who also have difficulties speaking the official language. This is very common in former colonies, where most of the native population has difficulties speaking Spanish or Portuguese in Latin America, and French or English in Africa for example. Moreover, generally the native population is among the poorest population in these regions.

Low performance in learning world-spoken language (i.e. English, French, Spanish, Chinese). The knowledge of a worldwide language such as English is considered one of the key features in improving the social situation of an individual, and in providing a chance for better job opportunities. As Kam et al (2007, pp. 1097) says *"World language fluency opens the door to further education, a larger or world marketplace, to new economy outsourced jobs, and often improves access to government health and legal services"*. This view is shared by both citizens and authorities of developing countries. Actually, governments in most of the developing countries include English as part of the official curricula in the schools. However, the effectiveness of the learning – especially in official schools – is very low. Children finish school unable to speak English. One probable reason is that children only have school time to be exposed to an English speaking environment, and outside classes there is no medium to practice what they have learnt (listen/speak). Some English schoolbooks offer audio material to practice at home, but how can a child with no access to computers or a CD stereo play them?

Very poor families. In the more underprivileged regions, where families are very poor, it is usual that children support their families, earning money to bring home. Therefore after elementary school, many children leave school.

No access to computers or the Internet. Only a very low percentage of schools in developing countries have access to computers (Valderrama et al., 2009; Betane, 2006). Those schools which do have computer labs are mainly located within or near the main cities.

BACKGROUND

Nowadays, in the digital era, the knowledge and skills in the use of computer technologies, including access to the Internet, are crucial to the improvement and development of a country. At the same time, the added value that technology could give to education is well documented (Bentley, 2007). The use of ICT can contribute enormously (1) to extending and to strengthening the content of all subjects, through accessing general related content on the web, and (2) enhancing the learning process, for example with the use of learning software to complement the teaching process, like English Learning software, etc.

Access to information and communication technology (ICT) is one of the measures used to classify developed and non-developed countries. The more citizens have access to free information and technology resources, the more developed is the country. Around one half of the population of our world has access to and skills in ICT, but the other half barely survive (Dias & Brewer, 2009). Many developing countries are aware of this digital divide and have made efforts to improve the use of computer technology in schools.

The Computer Lab Approach

This is the most traditional approach to provide computer access to children in schools. It consists of children visiting a classroom equipped with computers. Most developing countries, and some developed countries, have implemented this computer lab model in schools.

Despite the efforts made by many developing regions, this approach has not completely fulfilled the needs of acquisition and access to technology for the children. Among the limitations of the computer lab approach used by most developing countries are:

High number of students and low number of computers available. The majority of the population in developing countries is young (of school age), therefore the government needs to invest substantially to satisfy the basic educational demands of the children. Due to limited budgets, it is often necessary to prioritize equipment for the basic functionality of schools, hence the acquisition of enough equipment for the computer-labs in schools is lost. As a result, children have to share one computer, with two, three and even more students. The computer lab sessions are also very limited, varying from once a week to once a month. Considering that for many children in developing regions, this is the only opportunity for interaction with computers, these conditions make it difficult for them to advance significantly in the acquisition of the skills and also discourage teachers (Batane, 2006) in the use of computers as a support tool for teaching.

Old and badly equipped labs. The equipment –software and hardware– in computer labs is generally outdated and obsolete. As well as this, it is usual to find damaged equipment in the computer labs, which have not been operating from months. The small budgets of governments in developing countries make it difficult/impossible to pay the high administration costs of software and hardware.

Limited access to the Internet. It is very frequent, that most of the schools have no access to the Internet–on average less than 20% of schools in developing countries have Internet access (UN, 2009). Some of the reasons are (1) lack of adequate infrastructure for providing Internet access, especially in rural areas, (2) limited budget to provide the computer-lab with the Internet, and (3) in some cases internal policies such as security or concerns about distractions limit student access to the Internet (Valderrama, et al., 2009). In developed countries the latter can be a concern too. Hence, free access to the Internet is often restricted for students.

Not all schools have computer-labs. Only a low percentage of schools have computer-labs, the majority of which are located in urban areas. Schools located in rural areas, generally do not have any computer-lab.

In relation to the use of computers for education, usually there is no integration between the use of the computers and the subjects taught in the classrooms. Sessions in the computer-labs are mainly used to teach subjects relating to informatics/technology or the use of the computer itself. Teachers do not take better advantage of using the computer-labs to improve or complement their teaching lessons.

SHARING A COMPUTER IN THE CLASSROOM: MULTIPLE INPUT DEVICES AND SINGLE SHARED DISPLAY

As we discussed before, the scarcity of computers in schools in developing regions results in children having to share a computer with 2 or more peers. Sharing the computer implies sharing one single mouse. This situation leads to a limited experience for children in their interactions with computers. As an alternative and to overcome this situation, several researchers have developed different approaches based on the use of several input devices over one shared output device. The concept is based on the Single Display Groupware (SDG) for children. In general, a few children are each assigned a unique input device, i.e. mouse, connected to the same screen. On the screen, each mouse-cursor is identified through the assignation of different colors or shapes to each cursor.

Inkpen, Booth, Gribble & Klawe (1995) and Inkpen, Ching, Kuederle & Scott (1999) propose the utilization of two mice on one shared computer screen to enhance the collaborative learning between children. Pawar, Pal & Toyama (2006) presents the use of a single PC connected to two to five mice. The Mischief application developed by Moraveji et al. (2008) consists of the use of one large shared display, i.e. overhead projector, for the whole classroom with 10 to 32 children (Moraveji et al, 2009).

Based on the research above, several interesting conclusions have been drawn and we summarize them below:

1. it strongly supports collaborative learning between children
2. the level of learning achieved when two children are sharing a computer with a mouse each, is similar to the level achieved when each child has his/her own computer (Inkpen et al, 1995; Inkpen et al.,1999; Pawar et al., 2007, Moraveji et al. 2008)
3. the learning activities chosen directly affect the levels of learning, for example, competitive activities lead to a strong rivalry between students, specially between boys, in which case the levels of learning achieved are lower (Pawar et al, 2007).

One important concern relates to the feasibility of the number of mice connected to the output display. As the number of mice connected to the screen increases, the chances of the screen becoming messier increase, making it difficult to distinguish cursors from one another. The Mishief system (Moraveji et al., 2009) showed that as long as the targets on the screen are not too small, the performance of the children is similar to the performance they would achieve if fewer mice were connected. Nonetheless, they also conclude that due to the swan effect on the screen, caused by the connection of a large set of mice, it can be almost impossible to provide visual feedback to the children after the task has been completed. Sound/audio feedback is also discouraged because of the high number of the students in the classroom.

The utilization of multiple mice over one single shared display is a suitable alternative to one computer per child to overcome the scarcity of computers in schools of developing countries. The acquisition of several mice is far less costly than the acquisition of computers for each child. Its low cost for implementation as well as it facilitating collaborative learning make this approach a good solution for traditional schools, where all the children share similar knowledge backgrounds and have to learn the same curricula. However, in rural areas of developing regions, multigrade schools are the usual school model implemented; this approach could have some limitations (Valderrama et al., 2009). In multigrade schools, where children with diverse educational backgrounds share the same classroom, teachers have to prepare different learning material for each child or group of children within the classroom. Implementing different learning activities for each child while they are sharing a single display as within the Mischief approach (Morajevi et al., 2008) would be very difficult to implement.

Low Cost Laptop Designed for Children

A second approach to overcome the lack of access to computers in schools in developing regions is the creation of low-cost laptops specially designed for children and equipped with learning software. The most outstanding example of this approach is the One-Laptop-per-Child (OLPC) project.

The OLPC project was initiated by the MIT Media Lab with the aim of providing children in developing countries with a highly capable, low-cost laptop. As result, MIT Media lab created the XO-laptop (see Figure 1); in partnership with other organizations. The XO-laptop brings many innovations in terms of hardware and software.

With its striking outside design and the hard resistance of the material used, they are ideal for working with children. The main novelty of the XO laptops is the user interface Sugar. Sugar is a graphical interface based in Linux that strongly supports collaboration between XO-laptop owners. In contrast to the Window GUI concept, Sugar uses the concept of a diary book. All the documents and programs that children interact with are recorded in this diary, and afterward, the children can access the diary to review, recover or delete work done. The programs on Sugar are called activities, which are game-based learning applications. The activities allow for collaborative learning between other peers nearby through a wireless connection. For network connections XO laptops support both wireless and network meshed connections. A complete description can be found on the OLPC official site (http://laptop.org/en/laptop/index.shtml).

Most of the activities conceived in the OLPC project and implemented in the XO-laptop are based on the Constructionism theory – an extension of the Constructivist Theory, which states that *"learning can happen most effectively when people are actively making things in the real world"* (Constructionist learning theory, n.d.). Therefore the activities are oriented in such a way that children construct meaning by themselves more from collaborating with peers and less from their teacher's guidance. The activities on the XO laptops are not based on any learning curriculum content, rather on activities oriented towards constructing, designing or creating collaboratively. However, the learning theory implemented in the XO-laptops received some criticisms. The way children learn differs from child to child (Bentley, 2007), for example some children may be more talented in Math, other children in language arts. Similarly, the preferences of children differ, some children may like Math and other children may not. Therefore some children need to be more guided and motivated by the teacher than others. It could be inappropriate to allow children to learn only through creating and making by themselves without working within a structured learning curriculum due to the difficult conditions that children live in in underprivileged areas of developing countries. For this reason, governments have to ensure that children have learnt basic knowledge and skills, e.g. language skills and basic Math, by the end of elementary school. For these children, fulfillment of the learning curriculum is core and the OLPC approach provides little guidance for that.

Figure 1. XO-Laptop; the hardware created in the OLPC project.

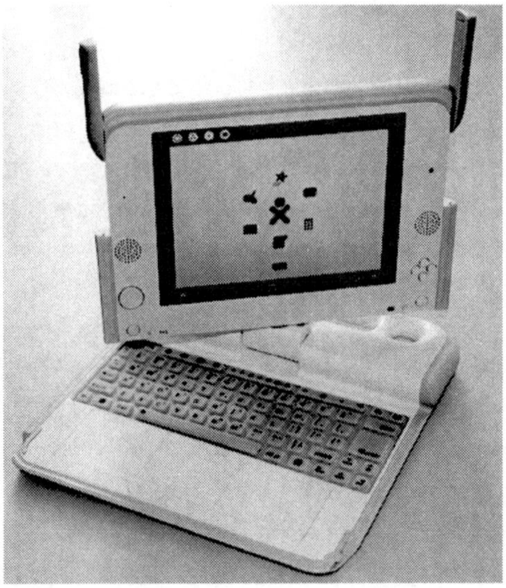

In addition, teaching in developing countries often includes teaching to large classes in an "ex-cathedra teaching" setup, where the hierarchy of teachers is strongly respected by students. This teaching approach is supported by the curriculum and to change it would require changing the entire educational culture–which would be a longer term project and it is doubtful that forcing the change through technology is a good approach.

Finally, one other criticism relates the use of Sugar as the GUI. Some governments of developing countries consider that using Sugar as the GUI could leave children at a later disadvantage, because they would be unfamiliar with a typical

Window based GUI (Kraemer, Dedrick, & Sharma, 2009), which is the dominant graphical interface in most known OS such as Windows, Linux, Mac or Solaris.

Current Learning Projects for Mobile Phones in Developing Countries

As discussed previously, mobile phones represent the main point of access to ICT for people in developing regions. Several research projects have been oriented in the utilization of mobile phones for supporting the education of children and adults in developing countries. Here we will present some current examples found in the literature about the use of mobile phones to support education in developing countries. Specifically, mobile learning experiences and research results from India and South Africa will be described. Most of the mobile learning applications illustrated here are part of different pilot research projects.

Supporting English Learning as a Second Language

In general, there are four major areas that should be achieved in order to be able to communicate in any language: reading, writing, speaking and listening comprehension. As one develops skills in each of these aspects/areas, the fluency and skills of the entire language improve as well. The realization of written and reading exercises, including grammar tasks, support the writing/reading skills, and the realization of oral exercises support the listening/speaking skills.

Books are the most commonly used tool for learning a second language in many developing countries, and many developed ones too. Language learning books include grammar lessons, written and oral activities; nevertheless, the written and grammar exercises make up the majority of exercises. For oral activities, some books include external audio resources such as CDs or cassettes. In contrast to developed countries, where most of children have easy access to audio players (MP3, or CD players), children in developing countries have limited access to such technologies; therefore the practice of listening is limited to schools sessions. Similarly, other books include online multimedia support or software CD to be played on computers, which again is a very limited resource for children from schools in developing countries, as discussed at the beginning of this chapter. Besides books, access to the Internet plays an important role here because with the Internet many resources are available to improve the learning of a language, from listening to international online broadcast e.g. radio, TV, to free online resources to learn the language, from dictionaries and access to journals for improving reading, to informal ways of learning a language such as chat or blog with other children who speak English. There are definitely a lot of advantages over children who have no computers and/or access to the Internet. There has been some initial research focusing on the utilization of mobile phones to improve the acquisition of second language skills of children in developing countries, one of these approaches is PACE Framework (Kam et al., 2007).

Kam et al. (2007) present the Pattern-Activity-Curriculum-Exercise PACE. PACE is a framework that supports the creation of learning games. Currently, the PACE framework implements learning games for the teaching of English as a second language (ESL) for children in rural areas in India (Kam et al., 2007). During the development of the PACE framework, the authors started their analysis with two main sources: computer games and English learning software. During the process of the implementation of the language learning pattern, Kam et al. (2007), faced some design problems in adapting computer games as well as ESL software for use in developing countries: (1) the user interface of the games had some Western biases and were not based in the context where the children would be using them (Kam, M., et al., 2009), (2) unfamiliarity of the design team with the English learning needs in India, (3) most children in the rural and poorer areas of developing countries are unfamiliar with video or electronic games, therefore (4) it was difficult to find a balance between the interface of the game and learning, either the games design was very challenging and had a high quality graphical interface but low learning content or there was high learning content with boring games. The games designed for PACE followed the Pimsleur language learning principles (Pimsleur Principle, 2009), and are composed of two phases, a receptive phase and an activation phase. During the receptive phase the players learn about specific content of ESL and in the activation phase, the game tests what has been learned during the receptive phase.

Based on the PACE framework, Kam et al. (2007) implemented two ESL learning games, one targeted at teaching the English alphabet and the second one aimed at teaching new vocabulary. To validate both games, they organized two user tests with children living in poor/rural areas in India. The first game was tested with fourteen children attending kindergarten aged from 4 to 6 years old and the second game with eleven children attending the 6th grade, with ages between 11 and 15 years old, with some basic English lectures (but only able to do simple greetings). In the first test, a bow-and-archery game was implemented, where the children had to match the phoneme with the corresponding grapheme of the letter. Only two thirds of the children knew the English alphabet prior to the user test. For this user test three exercises were designed which covered the whole English alphabet. Most of the children completed 2 of the 3 exercises, which reflects indirectly that the children had learnt the letters (Kam et al., 2007). However, a post learning test to bear out these indirect findings was not realized.

The second user test consisted of learning new vocabulary using pictures complemented with audio. Thus the children learnt the grapheme of the word with the picture (receptive phase) and then the game presented the picture and she/he had to match the correct word (activation phase). Kam et al. (20007) applied two post tests. In the first test children had to explain and read the words, and in the second post-test children had to write a small story using the words learnt in a team. In the first test, children knew an average of 4.3 (from 12) new words. In respect to the second test, almost half of the participants, 5 children (distributed in two teams), could write a small story using the correct context of the words learnt, although in some cases, because of their limited English vocabulary, Hindi was used to complete some phrases.

Even though it was a first experience with a mobile phone for all of the participants, they quickly learnt how to interact with the mobile phones, including how to enter text –as in the second user test. Only using the joystick button was difficult for them. In summary Kam et al. (2007, p. 1105) concluded that *"for unattended areas, such as rural or poor areas in developing countries, an iterative design with the users is essential in developing applications that meet their unique language learning needs"*.

Mathematics Learning: MOBI and M4Girls

MOBI (MOBI 2009) is a project in South Africa which provides Mathematics lessons to mobile phone users (Tectonic, 2007; Matthee & Liebenberg, 2007). The Maths lessons provided by MOBI match the 8th to 12th grade Mathematics curriculum of South Africa education. MOBI is based in the concept of edutainment, but not with the implementation of a game-learning approach. In MOBI the pupils do not learn while playing. Instead MOBI provides separate components for learning and for entertainment (Matthee & Liebenberg, 2007). For entertainment, children could listen to music or the radio, and a chat platform is provided which students could use for both collaborative learning or for socialization. MOBI-Maths is available in English as well as in Afrikaans language.

The learning component of MOBI is composed of five sections (MOBI, 2009, Matthee & Liebenberg, 2007): *Mobile Assessor, Quick Find, Background Material, Study Theory, Examples, Exercises, Past Papers,* and *Mobi Matric*. With the *Mobile Assessor* module, the student realizes a placement test to find out his/her proficiency in Maths as well as to determine his/her level of Maths knowledge according to the South African Math curriculum. The *Quick Find* module allows students to quickly search for a specific topic within other modules. The *Background Material* module offers students the possibility to review basic Maths content and thus fill gaps in their Maths knowledge from previous years. The *Study Theory, Examples* and *Exercise* modules provide grade-specific Mathematics background knowledge, corresponding examples and exercises respectively. In addition, the *Exercise* module prepares children for the realization of tests and exams, as well as to self-assess their progress. The *Past Paper* module presents students with the complete solutions including theory explanations of matriculation exams from previous years. The *Mobi Matric* module prepares students for the realization of the matriculation exam by providing examples of matriculation exams to write down.

The MOBI project focuses mainly on students in the three senior high school grades (10th to 12th grade). Currently (MOBI, 2009) MOBI-Maths offers, besides mathematics lessons, learning material in Physical Science, Geography, Business Studies, Economies, Life Sciences, English, Afrikaans, History, and Accounting. There is a monthly fee for the use of some services of MOBI (Tectonic, 2007).

Similarly to other mobile learning approaches, MOBI is not aimed at replacing the live classroom experience, but enhancing it. As Matthee & Liebenberg (2007, p. 153) said, MOBI-Math can *"assist learners in a homework environment as a tutor to assist in the understanding of maths concepts and as a revision tool during test and exam seasons"*.

Another project focused on learning Mathematics is M4Girls (Nokia, 2009a). M4Girls is a pilot project developed by Nokia, the Department Education of South Africa and the non-profit organization Mindset Network (http://mindset.co.za). The project was realized between December 2007 and December 2008. M4Girls provides learning games and other attractive content that matches the official mathematics syllabus (ITWEB, 2008). M4Girls was tested by 20 girls in the 10th grade of two public technical high schools located in South Africa (Nokia, 2009a). For the project, the girls used Nokia 6300 model. According to (Nokia, 2009b), the girls using M4Girls application improved their proficiency in mathematics, and M4Girls can be considered as a feasible means for learning.

Audio Wiki: MobilED

MobilED (MobilED, 2009) is a project of the Meraka Institute of the Council for Scientific and Industrial Research (CSIR) in South Africa and the Media Lab of the University of Art and Design Helsinki (UIAH) in Finland. The main objective of MobilED is *"to design learning environments that are meaningfully enhance with mobile technologies and services"* (MobilED, 2009).

Since most of the population in South Africa – and Africa as well – has no access to computers or the Internet, there is no opportunity to take part in online communities (Leinonen, Sari & Aucamp, 2006). Having said this, the government in South Africa has made efforts to improve skills in computer use, with the implementation of computer labs in schools. This model is far from being a success, due mainly, to a lack of infrastructure and ICT illiterate teachers (Ford & Botha, 2007). On the other hand, even though current mobile phones have capabilities to access to Internet, users are not familiar with this model (Leinonen, Sari & Aucamp, 2006). They only use mobile phones for making calls and sending SMS. Besides, most of the population there has low-profile mobile phones and the connection prices are unaffordable for them.

MobilED Audio Wiki works as follows (Leinonen, Sari & Aucamp, 2006; Ford & Botha, 2007): first, the mobile phone user sends an SMS-request with the concept that he want to search for information on, then the MobilED server calls back the user's mobile phone and a speech synthesizer reads the article found in Wikipedia about the concept. Similarly to accessing online (written) Wikipedia, users can navigate through the audio (next section, previous section). If the article does not exist, users can add their own article by dictating it to the system or if an article already exists, they can add their own annotations, also by dictating it to the system (Ford & Botha, 2007).

To validate MobilED, two user tests were done (Ford & Botha, 2007). The task to be completed in both tests was to search for information about HIV/AIDS and to add their own oral explanations using the MobilED system. The users were divided in teams, the audiocasting team and the thematic-expert team. This latter team was composed of one member of all of the audiocasting teams. The thematic-expert group had to search information about HIV/AIDS using the MobilED platform, then, based on the information retrieved and transmitted by the thematic-experts, the audiocasting-teams discussed the topic and added their own oral annotation to the original audio about HIV/AIDS. In both user tests students had to share mobile phones. There were 51 participants in the first user tests, who were between 15 and 17 years from the 11th grade of a private school in South Africa. All the pupils spoke English as their mother language. All the participants were fully skilled in the use of computers and the Internet. The second user test was carried out with pupils from 9th grade in a public school in a rural area of South Africa aged between 14 and 17. The students spoke different native languages, and had been learning English since 6th grade. In addition, all the students were computer-illiterate and few of them had used a mobile phone before. In both user tests, learners and teachers strongly supported the concept of Audio Wiki.

Some of the more outstanding findings (Ford &Botha, 2007) were that

1. Although many participants of the second user test were ICT-illiterate and many of them had no experience with mobile phones before, they quickly figured out how to interact with them. This finding is also shared in the experience with PACE framework (Kam et al., 2007).
2. While the users of the first user test had problems sharing mobile phones, the participants of the second user test had no problems sharing the mobile phones and working collaboratively. In many regions of Africa, especially to rural villages, the rule is to share the use of mobile phone between the family members or even between the communities.
3. Even though the students of the second user test were not fully skilled in English, most of the audiocasting-teams created the oral annotation in English, as the authors (Ford & Botha, 2007) stated that children are conscious of the key role of speaking English to have a better future.

Currently the MobilED system also includes retrieval information as an SMS as well.

Mxit Learning Experience: BOT and ImfundoYami/ImfundoYetho

MXit (http://mxit.com) is a Java-based free instant messaging software application for mobile phones. In order to install MXit users should be GPRS/3G mobile phones and support java. MXit also supports other instant messages applications such as MSN Messanger, ICQ, AOL Messanger and GoogleTalk. MXit was developed in South Africa, where it is also very popular among young people (Bernard & Moolman, 2007; Tang-

Figure 2. Screenshot of MXit running in mobile phone

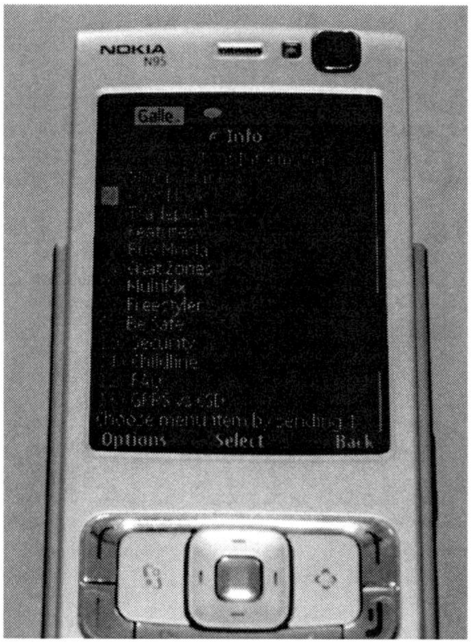

kuampien, 2009). Sending messages by MXit is cheaper than sending standard SMS.

Currently, MXit offer a version that runs over PC. Several researchers are interested in taking advantage of the popularity of MXit among young people and are looking for a means to include learning experiences using MXit.

Tangkuampien (2009) observed that children using MXit interact in multiple chat sessions leading to a long waiting time for the replies. Therefore he proposed Bot (Tangkuampien, 2009). Bot was implemented in a high school in South Africa. Bot works as follows: while children are waiting for replies from their peers, they can send a blank message to Bot, and then it answers them with information about material taught in class. Although there was no concluding proof of improved learning by the pupils using Bot, the service was popular and students were motivated to learn. With feedback from the first study, Tangkuampien (2009) enhanced Bot by adding a multi-choice questionnaire which provided feedback. When students answered correctly, Bot sent back additional information about the topic, and when students answered incorrectly Bot explained what was wrong with their response. In addition, an equation-solver, a dictionary and Wikipedia were added to Bot. The equation-solver guides students step-by-step to solve quadratic equations. The dictionary and Wikipedia support the students in looking up terms or concepts online using the MXit interface. Both the initial version and the enhanced version of Bot covered the Mathematics and IT classes.

Nokia and Nokia Siemens Networks (Nokia, 2009b) jointly developed the "ImfundoYami/ImfundoYetho" project with MXit. It is a pilot project to provide mathematics learning for mobile phone and MXit users. The project ran from February 2009 to June 2009, and 260 students from grade 10th participated in it. The participants used their own mobile phones during the study. They had to solve around 10500 Math-exercises available on MXit. The project is still ongoing and no results are published yet.

Overview of the Examples Presented

Most of the examples published over the last years were developed in South Africa and almost all of the educational applications described, with the exception of MOBI, were/are part of several research and pilot projects. South Africa was among the worst performing country in Mathematics in high school (Nokia, 2009a). Due to this, the South African government has been piloting several research projects to improve the low performance of students in Mathematics. M4Girls, ImfundoYami/ImfundoYetho and MOBI were developed for this reason. An overview of the examples presented in this section is shown in Table 2.

The applications presented explore many of the features of mobile phones. Two of the applications presented were game based, one was audio-based, one included video for explaining the content material, and the rest of them were text based. Only PACE framework included fully interactive games, where the children learnt while

Table 2. Example of Learning Applications with Mobile Phones for Developing Countries.

Learning Application	*Example*	*Description*	*Teaching resource*	*Mobile Phone features used*	*Country*
English	PACE Frame-work	-Application that supports English alphabet and vocabulary learning -Research project	-Game	-Audio -Pictures	India
Mathemat-ics and other subjects	MOBI	-Edutainment application that provides mathematics (mainly) and other subject content. -Based on the official curriculum	-Informative capsules -Multiple choice questionnaires -Feedback	-Video player -Text -Radio -Chat	South Africa
Mathematics	M4Girls	-Application that supports math-ematics learning -Based on the official curriculum -Pilot project	-Game and other content mean	n/a	South Africa
Mathematics	Imfundo Yami/ Imfundo Yetho	-Provided Mathematics exercises -Based on the official curriculum -Pilot project	-Solve exercises	-Mobile Instant Messag-ing (MXit) -Internet Access	South Africa
Mathematics, IT, information gathering	Bot	-Provided Mathematics and IT information based on the material taught in the course -Provided access to Wikipedia and online Dictionary -Research project	-Informative capsules -Multiple-choice questionnaires -Feedback	-Mobile Instant Messag-ing (MXit) -Internet Access	South Africa
Information gathering	MobiLED	-Provided access to an Audio Wikipedia -Research project	-Information gathering	-SMS -Phone Call -MMS	South Africa

playing. The mobile applications supporting mathematics were mainly text-based, with the exception of MOBI. MOBI presented an additionally video which explained, for example, the steps involved in solving an equation. In contrast to the rest of the applications described, and very interestingly, MobiLED explores the use of the two basic features of mobile phones: phone calls and SMS.

Reflection and Discussion

We have discussed that in the underprivileged areas of developing countries, the access to computers and Internet is very limited and that the available infrastructure is often unstable. In contrast mobile phones present a real and feasible alternative providing a similar potential for access to information and learning tools as classical ICT. Even though there is this potential it does not mean that mobile phones will solve the global digital divide and the unequal access for rural and urban areas in the short term.

So far most of the projects that explored novel approaches for access to ICT and to ICT for development have been small scale pilot projects often done by students and researchers. Often these projects are initiated by academic institutions and in many cases supported by ICT and mobile communication companies. The major problem with these projects and the validity of the findings is that the projects typically only run for a short period of time. Most of them are not sustained or put to a major scale deployment after the completion of the project. Two main reasons for this situation are described by (Dies & Brewer, 2009): (1) researchers/students involved in the projects look for better job opportunities after finishing their projects, and (2) little or no involvement of government or local known NGOs

(Non- Governmental Organizations) during the project that could promote them later. The support of government and/or NGOs is one key to ensure the continuation of the projects after the research part has finished. In most environments only in a close collaboration with the government will it be possible to initiate a stable, large scale, and long lasting project. Making educational tools a commercial success and achieving a sustained service as people earn money with it is an alternative direction. However there have not been larger scale examples in developing countries where this has worked – at least not so far.

Cultural and social practices play an important role in the acceptance and use of mobile applications in developing countries. The socio-cultural context differs from country to country and even within the same country. Especially in countries where there is significant ethnographic diversity and where a variety of languages is spoken - this is a central issue. Understanding the socio-cultural background as well as understanding local needs is the key to developing appropriate applications that can be used by people in developing regions (Leinonen, et al., 2006; Ford & Botha, 2007; Kam et al., 2007). A positive example is the PACE-framework. Its developers considered the local games practices in rural India, which differ strongly from game playing in USA or Germany. Similarly MobilED was inspired by the high number of illiterate and semiliterate people in the rural villages of Africa who used mainly voice features (phone) of mobile phones. Overall it is essential to consider the local practices of technology use in the design of applications. The central question and guideline is: Why should you make students adapt to using unknown technologies rather than taking advantage of the technologies they already know and are using? (Leinonen, et al., 2006, Sharples et al., 2005). One example is the implementation of learning applications included with MXit for kids in South Africa, where MXit is very popular.

Even though there are a very high number of mobile phone users in developing countries, it is essential to understand that most of them are not smartphones. Many of these phones have only basic capabilities. While researchers in developed regions are working with high-end mobile phones (e.g. 3G mobile phones or iPhones) such devices are unknown for citizens in underprivileged regions of developing countries. The successful delivery of mobile learning applications will depend on the use of mobile phone features which are readily available for these people. Nonetheless, as happens with other technologies, prices tend to decrease when there is a high demand for a technology. For this reason, today's smartphones in developed regions could be accessible widely in developing regions in the medium term. However it seems that the technology replacement cycles we see in the developed world (e.g. users have a new phone every 18 month) are not seen in the developing world, where technology is longer in use.

The involvement of teachers to create their own content is important for the long-term realization of mobile learning, especially in developing countries. Teachers are best-placed to know which material is best suited for children in the given learning situation. Mobile learning applications should provide an easy means for teachers to add/edit content (Valderrama et al., 2009) – which ideally should not be more complicated than preparing student material on paper. We see this as a central requirement for the future sustainability of mobile learning applications. If the job of the teachers could be eased by mobile technology and if they would get a direct benefit from preparing materials for use on mobile devices they would advocate the usage model.

We also think that this may go beyond content creation. If educational software for mobile devices could be developed by people in the respective regions and potentially using mobile devices as development tools this would solve many of the stated difficulties. Software developed locally would most likely fit the socio-cultural context without any additional effort. Additionally the cost structure would be very different, compared to having the software being developed in areas of the world that have much higher costs.

In summary, we consider the following points as central for creating successful mobile learning applications in developing countries:

- consider cultural and social local background
- reflect current practices of technology use in the design
- involve the teachers and educators in the creation and usage of materials
- consider the scholarly curriculum (if formal learning is the goal)
- ensure support of the government and of NGOs
- plan beyond the initial research stage of the project

In addition to these points, it is necessary to point out that the main stumbling block for the use of mobile or any other computer technology in underprivileged regions of developing countries is still the lack of electricity supply. Although situations differ from one developing country to another, many rural villages and many difficult-to-access areas in developing regions have unstable or no electricity supply. Nevertheless here mobile devices have a clear advantage as they can operate on rechargeable batteries for a significant amount of time. Furthermore this has lead to people establishing specific local practices to deal with these problems. For example, in Subsaharian African villages, there are more households with mobile phones than with electricity (STAAT, 2007), and those who have electricity supply offer the service of charging mobile phones to those who do not have an electricity supply. Solar powered chargers and solar panel phones, e.g. Samsung E1107 Crest, are the first step forward towards sustainable energy provision.

Finally, it is important to recall that mobile learning models that work in the research context or even in the urban/rich population of developing countries may not fit the culture and reality of underprivileged areas.

CONCLUSION

Access to the Internet and the use of computers is common, and, almost ubiquitous, in developed regions. As a result, citizens there have developed skills in the use of computers and the Internet. A very well established computer infrastructure exists for learning and in businesses. In contrast, in developing countries where computers are scare and Internet access is alien for many citizens, there is a lack of skills in using computers and the Internet. Mobile phones are the main type of ICT that people in developing countries get in contact and operate with. With the massive proliferation of mobile phones it is for the first time that citizens of underprivileged regions of developing countries may have access to computing and communication resources on a large scale. There may now be a real possibility that the digital divide between developed and undeveloped regions will become smaller.

Multimedia and network capabilities of current mobile phones (even low end phones) provide a high potential to support education. These features together with full portability of mobile phones makes them useful as learning platform, which can be used within or outside classrooms, during excursions or at home. It creates evidently the possibility of providing access to educational material everywhere students go. There are many examples of how mobile phones can contribute to both formal and informal learning environments (Sharples et al., 2007; Pierroux, 2008) and education. In developed countries we see a strong research focus on interaction with smartphones. However it is important to understand that for creating successful mobile learning applications for developing countries the requirements for mobile phones should be as low as possible.

In order to realize a useful and appropriate mobile learning experience for developing countries, it is important first to analyze the local cultural and social background of the target region, as well as the education needs. Direct implementation of successful mobile learning experiences from developed regions or from other developing regions, without adaptation to the local context and needs, will most certainly fail. In addition, teacher involvement is core; mobile learning applications should include easy mechanisms for teachers to create, edit and structure educational content. In cases where learning in formal settings (e.g. schools) should be supported, it is essential that the mobile learning applications fit the educational curriculum and the educational approach of that country. Finally, the participation of governments and NGOs is crucial to ensure the long term success of any project that involves learning and ICT for developing countries.

Due to the lack of computers at home as well as at schools, and the high number of users of cell phones, we believe that mobile phones can play an important role in education in developing countries and offer a potential to narrow the digital divide. Mobile learning in developing regions, tries to fulfill two goals, first, to enhance the learning experience inside or beyond classrooms and second, to help citizens to acquire skills in the use of information communication technology (ICT).

REFERENCES

Archibald, J. (2007). *Cellphones as an Educational Tool.* Retrieved July 9, 2009 from http://www.tectonic.co.za/?p=1396

Batane, T. (2006). When computers are Scarce: a case of Botswana Schools. In *Proceedings of the 4th IEEE International Workshop on Technology for Education in Developing Countries TEDC '06* (pp. 93-98).

Benteley, C. (2007). The OLPC Laptop: Educational Revolution or Devolution? *Proceedings of World Conference on ELearning in Corporate, Government, Healthcare and Higher Education* (pp. 647-652). Quebec, Canada

Constructionist learning theory. (n.d.). Retrieved June 18, 2009, from Wikipedia: http://en.wikipedia.org/wiki/Constructionist_learning

Dias, B., & Brewer, Eric. (2009). How Computer Science Serves the Developing World. *Communications of the ACM, 52*(8), 74–80. doi:10.1145/1516046.1516064

Ford, M., & Botha, A. (2007). MobilED-An Accessible Mobile learning Platform for Africa? *Proceedings of the IST-Africa Conference. Maputo Mozambique.* Retrieved December 1, 2009 from http://researchspace.csir.co.za/dspace/bitstream/10204/1777/1/Ford1_2007.pdf

Inkpen, K., & Booth, K. Gribble, S., & Klawe, M. (1995). Give and Take: Children collaborating on One Computer. *Proceedings of the International Conference on Computer Human Interaction CHI* Denver, CO. (pp. 258-259).

Inkpen, K. M., Ho-Ching, W., Kuederle, O., Scott, S. D., & Shoemaker, G. B. D. (1999). "This is fun! We're all best friends and we're all playing.": Supporting children's synchronous collaboration. In *Proceedings of Computer Supported Collaborative Learning (CSCL) '99*. Stanford, CA, (pp. 252-259).

ITU. (2009). *Measuring the Information Society. The ICT Development Index*. Geneve, Switzerland: International Telecommunication Union ITU.

ITWeb. (2008, April). *Mobile phones to teach Maths.* Retrieved July 13h, 2009 from http://www.itweb.co.za/sections/telecoms/2008/0804081343.asp?A=CEL&S=Cellular&O=FPPN

Jones, M., & Marsden, G. (2006). *Mobile Interaction Design*. London: John Wiley and Sons, Ltd.

Kam, M., Mathur, A., Kumar, A., & Canny, J. (2009). *Designing Digital Games for Rural Children: A Study of Traditional Village Games in India.* Paper presented at Computer Human Interaction CHI. Boston, MA. Retrieved December 1, 2009 from http://www.cs.cmu.edu/~mattkam/publications/CHI2009.pdf

Kam, M., Ramachandran, D., Devanathan, V., Tewari, A., & Canny, J. (2007). *Localized Iterative Design for Language Learning in Underveloped Regions: The PACE Framework.* ACM Conference in Human Interaction: Designing for Specific Cultures, San Jose, CA. Retrieved December 1, 2009 from http://www.eecs.berkeley.edu/~divya/CHI2007_pace.pdf

Kessy, D., Kaemba, M., & Gachoka, M. (2006). The Reasons for Under Use of ICT in Education: in the context of Kenya, Tanzania and Zambia. In *Proceedings of the 4th IEEE International Workshop on Technology for education in Developing Countries TEDC.*

Kraemer, K., Dedrick, J., & Sharma, P. (2009). One Laptop Per Child: Vision vs. Reality. *Communications of the ACM, 52*(6), 66–74. doi:10.1145/1516046.1516063

Leinonen, T., Sari, E., & Aucamp, F. (2006). *Audio Wiki for Mobile Communities: Information System for the Rest of Us.* Paper presented at the Workshop on Speech in Mobile and Pervasive Environments, during the ACM International Conference on Human Computer Interaction with Mobile Devices and Services MobileHCI. Espoo, Finland.

Matthee, M., & Liebenberg, J. (2007). Mathematics on the Move: Supporting Mathematics Learners through Mobile Technology in South Africa. In *Proceeding of the 6th conference on Mobile Learning Mlearn.* Melbourne, Australia. (pp. 156-161).

Mobi Math Project. (n.d.). *Mobi Math Project.* Retrieved July 8h, 2009 from http://www.mymobi.co.za/mobi_signin.php

Mobil, E. D. (2009). *MobilED Initiative*. Retrieved on July 10, 2009 from http://mobiled.uiah.fi/

Moraveji, N., Inkpen, K., Cutrell, E., & Balakrishnan, R. (2009). A Mischief of Mice: Examining Children's Performance in Single Display groupware Systems with 1 to 32 Mice. In *Proceedings of the International Conference on Computer Human Interaction CHI.* Boston.

Moraveji, N., & Kim, T. Ge, J., Pawar, U., Inkpen, K., & Mulcahy, K. (2008). Mischief: Supporting Remote Teaching in Developing Regions. In *Proceedings of the Computer Human Interaction Conference CHI.* Florence, Italy.

MXit. (2009). *MXit.* Retrieved July 8, 2009 from http://www.mxit.co.za

Nokia (2009a). M4Girls. *Exploring the Use of Mobile Phones for Learning among High School Girl Learners; Improving the Learning Condition of 2 High Schools.* Retrieved July 10, 2009 from http://www.nokia.com/NOKIA_COM_1/Corporate_Responsibility/World_map/pdf/MEA/CI_OnePager_South_Africa_m4girls.pdf

Nokia (2009b). *Education – Youth Development – Community – Corporate responsibility.* Retrieved July 13, 2009 from http://www.nokia.com/corporate-responsibility/society/youth-development/education

Pawar, S., Pal, J., Gupta, R., & Toyama, K. (2007). Multiple Mice for Retention Tasks in Disadvataged Schools. *Proceedings of the International Conference on Computer Human Interaction*. San José, CA. (pp. 1581-1590).

Pawar, S., Pal, J., & Toyama, K. (2006). Multiple Mice for computers in Education in Developing Countries. In *Proceedings of International Conference on Information and Communication Technologies and Development ICTD.* Berkley, CA. (pp. 64-71).

Pierroux, P. (2008) Extending Meaning from Museum Visits through the Use of Wikis and Mobile Blogging. *Paper presented at ICLS 2008: Internation Perspectives in the Learning Sciences, Utrecht, the Netherlands.*

Pimsleur Language Learning System. (n.d.). *Pimsleur Language Learning System*. Retrieved July 8h, 2009 from, http://en.wikipedia.org/wiki/Pimsleur_language_learning_system

Schwartz, M. (2008, April). Maths via mobile is South African pupil's to learning. *Developing Telecoms*, Retrieved July 20, 2009, from http://www.developingtelecoms.com

Sharples, M., Lonsdale, P., Meek, J., Rudman, P., & Vavoula, G. (2007). An Evaluation of My ArtSpace: A mobile Learning Service for School Museum Trips. Norma, A., & Pearce, J. (eds). In *Proceedings of the 6th Annual Conference on Mobile Learning mlearn 2007*. Melbourne: University of Melbourne, 238-244

Sharples, M., Taylor, J., & Vavoula, (2005). *Towards a Theory of Mobile Learning.* Paper presented at the 4th World Conference on mLearning, Cape Town, South Africa.

STAAT (2007) *Measuring Village ICT in Sub-Saharan Africa.* Markets and Statistics Unit STAAT. ITU's Development Sector.

Tangkuampien, J. (2009). The Potential for Technology-Enabled Connections: Kids, Education and Cellular Handsets. *Interaction, 16*(1).

Tectonic (2007). South Africa High School Math Syllabus goes Mobile. *Tectonic*. Retrieved July 8, 2009, from http://www.tectonic.co.za/wordpress/?p=1589

UN. (2008). *The Global Information Society. Partnership on Measuring ICT for Development.* Santiago, Chile: United Nation UN Publications.

Chapter 9
Historic Monument Education:
The Impact of a Collaborative Inquiry-Based Mobile Learning Strategy on Social Relationship Development

Ju-Ling Shih
National University of Tainan, Taiwan

Chien-Wen Chuang
National University of Tainan, Taiwan

Gwo-Jen Hwang
National Taiwan University of Science and Technology, Taiwan

ABSTRACT

The aim of this research was to design an inquiry-based mobile learning strategy that integrated the physical environment and digital resources to enhance elementary students' learning of historic monuments. The students were situated in both the real world and the virtual world to extend their learning experiences. Based on collaborative learning principles, students used the handheld device, PDA, to do their investigations. This research utilized the Peace Temple in Taiwan as the teaching and learning site and invited 32 fifth graders to participate in the learning activities. Through pre- and post-class questionnaires, as well as observations and focus group interviews, both qualitative and quantitative data about students' social relationships were collected and analyzed. The results showed that students' social relationships varied at different stages with different tasks. Inquiry-based mobile learning is shown to have positive effect on students' learning performances and learning satisfactions.

INTRODUCTION

In traditional classroom teaching where teachers conduct transmissive instructions, students are often not involved in sufficient learning activities. This results in a more passive learning attitude and less learning transfers. For this reason, we aimed to use an inquiry-based mobile learning strategy to supplement classroom learning. The strategy integrated both the physical (real world) and digital environments (virtual world) to allow for active knowledge construction to take place with the students.

DOI: 10.4018/978-1-61692-849-0.ch009

In order to conduct inquiry-based learning, the students were situated in an authentic environment to interact with real objects in the real world. They go through problem-solving processes, explore the learning context, conduct observations, and gain experiences in the everyday life environment. Situated learning theories suggest that knowledge construction and learning can only happen *in situ* and through the interaction of people and their social cultural environments. However, the teacher is unable to attend to many students in situated learning contexts, thus providing individual support is weak. With digital systems and tools, such as the PDA, students can get individual guidance when they are distributed in the field. In other words, the PDA is able to substitute the teacher's role and the pre-designed digital resources in the system work as learning scaffolds for the students.

Mobile devices with wireless communication services have enabled students to retrieve learning resources disregard of their locations. Due to the powerful functions and high portability of mobile devices they are used in numerous formal and informal learning situations to enhance learning effectiveness. The many formal learning situations from different fields around the world that widely use mobile devices include the natural sciences (e.g. Lai, Yang, Chen, Ho, & Chan, 2007), social sciences (e.g. Fischer & Konomi, 2007), mathematics (e.g. Zurita & Nussbaum, 2004) and languages (e.g. Cui & Bull, 2005), just to list a few. Informal learning situations include those in indoor settings, for example, museums, learning centers, laboratories and home, and outdoor settings, for example, parks, city centers and woodlands (Rogers, Price, Randell, Fraser, Weal, & Fitzpatrick, 2005). Other informal learning situations include bird-watching (Chen, Kao, & Sheu, 2003) and museum learning (Burgard et al., 1999). Among these, the Ambient Wood project led by Yvonne Rogers in England (Rogers et al., 2004) and butterfly learning with expert system on PDA led by Gwo-Jen Hwang in Taiwan (Chu, Hwang, Huang, & Wu, 2008) are eminent natural science mobile learning projects that led to positive learning results. Therefore, mobile technologies have fulfilled educational dreams by providing the possibility of creating innovative learning experiences.

In order to investigate the possibilities of applying mobile learning strategies to social science courses in elementary education, this research utilized the Peace Temple in Tainan as an example and invited 32 fifth graders to participate in the instructional activity. In this research, students were situated in the authentic learning environment where they could interact with the physical environment, their peers, and the digital devices. The objectives of learning local historic monuments were to let students to get familiar with local environments, cultures, people, and activities. The investigation was done by physically participating in it, and through observations and interview activities. Field inquiry instructional strategies were considered appropriate for social science learning. When students were situated in the learning context where they were interacting with the environment and their mobile devices, they were immersed in individual learning. In order to increase students' social interactions as well as to allow peer support to happen, collaborative learning strategies were implemented. Meanwhile, teachers could detect individual difficulties more easily.

Collaborative learning has also been proven to have positive learning effects on information-mediated instructions (e.g. Gokhale, 1995; Dillenbourg, Baker, Blaye, & O'Malley, 1996). When students are placed in a group, they have common goals, resources, and tasks to work toward. Hence, peer-interaction frequencies and learning motivation increases more than individual learning. In this research, we define students' interactions with peers as their social relationships.

This research first situated students in both the real world and the virtual world to extend their learning experiences; secondly, it designed educational activities between the field and the digital system to demonstrate the practices of mobile learning which emphasized learning happening close to real life; thirdly, it developed a digital system with learning content to facilitate students' field studies; and fourthly, it provided students with an opportunity to collaborate with peers in the learning process and finally, it used a comprehensive evaluation method to analyze students' social relationships in terms of learning performances and learning satisfaction.

In the following sections of this chapter, we will begin with the literature review that discusses the purpose and design strategies of inquiry-based learning, the features and implementation issues of mobile and ubiquitous learning, and the design and evaluation of collaborative learning. In the research design section, we describe the instructional design and the research methods. Then, in the system design section, learning guidance and the development of system interface are described. In the research results section, the social relationship analysis from pre- and post-class questionnaires, observations, and focus group interviews are presented. Students' feedback about their overall satisfaction is also evaluated at the end of this section. The last section is the conclusion section with summary of findings and future trends.

Literature Review

The literature is discussed from three perspectives, inquiry-based learning, mobile and ubiquitous learning, and collaborative learning.

Inquiry-Based Learning

Students' observation, classification, manipulation, and problem-solving abilities are necessary skills for the 21st century. Knowledge and skills that are closely related to real-life situations can only be retrieved and built in similar situations. Constructivists believe knowledge comes from the learning context and its construction done by the learner (von Glasersfeld, 1989). Learning meaningful knowledge has a strong relationship with the situation where students are placed in (Duffy & Jonassen, 1992). Practices should happen in real sites to gain the most impact since hands-on manipulations help to generate authentic experiences (Resnick, 1987; Montague & Knirk, 1993; Brown, Collins, & Duguid, 1989). McLellan (1993) claimed that it is only when teaching and assessment happens in real environments can they have high educational values. The connection to meaningful knowledge building also depends on appropriate contexts being available for it to happen (Brown et al., 1989). Since the knowledge cannot be separated from the authentic environment, learning situations must be unintentional and able to allow students to mentally connect concepts under study in the real-world environment (Lave & Wenger, 1991; Albanese & Mitchell, 1993). Some empirical researches (e.g. Artemeva, Logie, & St. Martin, 1999; Wolfson & Willinsky, 1998) have also pointed out that situated learning has positive effects on assisting learners to transfer knowledge and skills into their future work in the real world.

The teaching profiled in this research was based on inquiry-based learning (IBL) where students were involved in investigations and observations to obtain experiences. IBL is a concept which encourages teachers to allow learners to get in touch with authentic situations, explore and solve problems that are analogs to real life. By means of exploration, investigation, and observation, students become involved in more rigorous social interactions as well as higher level thinking. It is a teaching and learning model that promotes students' active construction of knowledge about the natural and human world. In the learning, students develop communication skills and habits of mind. They learn about their roles in society and interact with the world around them to become self-directed and independent learners. In the process of IBL, students are the active knowledge constructors and teachers are the facilitators. In the teaching process, by asking investigative questions and providing relevant content information about the learning context, students will vigorously formulate hypothesis, conduct exploration, perform testing, synthesis new knowledge and provide explanations based on discussions (Looi, 1998). The goal of IBL is not to find the standard answers but appropriate solutions to problems.

There are many different models of IBL, some are more focused on concept exploration and some emphasize more on guided investigation. Three basic models are found:

1. Inquiry by discovery (e.g. Bravo, van Joolingen, & de Jong, 2006; Kirschner, Sweller, & Clark, 2006) is a model that has three learning stages: exploration of the learning environment, definition of professional terms, and concept application.
2. Rational inquiry (e.g. Franklin, 1986; Hofstadter, 1998; Stenhouse, 1997) is a model in which the teacher initiates the course by providing the introduction to the topic, telling stories, or using films to replace free exploration. Then in the discussion phase between the teacher and students, students are required to use their logical thinking skills for concept development. In this model, teachers need to have very good questioning skills to induce conclusions with open questions.
3. Inquiry by experimentation (e.g. Hackling, 1995; Wells, 2001) is a model in which students need to follow steps to conduct experiments and to find the best hypotheses and improve on the existing hypotheses. In this process, students go through science experiment practices where they conduct self-directed designing and testing. Because it requires students to solve complex problems, the model is more suitable for students with higher cognitive levels. Ellis (2002) adds that there is a generation process in which students have to find out the implied meaning of their findings.

Suchman (1961) presented an inquiry training model to assist students to identify facts, build science concepts, and explain phenomena to gain knowledge (Gillani & O'Guinn, 2004). Schwab (1962) generated a process inquiry model in which teachers should first acknowledge that training students' ability to gain knowledge is more important than delivering knowledge. The teachers are facilitators who provide assistance during discussions. Students' learning attitude and skills are developed in the process.

Hamelin (2004) thought that IBL is to stimulate students' curiosity and let them choose what they are about to research. Schwab (1962) defined three different levels of IBL that the teacher (i) gives students problems, methods, and procedures to follow, (ii) gives only problems to let students explore the rest, or (iii) gives nothing so that students need to start with the identification of problems. Heron (1993) added one more level between the first and the second that allows students to have more scaffolds before they are elevated to another level.

The three levels Schwab (1962) defined are similar to what Colburn (2000) categorized. Colburn defined inquiry-based instruction as "the creation of a classroom where students are engaged in essentially open-ended, student-centered, hands-on activities. (p. 42)" He thinks inquiry-based instruction embraces several different approaches including structured inquiry, guided inquiry, open inquiry, and learning cycle. In structured inquiry, the teacher provides students with a hands-on problem to investigate, as well as the procedures and materials, but does not inform them of expected outcomes. Students are to discover relationships between variables or to generalize from data collected. For guided inquiry, the teacher provides only the materials and problem to investigate. Students devise their own procedure to solve the problem. In open inquiry, students formulate their own problems to investigate, and in a learning cycle, students are engaged in an activity that introduces a new concept. After the teacher provides the concept, students take ownership of the concept by applying it in a different context.

The National Science Education Standards notes that scientific inquiry refers to the diverse ways in which scientists study the natural world and propose explanations based on the evidence derived from their work (Colburn, 2000). Along this line, the aim of inquiry-based learning refers to the activities in which students develop knowledge and understanding of scientific ideas, as well as understanding of how scientists study the natural world. Therefore, Edelson, Gordin, and Pea (1999) stated that inquiry experiences can provide valuable opportunities for students to improve their understanding of both science content and scientific practices.

In this research, we aimed to extend the students' experience to social science learning which also requires systematic investigation of the real world. According to Collingwood (1994), history is a kind of science which involves inquiry into the past. Inquiry-based learning opens a new way for social science learning. Educators believe that students should systematically employ processes of critical historical inquiry to reconstruct and reinterpret the past, such as using a variety of sources and checking their credibility, validating and weighing evidence for claims, and searching for causality (National Council for the Social Studies, 1994).

Lim (2004) stated that online IBL ensures students' positive participation, on-task behavior, and rich collaboration. It also empowers students' ownership and self-directed learning by increasing their involvement and responsibility for their own learning. However, unguided online historical inquiry does not guarantee meaningful learning (Molebash, 2004). In the research of Manlove, Lazonder, and de Jong's (2008), they found that cognitive regulation involving the recursive processes of planning, monitoring, and evaluation during learning, is generally helpful with enhancing learning gains for the students. A support tool of goal-lists, hints, prompts, cues, and templates to support the cognitive regulation skills of students is incorporated in the students' fluid dynamic tasks. In these tasks, students are asked to produce two learning outcomes of their investigations: system dynamics models and laboratory reports. The researchers find that task complexity, time, and student prior experience have impacted their findings.

A similar research model was adopted in our research in the consideration of appropriate tasks, a sequence of hints and clues, as well as the desired interactive dynamics during the inquiry activities. It was to ensure that learning happens accordingly. In this research, we used the inquiry by discovery model (refer to Figure 3) and designed mobile learning strategies for the learning activities. Students needed to physically observe the environment around them, find problems, search, organize and analyze materials before they came to the conclusion. Using this learning model, we hoped the students would become active learners.

However, when students are distributed in the field, one teacher is not able to pay attention to each and every student, guide his/her learning according to the individual's needs, and answer questions that vary from student to student. Therefore, providing every student with a digital learning tool that can facilitate individualistic learning can be helpful.

Mobile and Ubiquitous Learning

Mobile learning entails the kind of learning in which learners use mobile devices with digital content inside, to learn in anytime and anywhere situations. Such devices include PDA, laptop computers, cellular phones with WIFI or other Internet connection capability, and customized hardware. To avoid students aimlessly wandering around, instructors need to carefully arrange the learning environment and design an interactive learning model, along with preparing meaningful learning content. Seppälä and Alamäki (2003) said that a flexible learning environment should be more accurately described as learning happening at the right time and in the right place and to substitute the idea of anytime and anywhere learning. This allows learners to experience the most authentic and enjoyable learning at the right time based on the individual's needs.

Weiser (1991) introduced mobile, ubiquitous, and context-aware ubiquitous computing concepts, and explains that learning is no longer limited to our desktops but could take place in the real world. Hull and Reid (2003) also suggested that the integration of mobile learning technology has allowed students to interact with many learning resources around them which have supplemented the classroom learning to create situational experiences for the students.

Mobile learning is an innovative learning model which combines the computing ability of mobile devices, communication functions, and smart use. Quinn (2000) pointed out that mobile learning is to utilize mobile computing devices to conduct learning activities. Trifonova and Rochetti (2003) stated three functionalities of mobile devices - the interaction of equipment, reception of content, and connection to services, are necessary for mobile learning. Leung and Chan (2003) thought that mobile learning is using PDA or the mobile phone with wireless Internet connection in the learning activities.

Context-aware ubiquitous learning, one term that is often used along side with mobile learning, has slightly different connotations from mobile learning. Based on Hwang, Tsai, and Yang's (2008) definition, they pointed out a few features to address the differences. Firstly, ubiquitous learning equips context-aware and location-aware abilities so that learners' surroundings are prepared with the detection systems. Secondly, the environment can offer the most accurate personal assistance at the right time in the right place. The data include personal profiles, current locations, learning traces, and learning conditions, so that the learning environment becomes a smart environment where it can recognize personal needs in accordance with personal goals. Thirdly, it is seamless in that learning would not be intermittent due to the shifting locations of learners. Learners are always connected to the smart system and the learning activity is continuous as long as the learners are within the system's coverage. Fourthly, it is self-adjustable, and can adopt different mobile devices and variable functions. A context-aware system is flexible enough to receive inputs from various mobile devices. For the students, having an authentic environment which is smart and can automatically provide resources and support are very important. A context-aware ubiquitous learning environment can adaptively provide learning assistance and document learning processes with digital technologies. Pinkwart et al. (2003) pointed out that the use of mobile devices is not to allow learning situations to be controlled by information technology but to let it become a part of traditional education. It is a tool which provides learning resources. Chen et al. (2002) stated that in a ubiquitous learning environment, learning services should include urgency of learning needs, initiative of knowledge acquisition, mobility of learning setting, interactivity of the learning process, situating of instructional activity and integration of instructional content. The ubiquitous learning environment should include five features such as permanency, accessibility, immediacy, interactivity, and situativeness of instructional activities.

This research, however, utilizes the PDA to facilitate students' learning where the learning environment is not equipped with context-aware facilities. In this regard this research will be defined as mobile learning.

Although the basic understanding of mobile learning and its application in various fields is not uniform, we can identify a few of the most distinguished characteristics through the literature review and instructional research practices such as those previously mentioned. They include:

1. Learning in the field. Mobile learning shifts students away from their seats in the classrooms, and leaves them in an authentic learning environment. Students are immersed in real life situations and create their own learning experiences. The instructional models can influence the achievement of their learning goals.
2. Use of mobile devices. With mobile devices, students can log onto the Internet and enter the virtual world. The digital system facilitates students to upload and store data in various forms such as texts, graphics, audio and video clips. It can also achieve the documenting and tracing functions. Meanwhile, students need to be accustomed to the new gadgetry. Its readability, stability, connectivity, and feasibility can all vary its learning effectiveness.
3. Design of learning content. The multimedia content can be text-based, graphic-based, or animation-based instruction, and allows for different levels of computer-human interactions. The structure of the content is linear and procedural, which is more suitable for declarative and procedural knowledge learning, or non-linear and exploratory, which is more suitable for problem-solving and inquiry-based knowledge learning. Some developers even simply offer only tool-supported functions to be adopted in instructions without any content placed inside the devices. Some researchers engineer mind-tools to support student learning of various subjects. To successfully adapt the various models of content presentation, it is essential for students to have high information and technology literacy.
4. Arrangement of a wide variety of resources. Resources must be rich, meaningful, and closely related to learning themes. Since mobile learning is a more individualistic learning model, resources to support various learning needs are necessary. Materials include hard-copy texts, resources on the Internet, and all kinds of data collected in the field inquiry. Materials can be categorized into three kinds. Mandatory materials are used to establish fundamental knowledge, suggested materials are used to extend students' learning according to individual preferences, and extra materials are those to be found by participants in the inquiry process and shared with peers to enrich the range of discussion.
5. Work in collaboration. Even though the objective of mobile learning is to support individualistic learning, peer interaction and social networking can actually enhance students' sense of involvement and extend learning in more depth and breath.
6. Self paced learning. Students can learn at individualized learning pace and process. Since students are not grouped together to listen to the same lengthy lectures, they can place more focus on materials that arouse their personal interest and they can arrange the amount of time to engage with specific materials according to their own pace and ability.
7. Switching roles as learning director. Teacher and students are playing different roles in the digital era. In traditional lecture-type classroom, it is teacher-centered instruction. With personal digital assistants, it becomes student-centered instruction.
8. Varied kinds of Interactions. The interaction happens between person and machine, between teacher and students, as well as between peers. These interactions stimulate the transferring of knowledge which involves thinking and communications.

The meaning of mobility is not just to stimulate learning motivations by transferring location. Its true value is to allow students to go through proper channels to learn knowledge and skills. This concept has provided an opportunity to bring the students outdoors to the field. First, students are placed in both the real world and virtual world and the design of learning activities should enable connections between the field and the digital system, allowing students to get just in time resources and support. In this regard, the design of mobile learning illustrates the concept of student-centered learning. In this research, we designed a mobile learning system that was implemented in the mobile device, PDA, which was used in the inquiry-based learning activity of the research. Students could use the mobile learning system when they were in the field, received guidance, and used the Internet to search for related information.

Although mobile devices can be used to support individual learning needs, there is the likelihood of students feeling isolated during the learning, and the human interactions decrease over time. For social science courses, we wanted an increase in student interactions and an enhancement of their social relationship as one of the main goals of the learning. Therefore, we designed learning activities that required peer collaborations.

Collaborative Learning

The learning activity in this research emphasized the formation of learning partners which was to enhance students' social interactions. Collaborative learning refers to students who continuously receive experiences from others, thus empowering their personal abilities. By accepting comments, interacting with peers, reaching consensus through discussions and conducting observations, students learn from the behaviors and experiences of other people (Bermejo, 2005). Cooperation means letting students work in small groups, working to common goals in order to gain the most benefits (Johnson et al, 1991; Huber, 2003; Gillet et al., 2005). A good collaborative learning activity design is to let all group members be responsible for the outcomes and accomplish the tasks together (Smith, 1995; Lang et al., 2006).

Collaborative learning is a systematic and structural teaching and learning strategy. In the constructive educational environment, students are not only immersed in the environment but are also involved in communicating and sharing knowledge with each other in the process. Therefore, the process of knowledge construction is more important than the outcome that emerges from it (Roth & Roychoudhury, 1992). Thus, collaborative learning not only draws learners together but also fosters positive interaction and inter-dependency (Johnson & Johnson, 1998).

The cognitive developmental perspective of learning is to help individuals to construct knowledge. Cognitive elaboration (Bandura, 1989) and cognitive development theories (Vygotsky, 1978) are concerned with collaboration. Cognitive elaboration theory entails that teachers use heterogeneous grouping to assign higher-achievement students to be the information provider and give lower-achievement students support. Adam and Hamm (1990) have shown that among individual learning, competitive learning, and collaborative learning, the latter has the biggest impact on students' learning. Collaborative learning not only help the students to improve their learning of content subjects, it also enhances social and affective aspects of the learning. Kagan (1985) stated that working and social skills are required in collaborations. Working skills are those skills that enable students to understand the context of their learning, discuss with others, express their own opinions, ask questions and follow instructions provided to conduct the activities. Social skills are skills that enable the students to understand the inter-dependent relationships.

Over the last two decades, scholars (e.g. Smith, 1995; Keyser, 2000; Ramsay et al, 2000; Rachel & Irit, 2002; Veenman et al, 2002; Janssen et al., 2007) have suggested various methods and provided research results to support cooperative and competitive learning models. Research results show that collaborative students are able to achieve better outcomes and are more productive.

Collaboration fosters better social relationships, healthier mental states and better self respect (Johnson et al, 1991; Veenman et al, 2002). In this regard, in the last two decades, there is an abundance of research in the literature that have reported the positive effects of collaborative learning in the classroom (e.g. Mevarech, 1993; Ghaith &Yaghi, 1998; Klingner & Vaughn, 2000; Porto, 2001; Swain, 2001; Ghaith, 2002). However the literature has also reported many problems with collaborative learning, such as the 'free-rider' effect (Hooper, 1992). Adams and Hamm (1990) suggested that three or four people in a collaborative group would have the best impact. Slavin (1989) as well as Sun and Chou (1996) thought that clear group goals and awareness of personal responsibilities are key elements to the success of collaboration while Hooper (2003) has found that students' interacting with each other has positive effects on their learning achievement. Cortez et al. (2004) have found that mobile computing technology can be used to help teachers in their teaching to achieve collaborative learning with their students. Within the context of this research, PDAs were used for collaborative learning. Students were divided into groups of three, and each group assigned with a theme. The student groups were required to plan the learning tasks, conduct field investigations, prepare group reports, and engage in peer evaluations.

Research Design

The research design section described below is divided into two parts, the instructional design and research methods.

Instructional Design

This project was based on inquiry-based learning principles. We incorporated activities in historic

Figure 1. Peace Temple

Figure 2. Floor map of Peace Temple

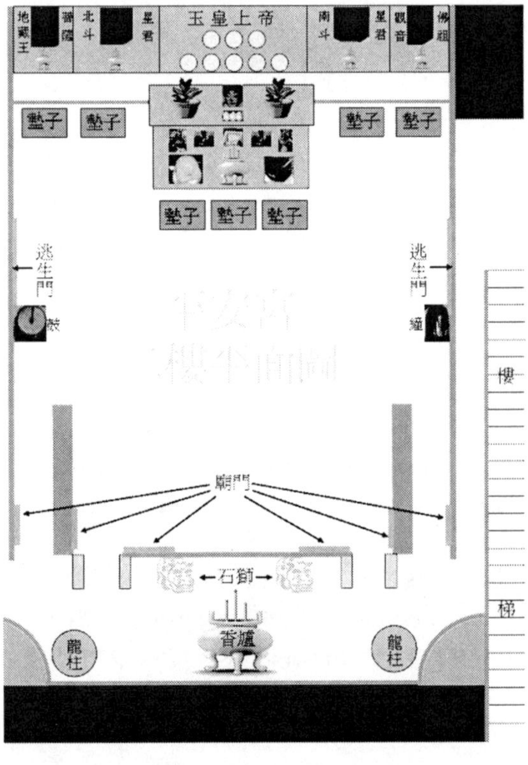

monument investigations using the handheld device, PDA. The objective of the instruction was to develop students' social relationships in their learning.

The instruction was designed for the fifth grade students in an elementary school. The learning content was adopted from their social science textbook Unit One, Lesson Four: Local Temples. The research took place in the Peace Temple in Tainan (Figure 1). Thirty-two fifth graders were invited to participate in the learning activity.

Two things were done in this research. The first was to design the learning content and activities according to the appearances of the historic monuments; in this case the design base was the floor map of Peace Temple (Figure 2). The second was to design the application model for mobile devices according to the learning themes, and to create a supportive learning system for the students' field investigation. Four levels of learning activities were designed (Figure 3).

To start the course, the teacher gave the course guidelines to the students. After creating students' learning motivation with interesting local stories and a series of interactive questions-and-answers, the teacher introduced the historic background of the temple, local cultures and religions, and other related information so that the students had the background information about the trip that they were going to undertake to the Peace Temple.

For the fieldtrip task, students were divided into groups of three and the groups were randomly assigned to different themes by the digital system, such as Gods, religious world, or cultures. Students then went on their fieldtrip courses with mobile devices in hand to explore the temple in person. They followed the digital guidance to conduct the field inquiries. Students visited spots of interests guided by the PDA. At the same time, they gath-

Figure 3. Concept map of instructional design

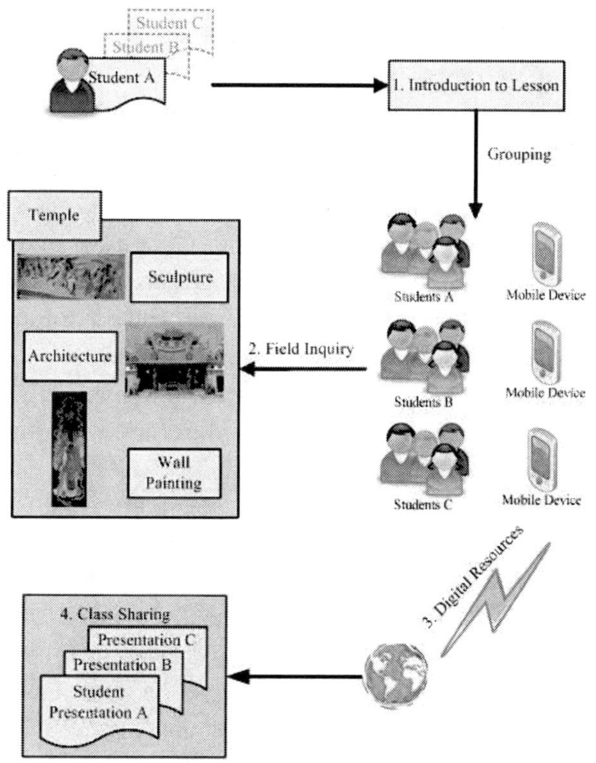

ered more information with the PDA by taking notes. They could conduct detailed observations of the architecture and placement of gods, and interviewed temple keepers and people living in the neighborhood regarding their thoughts and impressions about the temple.

With wireless connection, students could link up to the Internet to search for more information about what was not taught in class. In this regard, their learning resources were extended to the World Wide Web.

After they came back from the fieldtrip, students synthesized and categorized their collections, and constructed reports to share with their classmates. The discussions and feedback stimulated higher levels of thinking. Assessments of learning were conducted from various perspectives, including the teacher, their peers, and themselves. Hence multiple assessments were performed. Learning effectiveness was assessed based on perceptions and reviews of students' group discussions, collaboration and communication behaviors. Learning achievements were evaluated based on their oral and artifacts reports and the feedback they gave to others.

Research Methods

The research process was planned to correspond with the instructional design (Figure 4). Students were divided into two groups. In round one, the A group worked as the "collaborative group" in which students needed to work on the tasks collaboratively, and the B group worked as the "individual group" in which students needed to work on task individually. In round two, the A group became the "individual group" and the B group was the "collaborative group".

To initiate the learning process, the teacher provided the students with an orientation on the use of the PDA. Then, after students filled out the pre-class questionnaire, which was to enable the researchers to understand the students' preliminary understanding of the course materials, the students started to conduct their field inquiry in Peace Temple with two learning tasks assigned. The two learning tasks were designed to investigate if inquiry-based mobile learning can expand the breadth and depth of knowledge acquisition. One task was to ask students to tell an imagined story. It required students to creatively use their observations and interviews with the neighborhood people to conjure up a story. The purpose was to ask students to use their creativity to talk about what they have learned about the history of Peace Temple. The second task was a hands-on art work, in which we asked students to draw, cut, and paste to sketch a temple. They had to describe the special features of the temples they made to their peers. The purpose of this task was to encourage students to observe in detail the temple architecture, decoration and displays and to make impressions of them.

Throughout the whole process, researchers performed observations on students' learning attitudes and behavior by filling out observation forms. The learning process had three stages: physical exploration stage, presentation preparation stage, and presentation stage.

After completing the learning tasks, students returned to the classroom and filled out the learning sheets with regards to their thoughts about the learning process and outcomes. At the same time, a post-class questionnaire was distributed.

After researchers have obtained the preliminary results of the questionnaires, a follow-up focus group interview was conducted with students to get them to explain further on some parts of the thinking that were unclear in the questionnaire responses.

The research questionnaires were categorized according to the research objectives. The pre- questionnaire included personality, learning content, and social relationship questions; and the post- questionnaire included learning content, peer interaction, and satisfaction questions. Besides the content learning section, which was designed with yes/no type of questions to identify whether

Figure 4. Research process diagram

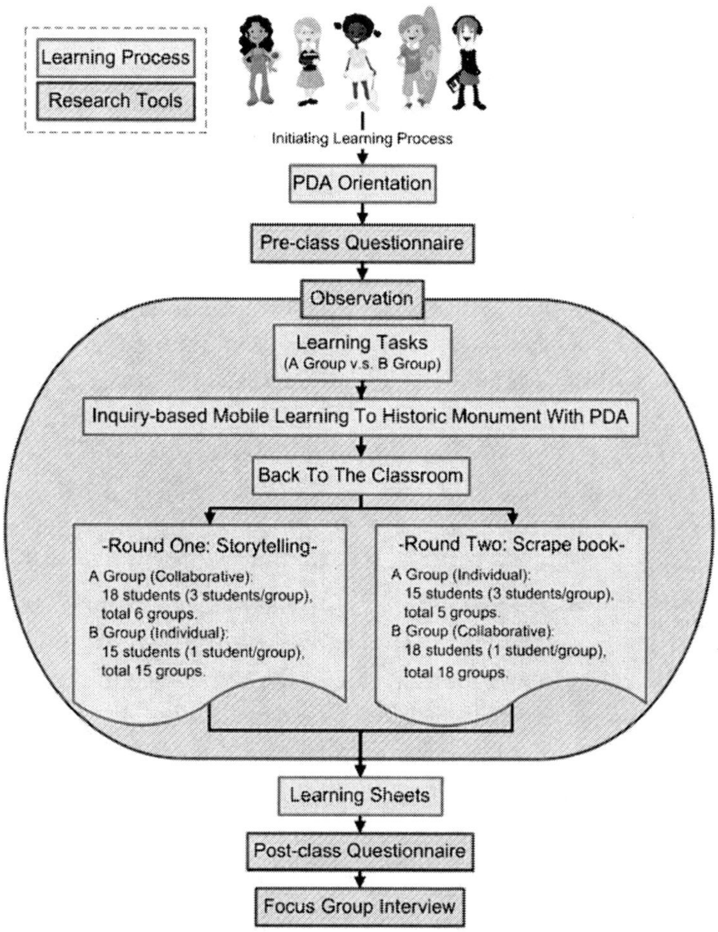

students were involved with specific mental activities, the other sections used six-point Likert scale from strongly agree to strongly disagree. The personality and social relationship sections were used to investigate the influence of the instructions on students' peer interactions.

The focus group interviews were conducted with students in groups of six. As mentioned above, the purpose was to clarify areas of uncertainty that researchers found in the research process. A semi-structured interview outline of questions was prepared to stimulate students' thinking in order to obtain feedback on all aspects of learning.

System Design

The system design for this research is presented in two parts, the learning guidance and system interface.

Learning Guidance

The design concept was based on inquiry-based investigations. Students were given two tasks, five themes, a number of hints in each theme, and selected choices for hints. Students could use those hints to collect the necessary information in order to complete the tasks. The hints were organized under themes, and were prompted consecutively and progressively. With each hint, students could look up more information from either the database or the Internet. Figure 5 shows the learning activity flow chart of the mobile system.

In the system, there were close-hints (CH) which provided selected choices for hints, and open-hints (OH) which only provided open-ended questions. Various investigation methods were also distributed into each theme so that different learning strategies and abilities could be used in the learning process. Method one (M1) referred to physical observation; Method two (M2) referred to physical visit and interviews; Method three (M3) referred to digital search; Method four (M4) referred to several styles of the mix of the M1 to M3. The complete learning content structure for task B is shown in Figure 6.

Task B was students' learning about the little temple architect. After students login to the system, the system randomly distributed themes to each group of students. Finished themes by each group would not be repeated. The themes included flat sculptures, 3-dimensional sculptures, gates, decorations, and architectural structures. Only when students finished with all the themes could they stop the digital system and end the learning activities.

System Interface

On the PDA, the interface design used few graphics in order to keep the information simple and easy to read on a small screen (Figure 7). On top of the screen, there was the dynamic menu of the themes. One hint was displayed at a time with the necessary selected related choices. They were either open hints or close hints to guide students to conduct physical observations or interviews. The system also provided suggested inquiry methods.

In the field, students could also conduct digital inquiry online. Whenever they felt the need to, they could click on the "online search" button to enter Google search engine. We chose Google because it had already a mobile version that could easily be adapted to the small screen of a handheld device. The display of the search results was similar to the PC version (Figure 8).

When the students returned to the classroom, they could retrieve their observation notes on the PDA (Figure 9) with their own account numbers

Figure 5. Learning activity flow chart of mobile system

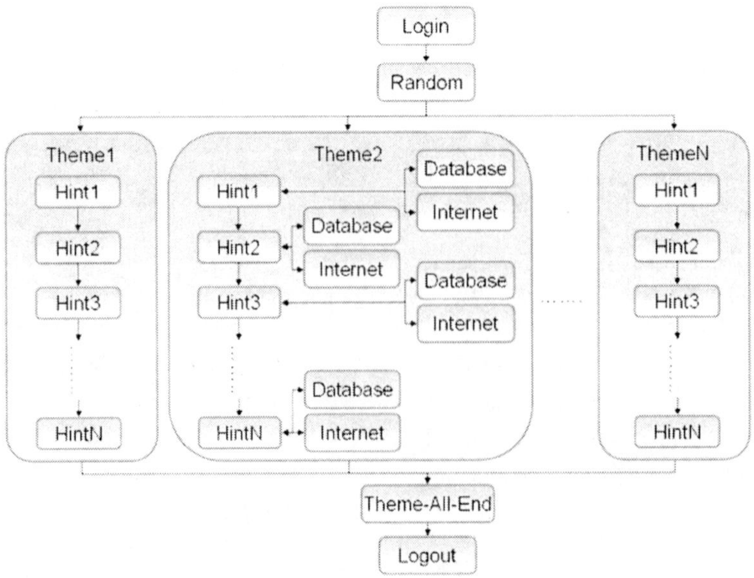

Figure 6. Learning content structure in the system

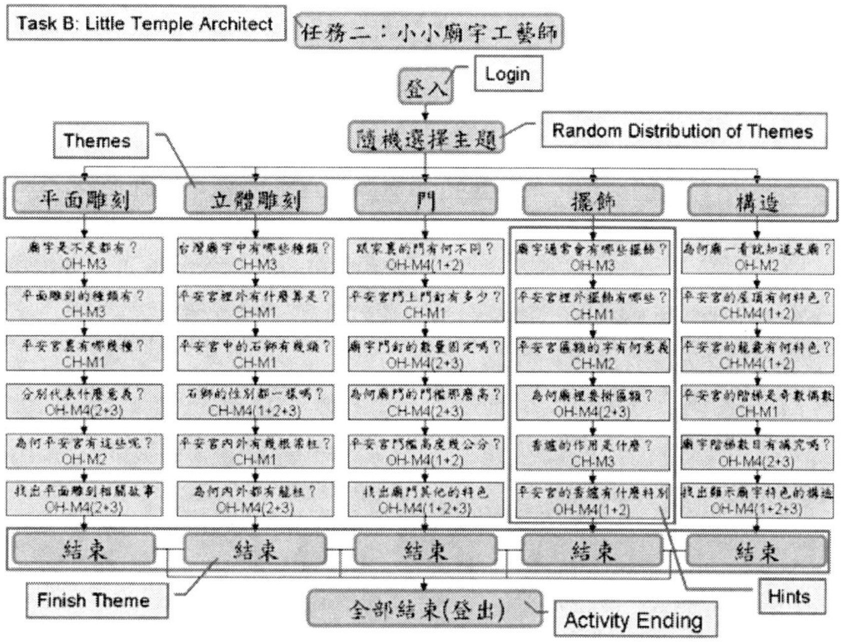

Figure 7. PDA interface for Inquiry

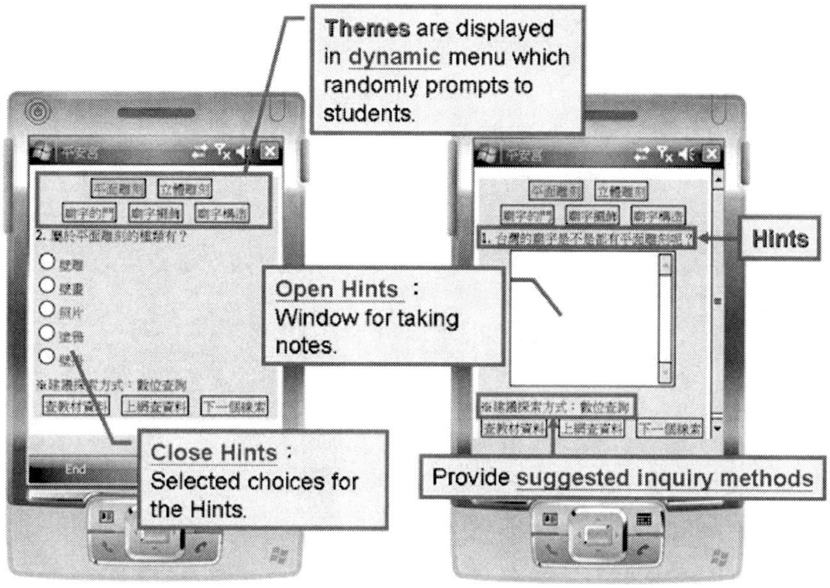

and passwords. Notes were documented according to hint orders. Students could then prepare for the task on reporting by comparing their notes with those of their peers'.

Research Results

The result of this research is presented in two parts, the social relationship and student satisfaction.

Figure 8. PDA interface for digital inquiry with Google

Social Relationship

Students' social relationships varied and changed when they were involved in either collaborative or individual learning modes. We compared the pre- and post-class questionnaires, learning process observations as well as focus group interviews, and reached some conclusions as follows.

Questionnaire

We used SPSS to conduct t-test for the results of the questionnaires in regards to the overall social relationship. It is found that students' overall social relationship (Table 1) changed from pre-test (mean=26.25) to post-test (mean=20.47), and has significantly dropped (p=.000). It showed that students preferred the individual learning mode in the inquiry-based learning situation.

In order to understand the reasons behind why students preferred individual learning that is contrary to our expectation, we looked at the details of the social relationship results by analyzing each of the question items and performed some paired comparisons.

Firstly, we investigated the preference for collaborative versus individual learning. From Table 2, we can see that the comparison of pre-test item 5 "I like to complete homework and reports by myself" (mean=3.69) and post-test item 4 "I like to learn by myself" (mean=3.56), there was no significant (p=.717) difference. We found that those students who preferred to learn individu-

Figure 9. PDA interface for observation notes

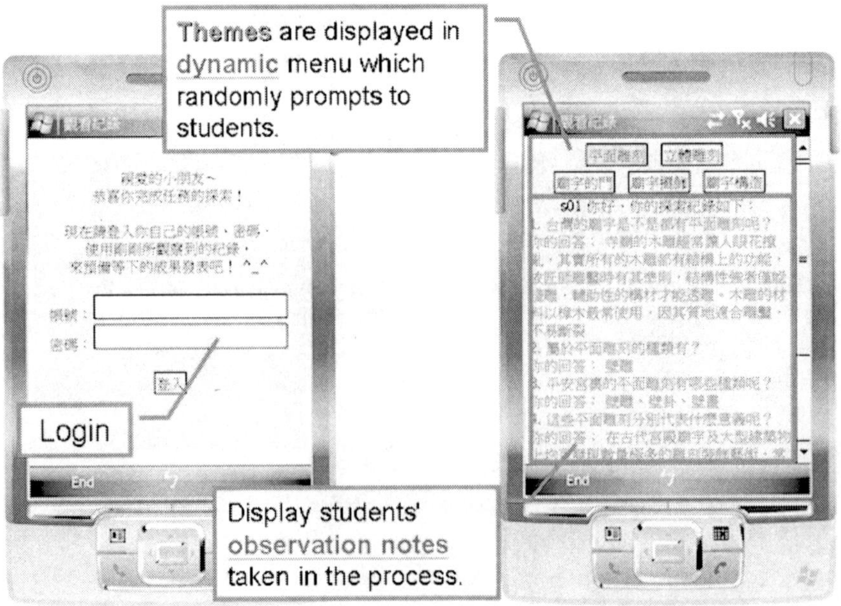

Table 1. Pre- and post-tests comparison of overall social relationship

	Mean	N	SD	Mean difference	*t*
Social relationship pre-test total	26.25	32	3.46410	5.78125	5.7373***
Social relationship post-test total	20.47	32	4.17353		

*** p<.001

ally already had the same tendency before they participated in this activity, therefore, showing no significant difference.

From the number counts of the above items (Table 3), we see that in the pre-test, half of the students preferred collaborative and the other half preferred individual learning. In the post-test, one student shifted his opinion from collaborative to individual learning, thus changing the statistics.

Secondly, we investigated whether students had leadership ability among group members. From Table 4, we see the comparison of pre-test item 2 "I have leadership ability among groups" (mean=4.19) and post-test item 3 "I feel I was leading the whole inquiry process in the activity" (mean=2.75), students leadership had significantly (p=.001) dropped. We assume that the design of the activity in this research was actually more appropriate for individual learning.

Thirdly, we investigated whether students were willing to discuss with classmates. From Table 5, we see the comparison of pre-test item 4 "I would discuss home work with classmates" (mean=4.53) and post-test item 2 "I discuss learning content with group members in the inquiry process" (mean=4.84). The results show that students' willingness to discuss with peers had increased but with no significant (p=.305) difference.

From the number counts of the above items (Table 6), we see that in the pre-test, there were 23 students who would discuss with peers, and after this inquiry learning exercise, only 3 more students would join the discussion, therefore no significant difference was found.

Lastly, we investigated whether students would accept suggestions from their peers. From Table 7, we see the comparison of pre-test item 6 "I would accept the suggestions from peers" (mean=4.78) and post-test item 5 "I accepted the suggestions from peers in the inquiry process" (mean=4.56). The results show that students preferred to depend on themselves hence causing no significant (p=.543) difference.

Table 2. Social relationship: pre-test item 5 vs. post-test item 4

	Mean	N	SD	Mean difference	*t*
Social relationship pre-test item 5	3.69	32	1.59510	.12500	.366
Social relationship post-test item 4	3.56	32	1.83052		

Table 3. Social relationship: Number counts of pre-test item 5 vs. post-test item 4

	agree	disagree
	Count (%)	Count (%)
Social relationship pre-test item 5	16 (50.0%)	16 (50.0%)
Social relationship post-test item 4	17 (53.1%)	15 (46.9%)

Table 4. Social relationship: pre-test item 2 vs. post-test item 3

	Mean	N	SD	Mean difference	*t*
Social relationship pre-test item 2	4.19	32	1.30600	1.43750	3.559***
Social relationship post-test item 3	2.75	32	1.52400		

*** p<.001

Table 5. Social relationship: pre-test item 4 vs. post-test item 2

	Mean	N	SD	Mean difference	*t*
Social relationship pre-test item 4	4.53	32	1.62609	-.31250	-1.044
Social relationship post-test item 2	4.84	32	1.58845		

Table 6. Social relationship: Number counts of pre-test item 4 vs. post-test item 2

	agree	disagree
	Count (%)	Count (%)
Social relationship pre-test item 4	23 (71.9%)	9 (28.1%)
Social relationship post-test item 2	26 (81.3%)	6 (18.8%)

From the number counts of the above items (Table 8), we see that in the pre-test, there were 29 students who would accept suggestions from peers, and after this inquiry learning, 4 students preferred to depend on themselves, therefore no significant difference was found.

Observation

In the learning process stage, researchers used observation forms to record students' behaviors during the learning process. We analyzed students' social relationships when they were in either col-

Table 7. Social relationship: pre-test item 6 vs. post-test item 5

	Mean	N	SD	Mean difference	*t*
Social relationship pre-test item 6	4.78	32	1.40814	.21875	.615
Social relationship post-test item 5	4.56	32	1.58496		

Table 8. Social relationship: Number counts of pre-test item 6 vs. post-test item 5

	agree	disagree
	Count (%)	Count (%)
Social relationship pre-test item 6	29 (90.6%)	3 (9.4%)
Social relationship post-test item 5	25 (78.1%)	7 (21.9%)

laborative or individual groups, conducting task A or B, and at the three different stages of the activity.

Table 9 shows the social relationships of both groups in the three stages of task A. Students' social relationship showed a significant difference (p=.045) in stage 3 "presentation stage" (Table 9). The mean of the collaborative group (mean=0.82) was significantly higher than the individual group (mean=0.33). In stage 1, the "physical observation stage" and stage 2, the "presentation preparation stage", even though individual group's respective means (2.27 and 2.00) were higher than those of the collaborative group's means (mean=2.18 and mean=1.77), they were not significantly different enough. Therefore, we could conclude that in the storytelling task, students tended to work individually in the first two stages, but tended to collaborate in the presentation stage.

Table 10 shows the social relationships of both groups in the three stages of task B. Students' social relationship was significantly different (p=.000 and p=.001) in stage 2 "presentation preparation stage" and stage 3 "presentation stage" respectively, and collaborative group means (mean=2.8 and mean=1.67) were higher than those of the individual group means (mean=1.29 and mean=0.41). In stage 1, the "physical observation stage", although collaborative group mean (mean=1.87) was higher than the individual group mean (mean=1.71), it did not achieve a significant level. Therefore, we could conclude that task B "Little Temple Architect" was more suitable for collaborative learning and could enhance students' social relationship especially in stage 2 "presentation preparation stage" and stage 3 "presentation stage".

Table 11 shows students' social relationship in collaborative group. There was a significant difference (p=.002 and p=.019) for stage 2 "presentation preparation stage" and stage 3 "presentation stage". Task B's respective means (2.8 and 1.67) were higher than task A's means (1.76 and

Table 9. Social relationships of both groups in three stages of task A

	Task 1	n	Mean	SD	*t*-value
Social relationship stage 1	Collaboration group	17	2.18	.72761	-.282
	Individual group	15	2.27	1.03280	
Social relationship stage 2	Collaboration group	17	1.77	1.14725	-.568
	Individual group	15	2.00	1.19523	
Social relationship stage 3	Collaboration group	17	.82	.80896	2.102*
	Individual group	15	.33	.48795	

* p<.05

Table 10. Social relationships of both groups in three stages of task B

	Task 2	n	Mean	SD	*t*-value
Social relationship stage 1	Collaboration group	15	1.87	1.18723	.492
	Individual group	17	1.71	.46967	
Social relationship stage 2	Collaboration group	15	2.80	.41404	6.986***
	Individual group	17	1.29	.77174	
Social relationship stage 3	Collaboration group	15	1.67	1.11270	3.872***
	Individual group	17	.41	.61835	

** p<.001

0.82). However, in stage 1 "physical observation stage", task A "Storytelling" (mean=2.18) was higher than that of task B "Little Temple Architect" without achieving a significant difference level. Therefore, we conclude that for collaborative group work, stage 2 and stage 3 could better enhance students' social relationship for task B, but for stage 1, task A was more appropriate for enhancing students' social relationship.

Table 12 shows students' social relationship in their individual group. There was no significant difference between the two tasks in all the three stages. But in stage 1 "physical observation stage" and stage 2 "presentation preparation stage", task A's "Storytelling" respective means (2.27 and 2) were higher than task B's means "Little Temple Architect" (1.71 and 1.29). For stage 3 "presentation stage", on the other hand, task B mean "Little Temple Architect" (mean=0.41) was a little higher than task A mean "Storytelling" (mean=0.33). Therefore, we conclude that for the individual group task, stage 1 and stage 2 would better enhance students' social relationship for task A, but for stage 3, task B was more appropriate for enhancing students' social relationship.

From the analysis above, we could conclude that for the different tasks and at different stages, students preferred different learning modes, that is, either individual or collaborative. Instructors should design learning activities using different models according to the learning content.

Focus Group Interviews

From students' discussions and feedback from the focus group interviews, students have shown different preferences for learning modes.

One student said, *"During the physical observation and inquiry process, I like individual learning. But when I am preparing for class presentations, I like to collaborate with peers"* (fg-20081124-001). Another student said, *"I don't like to work with classmates I dislike, but I want to work together with those I like"* (fg-20081124-

Table 11. Social relationships of collaborative group in three stages

	Collaboration group	n	Mean	SD	*t*-value
Social relationship stage 1	Task A	17	2.18	.72761	.876
	Task B	15	1.87	1.18723	
Social relationship stage 2	Task A	17	1.77	1.14752	-3.473**
	Task B	15	2.80	.41404	
Social relationship stage 3	Task A	17	.82	.80896	-2.472*
	Task B	15	1.67	1.11270	

* p<.05, ** p<.01

Table 12. Social relationships of individual group in three stages

	Individual group	n	Mean	SD	*t*-value
Social relationship stage 1	Task 1	15	2.27	1.03280	1.934
	Task 2	17	1.71	.46967	
Social relationship stage 2	Task 1	15	2.00	1.19523	1.956
	Task 2	17	1.29	.77174	
Social relationship stage 3	Task 1	15	.33	.48795	-.394
	Task 2	17	.41	.61835	

Table 13. Reasons for preferring collaborative or individual learning in terms of social relationships

	Collaborative Learning	Individual Learning
Ratio	77.8%	22.2%
Reasons	a. It is boring to do works individually. I am happier to work with classmates and have more fun. b. Friends can help each other. Works can be distributed between each other. It is more fun. Works can be finished faster. c. I can copy classmates' answers.	a. I trust myself more than others. My answers are more correct. b. Some classmates would copy my answers. c. I like to conduct observation by myself. d. Sometimes I have different opinions with classmates.

001). From opinions such as these, we conclude that students' preference for certain learning mode is also due to the choice of partners.

From Table 13, we could see that 22.2% of the students liked "individual learning" (Figure 11, 13, 15) and 77.8% of students liked "collaborative learning" (Figure 12, 14, 16). For those students who liked individual learning, they were mostly more confident than others. They believed their answers were more correct than others, and were sometimes in disagreement and disliked their peers copying their answers. They were also more independent and preferred to learn by themselves. However, most students said that they liked collaborative learning because they felt bored when they worked alone and they felt happier and had more fun when they were with friends. They said friends could help each other and they distributed work among themselves. While some found materials, others discussed the answers; it was more fun and they could finish the work faster when learning collaboratively. Some students said they liked collaborative learning because they could copy their classmates' home work.

In addition, we conducted interviews with teachers and learned what the teachers saw in students' changes in both behaviors and attitudes. The teacher expressed opinions such as:

Students learned more when they collaborated in the discussion session. ... If students were learning subjects that required attentions such as analysis to natural science or material memorization about social science, collaboration would cause fewer benefits. If collaborative strategies were used in discussion sessions and with more meaningful content, students could hear information from many different aspects. For subjects that required extensive readings and physical explorations, they would give the class lots of constructive feedbacks and feel very interesting. They learned

Figure 10. Individual learning (Physical Exploration Stage)

Figure 11. Collaborative learning (Physical Exploration Stage)

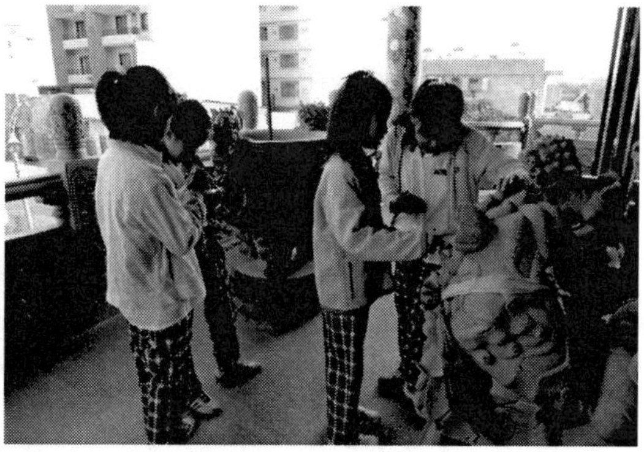

Figure 12. Individual learning (Presentation Preparation Stage)

Figure 13. Collaborative learning (Presentation Preparation Stage)

Figure 14. Individual learning (Presentation Stage)

Figure 15. Collaborative learning (Presentation Preparation Stage)

Figure 16. Focus group interviews

Figure 17. Interview with teachers

knowledge that was outside of textbooks, such as living experiences of grandparents, the Internet, and their own observations. Information and knowledge were accumulated throughout the process. Although that knowledge would not be tested in the academic examinations, they were very useful to their growth. In this inquiry-based mobile learning research, I think collaboration could work better than individual learning because they could discuss about what they saw and heard, and the discussion had more depth than perceived by themselves (ti-20081125-001).

High-achievement students had higher learning standards. No matter they were learning individually or collaboratively, they learned about the same things. Yet, when students were placed in groups, high-achievement students could help mid- and low-achievement students. Slow learners needed more hints and reminders that made their learning more meaningful and more effective (ti-20081125-001).

From Table 14, we see that 77.8% of students liked to discuss with classmates and 22.2% of students did not like to discuss with classmates. In this regard, the gender of group members and whether the group members were in friendship groups could influence the students' attitudes in participating in discussions. The students thought that in collaborative work, they spent less time and felt more comfortable with finishing the work. For those who refused to discuss with their classmates, there were three different reasons. For students with high confidence, they thought that their answers were correct and did not want to discuss with others; for students who were shy, they tended not to express their own opinions hence they did not discuss with their classmates; and for the rest of the students, since they learned at different speeds, they were gradually separated from others when they were conducting field inquiries.

We further investigated whether students would accept suggestions from their peers. From Table 15, we could see that 61.1% students said they would accept suggestions from peers, 22.2% of students said it depended on the condition, and 16.7% of students said they would not accept suggestions from their peers. From the reasons the students provided in the focus group interviews, we conclude that those who would accept suggestions were less confident since they often doubted the correctness of their own answers; those who accepted suggestions conditionally were thinkers who judged by comparing the answers from peers and what they saw in person before they made decisions; and those who would not accept suggestions were those with strong confidence in that no matter what answers other had, they believed in their own answers.

Satisfaction

Table 16 shows students' satisfaction level to the overall mobile learning activity. The learning satisfaction questionnaire was designed with six-point Likert's Scale. From the statistics, we see that most students felt satisfied with the overall

Table 14. Reasons for the willingness to discussion in collaboration groups

	Would discuss with classmates	Would not discuss with classmates
Ratio	77.8%	22.2%
Reasons	a. If they are the same gender as I am, I do not care if they are good friends of mine. b. I only discuss with good friends. c. Working with classmates is faster and more convenient.	a. I do not need to discuss when I am in the field inquiry. b. It is troublesome to having to walk with classmates. c. I am shy; I do not like to ask other people. d. We have different learning speed. We often walk separately when in the field.

Table 15. Reasons for whether accepting suggestions from peers

	Accept	Accept conditionally	Not accept
Ratio	61.1%	22.2%	16.7%
Reasons	a. As long as there is an answer. b. Other answers sound more reasonable. c. Sometimes others are correct so I should consider their answers.	a. Classmates' suggestions are only as reference. b. Only accept answers from girls. c. Not accept answers from those who are bad.	a. My answers are correct.

learning activity (mean=5.552). Among which, we found that the average of students' answer to every item was above 5. Except for item 1, 3, and 8, the average of all other items were between 5.5 and 6. Item 4 was especially high (mean=5.906) which showed that students felt inquiry-based learning field trip was more liberating and motivating.

CONCLUSION

This research emphasized on the inquiry-based learning practices which integrated digital and physical environments. By using mobile devices, students could have more customized learning pace and process, and could get individual attention when they were distributed in the fields. This research was aimed at examining whether such teaching strategies could effectively promote students' social relationship. In the research, we investigated students' social relationship from their learning performances and learning satisfaction.

The results of this research showed that students' social relationships vary in the different stages of their learning and with different types of tasks. Students preferred individual mode of learning when they were conducting field inquiries, but preferred collaboration when preparing for presentations. Their social relationships and learning satisfaction showed positive results. From

Table 16. Satisfaction of Inquiry-based mobile learning activity

	Strongly agree	Agree	Slightly agree	Slightly disagree	Disagree	Strongly disagree	Mean
1. I think using PDA in field learning is more interesting then teacher-guided field trip.	90.6%	3.1%		3.1%		3.1%	5.498
2. I like to learn with physical object manipulation and observation	78.1%	15.6%	6.3%				5.718
3. I like to conduct inquiry and exploration in groups.	65.6%	12.5%	6.3%	9.4%	6.3%		5.221
4. I feel more free and relaxed in inquiry-based learning field trip than in the traditional lectures.	90.6%	9.4%					5.906
5. I think I learn more in this inquiry-based learning field trip than listen to teacher's explanation in the classroom.	68.8%	25.0%	3.1%		3.1%		5.564
6. Using PDA can increase my learning motivation.	75.0%	18.8%	3.1%	3.1%			5.657
I wish I can have more chances to use PDA to learn in the future.	78.1%	9.4%	9.4%			3.1%	5.563
8. I will recommend inquiry-based learning field trip to my classmates.	59.4%	21.9%	6.3%	9.4%		3.1%	5.224
9. If I have to give grade to this learning experience, I would give it …	78.1%	15.6%	3.1%			3.1%	5.621

their learning performances, both their presentations of task results gained high appraisals from their teachers, and their physical participations and affective involvements were both significantly better than those displayed in traditional classroom teaching. Through pre-designed materials, field investigations in authentic learning environment, as well as in-depth community visits and interviews, students' understanding about historical monuments was increased. Social science lessons such as this, especially with geographical and historical content, inquiry-based learning strategies could be useful. When we integrated mobile learning by giving each student a PDA to assist them with their field inquiry, the learning effectiveness increased significantly.

What makes this research unique is that it is one of a few mobile learning researches that have been done with a social science course while most research focused on natural science content. The instructional design does not offer pre-defined answers to students but encouraged them to conduct inquiries that may extend their learning experiences to outside of textbooks and teachers' knowledge. At the same time, collaborations are encouraged in the learning process. The research found that neither individualistic learning nor collaborative learning brought definitive positive effects to the students' learning. The results showed that instructional strategies have to be used wisely according to the subject under study to consider activities, amount of time and scaffolding the learning to bring about positive impact.

To conclude, this study's inquiry-based mobile learning model can be further extended to other courses and subjects. The results of this research concord with other similar research that students' achievements with a mobile-learning strategy can be improved. For our future research, we will conduct research to understand other aspects of mobile-mediated learning, such as the development and use of critical and creative thinking. We hope this research has provided some fundamental understanding about inquiry-based mobile learning for researchers.

ACKNOWLEDGMENT

This research was supported by the National Science Council in Taiwan under projects NSC 97-2511-S-024-009-MY2, NSC 98-2511-S-024-006-MY2, and NSC 98-2631-S-024-001.

REFERENCES

Adam, D. M., & Hamm, M. E. (1990). *Cooperative learning critical thinking and collaboration across the curriculum*. Charles C. Thomas.

Albanese, M. A., & Mitchell, S. (1993). Problem-based learning: A review of literature on its outcomes and implementation issues. *Academic Medicine*, *68*, 52–81. doi:10.1097/00001888-199301000-00012

Artemeva, N., Logie, S., & St-Martin, J. (1999). From page to stage: How theories of genre and situated learning help introduce engineering students to discipline-specific communication. *Technical Communication Quarterly*, *8*(3), 301–316.

Bandura, A. (1989). Human agency in social cognitive theory. *The American Psychologist*, *44*(9), 1175–1184. doi:10.1037/0003-066X.44.9.1175

Bermejo, S. (2005). Cooperative electronic learning in virtual laboratories through forums. *IEEE Transactions on Education*, *48*(1), 140–149. doi:10.1109/TE.2004.837045

Bravo, C., van Joolingen, W. R., & de Jong, T. (2006). Modeling and Simulation in Inquiry Learning: Checking Solutions and Giving Intelligent Advice. *Simulation*, *82*(11), 769–784. doi:10.1177/0037549706074190

Brown, J. S., Collins, A., & Duguid, P. (1989). Situated cognition and the cultural of learning. *Educational Researcher*, *18*(1), 32–42.

Burgard, W., Cremers, A. B., Fox, D., Hähnel, D., Lakemeyer, G., & Schulz, D. (1999). Experiences with an interactive museum tour-guide robot. *Artificial Intelligence*, *114*, 3–55. doi:10.1016/S0004-3702(99)00070-3

Chen, Y., Kao, T., Sheu, J., & Chang, C. (2002, August). A Mobile Scaffolding-Aid-Based Bird-Watching Learning System. In *Proceedings of the IEEE International Workshop on Wireless and Mobile Technologies in Education* (WMTE'02), Vaxjo, Sweden.

Chen, Y. S., Kao, T. C., & Sheu, J. P. (2003). A mobile learning system for scaffolding bird watching learning. *Journal of Computer Assisted Learning*, *19*(3), 347–359. doi:10.1046/j.0266-4909.2003.00036.x

Chu, H.-C., Hwang, G.-J., Huang, S.-X., & Wu, T.-T. (2008). A knowledge engineering approach to developing e-libraries for mobile learning. *The Electronic Library*, *26*(3), 303–317. doi:10.1108/02640470810879464

Colburn, A. (2000). An Inquiry Primer. *Science Scope*, *23*(6), 42–44.

Collingwood, R. G. (1994). In Jan van Der Dussen (Ed.), The idea of history [1946] revised edition with lectures 1926–1928. Oxford: Clarendon Press.

Cortez, C., Nussbaum, M., Santelices, R. A., Rodríguez, P., Zurita, G., Correa, M., & Cautivo, R. (2004). *Teaching Science with Mobile Computer Supported Collaborative Learning (MCSCL)*. Proceedings of IEEE International Workshop on Wireless and Mobile Technologies in Education (WMTE 2004), (pp. 67-74).

Cui, Y., & Bull, S. (2005). Context and learner modelling for the mobile foreign language learner. *System*, *33*, 353–367. doi:10.1016/j.system.2004.12.008

Dillenbourg, P., Baker, M., Blaye, A., & O'Malley, C. (1996). The evolution of research on collaborative learning. In Spada, E., & Reiman, P. (Eds.), *Learning in Humans and Machine: Towards an interdisciplinary learning science* (pp. 189–211). Oxford, UK: Elsevier.

Duffy, T. M., & Jonassen, D. H. (1992). Constructivism: New implications for instructional technology. In Duffy, T. M., & Jonassen, D. H. (Eds.), *Constructivism and the technology of instruction: a conversation* (pp. 1–16). Hillsdale, NJ: Lawrence Erlbaum Associates.

Edelson, D. C., Gordin, D. N., & Pea, R. D. (1999). Addressing the Challenges of Inquiry-Based Learning Through Technology and Curriculum Design. *Journal of the Learning Sciences*, *8*(3-4), 391–450. doi:10.1207/s15327809jls0803&4_3

Ellis, A. K. (2002). *Teaching and Learning Elementary Social Studies*. Boston: Allyn and Bacon, Inc.

Fischer, G., & Konomi, S. (2007). Innovative socio-technical environments in support of distributed intelligence and lifelong learning. *Journal of Computer Assisted Learning*, *23*, 338–350. doi:10.1111/j.1365-2729.2007.00238.x

Franklin, D. L. (1986). Mary Richmond and Jane Addams: From Moral Certainty to Rational Inquiry in Social Work Practice. *The Social Service Review*, *60*(4), 504–525. doi:10.1086/644396

Ghaith, G. M. (2002). The relationship between cooperative learning, perception of social support, and academic achievement. *System*, *30*, 263–273. doi:10.1016/S0346-251X(02)00014-3

Ghaith, G. M., & Yaghi, H. (1998). Effect of cooperative learning on the acquisition of second language rules and mechanics. *System, 26*, 223–234. doi:10.1016/S0346-251X(98)00005-0

Gillani, B., & O'Guinn, C. (2004). *Cognitive Theories and the Design of E-Learning Environments*. IOS Press.

Gillet, D., Ngoc, A., Nguyen, V., & Rekik, Y. (2005). Collaborative web-based experimentation in flexible engineering education. *IEEE Transactions on Education, 48*(4), 696–704. doi:10.1109/TE.2005.852592

Gokhale, A. A. (1995). Collaborative Learning Enhances Critical Thinking. *Journal of Technology Education, 7*(1), 22–30.

Hackling, M. W., & Garnett, Patrick J. (1995). The Development of Expertise in Science Investigation Skills. *Australian Science Teachers Journal, 41*, 80–85.

Hamelin, D. (2004, June), *Searching the Web to Develop Inquiry and Collaborative Skills*. Annual Joint Conference Integrating Technology into Computer Science Education. Working group reports from ITiCSE on Innovation and technology in computer science education, United Kingdom.

Heron, J. (1993). *Co-operative inquiry and the primacy of the practical; Adapted from Chapter 2, Co operative Inquiry*. London: Sage. Retrieved November 25, 2009 from http://www.humaninquiry.com/primprac.htm

Hofstadter, D. R. (1998). Popular Culture and the Threat to Rational Inquiry. *Science, 281*(3576), 512–513. doi:10.1126/science.281.5376.512

Hooper, S. (1992). Effects of peer interaction during computer-based mathematics instruction. *Journal of Educational Research and Development, 85*(3), 180–189.

Hooper, S. (2003). The effects of persistence and small group interaction during computer-based instruction. *Computers in Human Behavior, 19*, 211–220. doi:10.1016/S0747-5632(02)00033-X

Huber, G. L. (2003). Processes of decision-making in small learning groups. *Learning and Instruction, 13*, 255–269. doi:10.1016/S0959-4752(01)00039-1

Hull, R., & Reid, J. (2003). Designing engaging experiences with children and artists. In Blythe, M. A., Overbeeke, K., & Monk, A. F. (Eds.), *Funology: From usability to enjoyment* (pp. 108–116). Dordrecht, Netherlands: Kluwer.

Hwang, G.-J., Tsai, C.-C., & Yang, S. J.-H. (2008). Criteria, Strategies and Research Issues of Context Aware Ubiquitous Learning. *Journal of Educational Technology & Society, 11*(2), 81–91.

Janssen, J., Erkens, G., Kanselaar, G., & Jaspers, J. (2006). Visualization of participation: Does it contribute to successful computer-supported collaborative learning? *Computers & Education, 49*(4), 1037–1065. doi:10.1016/j.compedu.2006.01.004

Johnson, D. W., & Johnson, R. T. (1991). *Learning Together and Alone: Cooperative, competitive, and individualistic learning* (3rd ed.). Englewood, NJ: Prentice Hall.

Johnson, D. W., Johnson, R. T., & Smith, K. A. (1998). Cooperative learning returns to college: What evidence is there that it works? *Change, 30*(4), 26–35. doi:10.1080/00091389809602629

Johnson, D. W., Roger, T., & Smith, K. A. (1991). *Active Learning: Cooperation in the College Classroom*. Edina, MN: Interaction Book Company.

Kagan, S. (1985). Co-op Co-op: A flexible cooperative learning technique. In In Slavin, R. (Eds.), *Learning to cooperate, cooperating to learn* (pp. 437–462).

Keyser, M. W. (2000). Active learning and cooperative learning: understanding the difference and using both styles effectively. *Research Strategies*, *17*, 35–44. doi:10.1016/S0734-3310(00)00022-7

Kirschner, P. A., Sweller, J., & Clark, R. E. (2006). Why Minimal Guidance During Instruction Does Not Work: An Analysis of the Failure of Constructivist, Discovery, Problem-Based, Experiential, and Inquiry Based Teaching. *Educational Psychologist*, *41*(2), 75–86. doi:10.1207/s15326985ep4102_1

Klingner, J. K., & Vaughn, S. (2000). The helping behaviors of fifth graders while using collaborative strategic reading during ESL content classes. *TESOL Quarterly*, *34*, 69–98. doi:10.2307/3588097

Lai, C.-H., Yang, J.-C., Chen, F.-C., Ho, C.-W., & Chan, T.-W. (2007). Affordances of mobiletechnologies for experiential learning: the interplay of technology and pedagogical practices. *Journal of Computer Assisted Learning*, *23*, 326–337. doi:10.1111/j.1365-2729.2007.00237.x

Lang, B., Lu, Y., Zhang, Z., & Li, W. (2003). *A flexible access control mechanism supporting large scale distributed collaboration*. In The eighth international conference on computer supported cooperative work in design proceeding, 1, 500-504.

Lave, J., & Wenger, E. (1991). *Situated learning: Legitimate peripheral participation*. New York: Cambridge University Press.

Leung, C. H., & Chan, Y. Y. (2003). *Mobile learning: A new paradigm in electronic learning*. 3rd IEEE International Conference on Advanced Learning Technologies(ICALT'03), 76-80.

Lim, B. R. (2004). Challenges and issues in designing inquiry on the Web. *British Journal of Educational Technology*, *35*(5). doi:10.1111/j.0007-1013.2004.00419.x

Looi, C. K. (1998). Interactive learning environments for promoting inquiry learning. *Journal of Educational Technology Systems*, *27*(1), 3–22. doi:10.2190/L4Q4-8QMM-QPJ3-B5LJ

Manlove, S., Lazonder, A. W., & de Jong, T. (2008). Trends and issues of regulative support use during inquiry learning: Patterns from three studies. *Computers in Human Behavior*, *25*(4), 795–803. doi:10.1016/j.chb.2008.07.010

McLellan, H. (1993). Evaluation in a situated learning environment. *Educational Technology*, *33*(3), 39–45.

Mevarech, Z. A. (1993). Who benefits from cooperative computer-assisted instruction? *Journal of Educational Computing Research*, *9*(40), 451–464.

Molebash, P. (2004). Web historical inquiry projects. *Social Education*, *68*(3), 226–234.

Montague, W. E., & Knirk, F. G. (1993). What works in adult instruction: the management, design and delivery of instruction? *International Journal of Educational Research*, *19*, 329–331.

National Council for the Social Studies. (1994). *Expectations of excellence: Curriculum standards for social studies*. Washington, DC: National Council for the Social Studies.

Pinkwart, N., Hope, H., Milrad, M., & Perez, J. (2003). Educational scenarios for cooperative use of Personal Digital Assistants. *Journal of Computer Assisted Learning*, *19*, 383–391. doi:10.1046/j.0266-4909.2003.00039.x

Porto, M. (2001). Cooperative writing response groups and self-evaluation. *ELT Journal*, *55*(1), 38–46. doi:10.1093/elt/55.1.38

Quinn, C. (2000). *mLearning: mobile, wireless and in-your-pocket learning*. Line Zine Magazine. Retrieved June 3, 2009, from http://linezine.com/2.1/features/cqmmwiyp.htm

Rachel, H. L., & Irit, B. N. (2002). Writing development of Arab and Jewish students using cooperative learning (CL) and computer-mediated communication (CMC). *Computers & Education*, *39*, 19–36. doi:10.1016/S0360-1315(02)00019-2

Ramsay, A., Hanlon, D., & Smith, D. (2000). The association between cognitive style and accounting students' preference for cooperative learning: an empirical investigation. *Journal of Accounting Education*, *18*, 215–228. doi:10.1016/S0748-5751(00)00018-X

Resnick, L. (1987). Learning in school and out. *Educational Researcher*, *16*, 13–20.

Rogers, Y., Price, S., Fitzpatrick, G., Fleck, R., Harris, E., & Smith, H. (2004). Designing new forms of digital augmentation for learning outdoors. In *Proc. Interaction Design and Children, ACM* (pp. 3–10). New York: Ambient Wood.

Rogers, Y., Price, S., Randell, C., Fraser, D. S., Weal, M., & Fitzpatrick, G. (2005). Ubi-learning Integrates Indoor and Outdoor Experiences. *Communications of the ACM*, *48*(1), 55–59. doi:10.1145/1039539.1039570

Roth, W. M., & Roychoudhury, A. (1992). The social construction of scientific concepts or the concept map as conscription device and tool for social thinking in high school science. *Science Education*, *76*, 531–557. doi:10.1002/sce.3730760507

Schwab, J. J. (1962). The teaching of science as enquiry. In Schwab, J. J., & Brandwein, P. F. (Eds.), *The teaching of science*. Cambridge, MA: Harvard University Press.

Seppälä, P., & Alamäki, H. (2003). Mobile learning in teacher training. *Journal of Computer Assisted Learning*, *19*, 330–335. doi:10.1046/j.0266-4909.2003.00034.x

Slavin, R. E. (1989). Research on cooperative learning: consensus and controversy. *Journal of the Educational Leadership*, *47*(4), 52–54.

Smith, R. K. (1995). Emerging tools for team collaboration: desktop videoconferencing experience and insight. *FID News Bulletin*, *45*(4), 121–127.

Stenhouse, L. (1997). An Appeal for Evidence of the Effectiveness of an Alternative Role to that of Neutral Chairman in Promoting Rational Inquiry into Moral Issues: A Reply to John T. Hyland. *Journal of Further and Higher Education*, *1*(2), 50–54.

Suchman, J. R. (1961). Inquiry training: Building skills for autonomous discovery. *Merrill-Palmer Quarterly of Behavior and Development*, *7*(3), 147–169.

Sun, C. T., & Chou, C. (1996). Experiencing CORAL: design and implementation of distant cooperative learning. *IEEE Transactions on Education*, *39*(3), 357–366. doi:10.1109/13.538759

Swain, M. (2001). Integrating language and content teaching through collaborative tasks. *Canadian Modern Language Review*, *58*, 44–63. doi:10.3138/cmlr.58.1.44

Trifonova, A., & Ronchetti, M. (2003, November). Where is Mobile Learning Going? *Proceedings of the World Conference on E-learning in Corporate, Government, Healthcare, & Higher Education*, Phoenix, AZ, 1794-1801.

Veenman, S., Benthum, N. V., Bootsma, D., Dieren, J. V., & Kemp, N. (2002). Cooperative learning and teacher education. *Teaching and Teacher Education*, *18*, 87–103. doi:10.1016/S0742-051X(01)00052-X

Von Glasersfeld, E. (1989). Cognition, construction of knowledge, and teaching. *Synthese*, *80*(1), 121–140. doi:10.1007/BF00869951

Vygotsky, L. S. (1978). *Mind in society: The development of higher psychological processes*. Cambridge, MA: Harvard University.

Weiser, M. (1991). The computer for the twenty-first century. *Scientific American*, 94–104. doi:10.1038/scientificamerican0991-94

Wells, G. (2001). *Action, Talk, and Text: Learning and Teaching Through Inquiry*. Teachers College Press.

Wolfson, L., & Willinsky, J. (1998). What service-learning can learn from situated learning. *Michigan Journal of Community Service Learning*, *5*, 22–31.

Zurita, G., & Nussbaum, M. (2004). Computer supported collaborative learning using wirelessly interconnected handheld computers. *Computers & Education*, *42*(3), 289–314. doi:10.1016/j.compedu.2003.08.005

APPENDIX A: PRE-CLASS QUESTIONNAIRE

Pre-class Questionnaire for Historic Monument Mobile Learning

Class: ________ School Number: ________ Name: ________ Group: ________

Thank you for taking your time on filling out this questionnaire. The purpose of the questionnaire is for us to understand your learning condition and feelings about Historic Monument Mobile Learning activity. Please carefully answer the following questions. There is no correct answer for any of the questions. Your input is very important for us. The information of this questionnaire will be confidential to the research group only so please feel free to write down your response.
Best Regards,
Department of Information and Learning Technology, National University of Tainan
Gwo-Jen Hwang, Professor
Ju-Ling Shih, Associate Professor
Chien-Wen Chuang, Graduate Student
November 2008

【Part I: Personality】

	Strongly Agree	Agree	Somehow Agree	Somehow Disagree	Disagree	Strongly Disagree
1. I like to try new thing.	□	□	□	□	□	□
2. I am sentimental.	□	□	□	□	□	□
3. I care how other people see and treat me.	□	□	□	□	□	□
4. I care when people know my failure.	□	□	□	□	□	□
5. I care if I give people good impression.	□	□	□	□	□	□
6. I am capable to deal with accidental conditions.	□	□	□	□	□	□
7. I can remain calm when dealing with difficulties.	□	□	□	□	□	□
8. It is easy for me to be faithful to the goal.	□	□	□	□	□	□
9. I can solve most problems when I put enough efforts.	□	□	□	□	□	□
10. I know how to solve problems.	□	□	□	□	□	□

【Part II: Social Relationship】

	Strong-ly Agree	Agree	Some-how Agree	Some-how Dis-agree	Dis-agree	Strong-ly Dis-agree
1. I like to discuss questions with peers.	□	□	□	□	□	□
2. I have leadership ability among groups.	□	□	□	□	□	□
3. I am shy in presenting myself among peers.	□	□	□	□	□	□
4. I would discuss home work with classmates.	□	□	□	□	□	□
5. I like to complete homework and reports by myself.	□	□	□	□	□	□
6. I would accept the suggestions from peers.	□	□	□	□	□	□

Appendix B: Post-Class Questionnaire

Post-class Questionnaire for Historic Monument Mobile Learning

Class: _________ School Number: _________ Name: _________ Group: _________

Thank you for taking your time on filling out this questionnaire. The purpose of the questionnaire is for us to understand your learning condition and feelings about Historic Monument Mobile Learning activity. Please carefully answer the following questions. There is no correct answer for any of the questions. Your input is very important for us. The information of this questionnaire will be confidential to the research group only so please feel free to write down your response.
Best Regards,
Department of Information and Learning Technology, National University of Tainan
Gwo-Jen Hwang, Professor
Ju-Ling Shih, Associate Professor
Chien-Wen Chuang, Graduate Student
November 2008

【Part I: Peer Interaction】

	Strong-ly Agree	Agree	Some-how Agree	Some-how Dis-agree	Dis-agree	Strong-ly Dis-agree
1. I feel I have similar ideas with peers.	□	□	□	□	□	□
2. I discuss learning content with group members in the inquiry process.	□	□	□	□	□	□
3. I feel I was leading the whole inquiry process in the activity.	□	□	□	□	□	□
4. I like to learn by myself.	□	□	□	□	□	□
5. I accepted the suggestions from peers in the inquiry process.	□	□	□	□	□	□

【Part II: Satisfaction】

	Strong-ly Agree	Agree	Some-how Agree	Some-how Dis-agree	Dis-agree	Strong-ly Dis-agree
1. I think using PDA in field learning is more interesting then teacher-guided field trip.	□	□	□	□	□	□
2. I like to learn with physical object manipulation and observation	□	□	□	□	□	□
3. I like to conduct inquiry and exploration in groups.	□	□	□	□	□	□
4. I feel more free and relaxed in inquiry-based learning field trip than in the traditional lectures.	□	□	□	□	□	□
5. I think I learn more in this inquiry-based learning field trip than listen to teacher's explanation in the classroom.	□	□	□	□	□	□
6. Using PDA can increase my learning motivation.	□	□	□	□	□	□
I wish I can have more chances to use PDA to learn in the future.	□	□	□	□	□	□
8. I will recommend inquiry-based learning field trip to my classmates.	□	□	□	□	□	□
9. If I have to give grade to this learning experience, I would give it …	6 □	5 □	4 □	3 □	2 □	1 □
1. Other Suggestions						

Appendix C: Focus Group Interview Outline

Focus Group Interview Outline:

1. After the two rounds of activities of this course, do you prefer collaborative or individual learning?
2. When you are in collaborative groups, do you discuss with other group members? Why or why not?
3. In the process of group discussion, do you accept other people's suggestions? Why or why not?
4. If there is another chance to participate such inquiry course, would you want to work collaboratively or individually? Why?

Teacher Interview Outline:

1. Do you often design collaborative works for students in your other courses?
2. What are the students' learning conditions when they are working collaboratively and individually?
3. From your point of view, what do you feel about students' learning conditions in this field inquiry activity?
4. If there is another chance to conduct such inquiry course, which kind of inquiry method do you think can help students learn better?

Chapter 10
Implementation of Mobile Learning at the Open University Malaysia

Chng Loi Peng
Open University Malaysia, Malaysia

Zoraini Wati Abas
Open University Malaysia, Malaysia

Norlia T. Goolamally
Open University Malaysia, Malaysia

Yuzery Yusoff
Open University Malaysia, Malaysia

Harvinder Kaur Dharam Singh
Open University Malaysia, Malaysia

ABSTRACT

As part of its ongoing effort to offer an effective blend of learning modes, the Open University Malaysia (OUM) recently incorporated mobile learning into its blended learning model. Having enrolled more than 86,000 learners over the last eight years, OUM strives to continually improve its delivery mode for the benefit of its learners. Mobile learning is seen to be the next wave of learning given the fact that almost all OUM learners have mobile phones. A study conducted in late 2008 to determine the readiness of its learners for mobile learning indicates that about 99% of OUM respondents have a mobile phone. The readiness survey also found that majority of respondents would be ready for mobile learning within six months of the study. The chapter will highlight, among others, the introduction of mobile learning at OUM, the initial finding with the pilot implementation and recommendations for the way forward.

DOI: 10.4018/978-1-61692-849-0.ch010

INTRODUCTION

Open University Malaysia (OUM) is established in August 2001 in response to the country's need to democratise education and to offer opportunities for higher education to the country's working adults. It is the pioneer in open and distance learning (ODL) in Malaysia. OUM believes that education should be made available to all, regardless of time, place, age and socioeconomic background and therefore has set its mission to widen access to quality flexible education and provide lifelong learning opportunities by leveraging on technology, adopting flexible modes of learning, and providing a conducive and engaging learning environment at competitive and affordable cost. Subsequently, OUM adopted the motto "University for All". OUM offers a blend of pedagogies comprising of face-to-face tutorials, self-managed learning and e-learning. This unique blend of pedagogies has been well received by the adult working population as testified by a total enrolment of 86,000 in May 2009 as compared to 753 learners enrolled in 2001. Internationally, OUM has learners in Bahrain, Yemen, Maldives, Indonesia and Singapore.

The use of computers and Internet facility is important to many ODL institutions like OUM. In OUM, the e-learning is supported by the learning management system known as My Virtual Learning Environment (MyVLE). MyVLE enables learners to communicate with their tutors and peers in the virtual world and is available 24/7. To enable e-learning to take place, learners need to have a computer with Internet access. In a report published by the Malaysian Communication and Multimedia Commission (MCMC, 2008), only 28.2% of Malaysians had access to personal computers in 2004 and the broadband penetration rate was only 15.2% in 2007 as compared to the 85.1% of the mobile phone penetration rate (per 100 households). In view of the latter, OUM decided to explore the possibility of disseminating learning materials through the mobile phone. After all, the mobile phone technology has advanced tremendously and has become reasonably affordable and most importantly, its use is widespread.

Soon after OUM's decision to embark on mobile learning, a study to determine learners' readiness was conducted late in 2008. The readiness factors were those from the aspects of technology readiness and psychological readiness of OUM learners for mobile learning. The study also aimed to determine how soon learners would be ready and what their expectations were in terms of learning materials that they would like to access through their mobile phones. This readiness survey also determined learners' attitude and acceptance of mobile learning.

MOBILE LEARNING

Definition

There are many definitions of mobile learning. Mobile learning means differently to different groups of people. It can be defined as either learning that happens across locations or learning opportunities that are offered by portable devices that typically overcomes physical and geographical barriers of learning. Those found in the literature often overlap.

Geddes (2004) defined mobile learning as the acquisition of any knowledge and skill by using mobile technology; anywhere, anytime that results in an alteration in behaviour. Lehner and Nosekabel (2002) refer mobile learning as a facility that supplies a learner with general electronic information and educational content that aids in acquisition of knowledge regardless of location and time. In another definition of mobile learning, O'Malley et al. (2003) defined it as any sort of learning that happens when the learner is not at a fixed, predetermined location, or learning that happens when the learner takes advantage of the learning opportunities offered by mobile technologies and Smith (2008) refers to it as learning activities made possible by lightweight handheld device that are easily transported from place to place. According to Nyiri (2002), mobile learning is fundamentally delivered through mobile devices such as Pocket Digital Assistance (PDA), smart phones and any other handheld devices. Mobile learning is seen as the next generation of e-learning and an important instrument for lifelong learning (Sharples, 2000). Traxler (2009) defined mobile learning in terms of its technology and hardware such as PDA, smart phones or wireless laptop computers, which are portable. If the characteristics of mobile learning were considered, words to describe mobile learning would include personal, spontaneous, opportunist, informal, pervasive, situated, private, context aware, bite-size, portable and ubiquitous learning.

Benefits of Mobile Learning

Brown and Parsons (2006) stated that mobile technology has the potential to offer many different levels of engagement, and most applications offered in mobile devices deliver content almost similar to e-learning. Ally (2009) highlighted that through the use of wireless networks, mobile learning allows anyone to access information and learning materials from anywhere and at anytime. This results in learners having control of when they want to learn and where they want to learn from. According to Savill-Smith and Kent (2003), handhelds or mobile devices are part of a new generation of technology that highlight mobility and connectivity; handheld computers are at the forefront of a new wave in the evolution of technology that involves very small computers and wireless connectivity that can provide for learning to anyone, anywhere, at anytime. These mobile devices allow the process of learning to take place everywhere, while commuting, while waiting for services being rendered, etc as opposed to the conventional education whereby learners are confined to their classrooms. Besides this, mobile devices also encourage learners to be independent in their study and promote collaborative learning. The functions and capabilities of powerful computers are now available in mobile devices and mobile phones at a minimum price.

Mobile Learning Pedagogies

For the mobile community, existing learning methodologies are not enough to satisfy their learning needs as their needs are changing (Oblinger, 2003). Mobile learning tools can be developed to enhance situated learning where students can experience, plan and also reflect on their actions (Chen, Kao and Yu, 2004; Proctor and Burton, 2003; Fortier, 2008). According to Gardner (2000), any new technology to be incorporated in the educational field must be inline with the educational goals. However there are issues that need to be addressed when implementing mobile learning pedagogies. According to the literature, meaningful learning takes place in authentic context or simulated environments where the learner can experience through his own experimentation and learn through his mistakes (Engestrom, 1994; McMahon, 1997). The art of learning involves manipulating the environment as a method to guide learners to improve their problem solving skills and acquiring new knowledge. Deep learning is a constructive process where students are able to work with the environment and accommodating or assimilating new knowledge (Sharples, 2000). Different methodologies of the most applicable learning theories in the teaching and learning process can be accommodated through mobile learning technology (Naismith et al., 2005). Learning material supported by the appropriate mobile technology tools will provide different activity based approaches that can either change the learners' future course of actions as in the behaviourist learning theory or the constructivist learning theory where learners will construct new knowledge based on already learnt concepts or previous experiences (Roschelle, 2003; Colella, 2000; Facer et al., 2004). Problem based learning activities and real scenario can be conducted in the application environment through the use of mobile devices equipped with advance technology.

In depth study of the learning processes incorporated in the mobile learning format revealed that teaching and learning processes are not far from the learning theorists approach. Operant conditioning in Skinner's learning theory which involved the association between stimulus and responses is characterised by the drill-and-practice and present-test-feedback activities. In this situation, reinforcement to the learners will be in the form of a problem or information, learners will then have to respond to questions and feedback will be sent to the learners upon receiving their responses. One can indicate that learning has taken place through observed behaviours, however, there is more to learning than just to consider a change in behaviour. Therefore the progression from behaviourism to cognitivism evolved. Cognitive psychologists believe that learning involves the use of memory, reflection, motivation and thinking which sometimes are not displayed as behaviours. Learning takes place internally and amount learned is dependent on the ability of learners to comprehend and process (Craik and Lockhart, 1972).

The theoretical framework by Bruner (1986, 1990, 1996), that is the constructivist learning theory, encompasses the socio-cultural aspects of learning where the three underlying principles are:

i. learning processes must be structured according to the experiences and actual contexts that the students are able and comfortable to learn.
ii. the elements in the teaching process should be designed and structured that it can be easily absorbed by students.
iii. instruction should be designed so that it can lead learners to go beyond and collect information.

The three learning theories - behaviourism learning theory, social cognitive theory and constructivist learning theory, can be combined to design appropriate online learning materials to match learners' learning styles. Increasing cognitive ability from just knowing the facts (behaviourist learning theory) to deeper learning (cognitive learning theory) and to the higher order thinking skills required in constructivism will be brought about by aligning these three major learning theories.

Learning through Mobile Devices

With the technological boom in the mobile devices, many forms of mobile devices have emerged, such as the PDA, music players (popularly known as MP3 players), multimedia players (such as the Apple's iPOD and Creative's Zen) and mobile phones. Mobile phones are currently being studied as a mode of learning by many researchers as the manufacturers of mobile phones have incorporated many applications which were previously available only to computers such as office applications, viewing of multimedia and animations, and many more. For example, mobile phones were being investigated to teach language subjects such as the teaching of English language in Japan (Kimura, 2006; Thornton and Houser, 2004) and China (Cui and Bull, 2005), and learning of Japanese kanji characters (Lin, Kajita, and Mase, 2007). Kimura (2006) found that mobile phones in preparing Japanese college students for the Test of English for International Communication and the video-on-mobile-phones are also effective for vocabulary learning and listening comprehension. Besides learning languages, Bradley, Haynes, and Boyle (2005) experimented with the teaching of Java programming using PDAs. The use of podcasting has been studied by Lee and Chan (2008) in disseminating supplementary materials for an information technology based subject to distance learners. It is interesting to also highlight that mobile science learning for the blind has also being investigated (Sánchez, Flores, and Sáenz, 2008).

Beside these, collaborative learning through mobile learning has also been investigated, for example collaborative learning using PDAs among the children visiting a museum (Koji Yatani et al., 2004; Schroyen, et al., 2008), collaborative learning using mobile phones in storytelling, outdoor games and activities (Stanton and Neale, 2002), collaborative learning using mobile devices in archaeological parks (Ardito, Costabile, and Lanzilotti, 2009) and in zoo (Ohashi, Ogawa, and Arisawa, 2008). In another context, the use of mobile gaming to solve science problems among the 8th grade science learners has also been explored (Sánchez and Salinas, 2008). The progress in the application of mobile learning technologies can be changed accordingly to suit the age of the learners or the subjects to be learnt and even the skills to be applied in that particular context (Tétard, Patokorpi, and Carlsson, 2008). Learning content can be added and customised to suit individual learners (Abfalter, Mirski, and Hitz, 2004).

DESIGN AND DEVELOPMENT OF MOBILE LEARNING AT OUM

The mobile learning research and development team was established in August 2008 with the purpose of conceptualising a mobile learning initiative for OUM's learners. The team researched on various possibilities for mobile learning ranging from texting via short message service (SMS) to multimedia rich content, and podcasting. It was also decided that the team should establish how ready OUM learners are for mobile learning; hence a learner readiness survey (Abas, Chng and Mansor, 2009) was conducted during the September 2008 semester. The findings are reported below.

Learner Readiness

The objective of this survey was to determine the level of readiness in terms of the learners' technology, financial and psychological readiness. A total of 6,000 questionnaires were randomly distributed to learners undertaking various academic programmes offered by OUM. A total of 2,837 questionnaires were received, translating to a response rate of 47.28%.

The respondents came from various regional learning centres in Malaysia. The majority of respondents came from Sabah and Sarawak (31.76%) and from the central learning centres (in Kuala Lumpur, Selangor and Negeri Sembilan) and in Peninsular Malaysia (26.01%). The number of female respondents (60.42%) outweighed the number of male respondents (39.58%). Majority of respondents were undertaking their bachelors' degree (91.05%), and about 5% were undertaking their postgraduate degrees. About 55% respondents were identified as the young working adults as their age were in the 26-35 year old category, which explains their relatively low monthly household income, between RM1,501 to RM3,500 per month (USD417 to USD972, 1 USD = RM3.60) (refer to Table 1).

Table 1. The age distribution and monthly household income of respondents

Profile of Respondents		n (%)
Age Distribution	≤ 25 years old	478 (16.85)
	26 - 30 years old	768 (27.07)
	31 - 35 years old	804 (28.34)
	36 - 40 years old	401 (14.13)
	> 41 years old	383 (13.50)
	Did not respond	3 (0.11)
Monthly household income	≤ RM1500	458 (16.14)
	RM1501-RM2500	775 (27.32)
	RM2501-RM3500	651 (22.95)
	RM3501-RM4500	377 (13.29)
	RM4501-RM5500	251 (8.85)
	RM5501-RM6500	136 (4.79)
	RM6501-RM7500	79 (2.78)
	RM7501-RM8500	44 (1.55)
	RM8501-RM10000	26 (0.92)
	≥ RM10001	35 (1.23)
	Did not respond	5 (0.18)

From the total of 2,837 respondents, 98% of the respondents owned at least one mobile phone and this corresponds to the findings of Karimuddin (2008). The year and purchase price of their mobile phones were also investigated as generally more recent mobile phones would have various multimedia features such as 3G, Bluetooth, audio and video playback, etc embedded and at an affordable price. A random market survey of the price of mobile phone conducted in 2007 revealed that mobile phones with the above features were generally priced from RM600 (USD167). The results also revealed that a large percentage of respondents purchased their mobile phones in year 2007 (30%) and in year 2008 (20.9%), and the price ranged between RM301-RM800 (USD84-USD222). The average amount spent on mobile phone services per-month was between RM30-RM59 (USD8-USD16) (37.86%), followed by RM60-RM89 (USD17-USD25) (19.07%) and less than RM30 (USD8) (14.17%).

To determine the psychological readiness, a scenario of mobile learning showing learning activities through mobile phones (see Figure 1) were given to the respondents. They were asked, "Can you imagine yourself learning in the given scenario?". Eighty-two percent of the respondents indicated that they would imagine themselves learning through this scenario. The respondents were then given a list of courses that they would like to learn through their mobile phones, and 52.90% of them would want to learn University courses such as languages, theological and entrepreneurship courses. Less than 10% indicated that they would like to study courses such as mathematics and science courses via their mobile phones. The respondents were then given a choice of possible mobile learning materials to select and 60% of the respondents would like to listen to the OUM learning materials and 53% of the respondents would like to receive reminders through SMS. As the majority of OUM learners are working adults, the results suggested that the respondents would like to utilise their learning time by listening to their courses while commuting or waiting for services being rendered to them (e.g. queuing in banks and post offices). Furthermore, working adults generally have busy schedules, the respondents felt that they needed to be reminded of the important events relating to their study such as submission of assignments, examination date and venue, etc. Table 2 summarises the mobile learning preferences among the respondents.

To enable mobile learning, there is a possibility that learners may need to purchase new mobile phones and subscribe to certain services. About 65% of the respondents were willing to purchase a new mobile phone between the prices of RM250-RM500 (USD69-USD139). About 56% of the respondents were willing to spend between RM100-RM200 (USD28-USD56) per-month to subscribe to relevant services that would enable mobile learning. The respondents were eager to experience mobile learning as 47% of the respondents would see themselves enhancing their learning through mobile phone within the next 6 months. Table 3 shows the response on the willingness to embark on mobile learning.

On the perceived benefits of mobile learning, the respondents agreed that mobile learning would help them in optimising their time (41%) and help them to stay focussed in their learning (40%). The respondents also agreed that mobile learning will increase their motivation to learn (42%), attract their interest in learning (44%). The respondents also felt that mobile learning will make their learning more flexible (49%).

DESIGNING THE CONTENT FOR MOBILE LEARNING

A model for mobile learning development was conceptualised for implementation. The model incorporates four elements, illustrated in Figure 2. The team felt that the content design and development would be influenced by pedagogical as well as the technological aspect. The design and

Figure 1. The visual used when respondents were asked "Can you imagine yourself learning in the given scenario?" (Abas, Chng and Mansor, 2009)

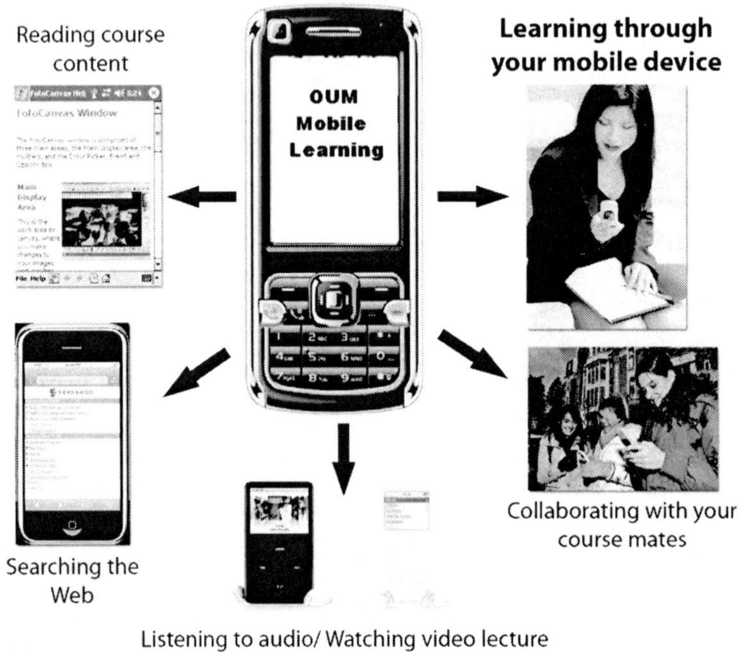

development of content would provide optimal learning and to achieve the learning outcomes of a selected course.

The mobile learning materials developed were made available to the learners via podcasting. The podcast is a multimedia file distributed over the

Table 2. The preferred mobile learning materials

Mobile Learning Materials		n (%)
Courses	University courses Information Technol- ogy Mathematics Science Social Science Others Did not response	1501 (52.90) 397 (13.99) 248 (8.74) 218 (7.68) 64 (2.26) 118 (4.16) 291 (10.26)
Mobile learning materials (multiple respons-es were allowed)	Listen to the OUM learning materials Reminders Tips Online tutorial (video) Quizzes Main concepts Graphical simulations	1701 (59.96) 1531 (53.97) 1382 (48.71) 1127 (39.73) 1115 (39.30) 1067 (37.61) 1068 (37.66)

Table 3. Willingness of respondents to embark on mobile learning

Items		n (%)
Willingness to pur-chase new mobile phone	≤ RM250 RM251-RM500 RM501-RM750 RM751-RM1000 RM1001- RM1250 RM1251- RM1500 > RM1501 Not willing Did not respond	758 (26.72) 389 (13.71) 200 (7.05) 237 (8.35) 146 (5.15) 70 (2.47) 65 (2.29) 963 (33.94) 9 (0.32)
Willingness to sub-scribe to relevant services to enable mobile learning (GPRS/WAP/3G)	RM100 RM200 RM300 RM400 RM500 Not willing Did not respond	1244 (43.85) 200 (7.05) 78 (2.75) 22 (0.78) 31 (1.09) 1242 (43.78) 20 (0.70)
I see myself en-hancing my learn-ing through mobile phone in the next ____ months	≤ 6 months 6-12 months 13-24 months Not willing Did not respond	1345 (47.41) 441 (15.54) 165 (5.82) 853 (30.03) 34 (1.20)

Figure 2. Mobile learning development model

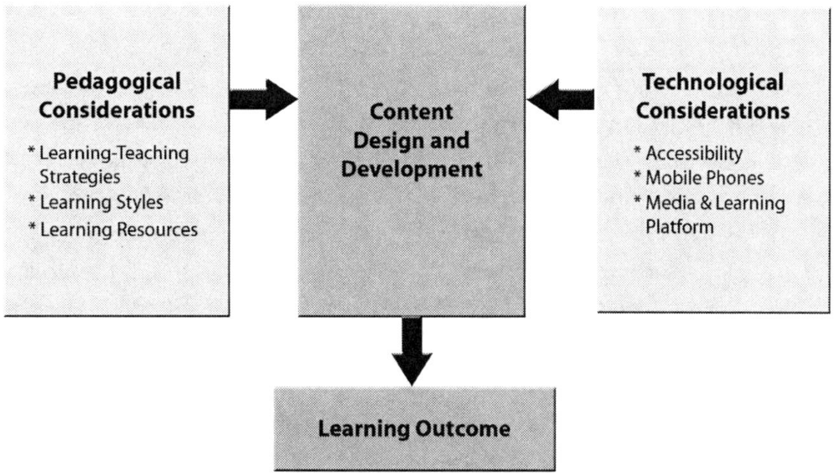

Internet which can be played on computers and mobile devices. Podcasting was selected in view of its increasing popularity from the emergence of the smartphones and portable media players (Andronache, Brust, and Rothkugel, 2007).

Pedagogical Considerations

The pedagogical framework for mobile learning at OUM consists of three major components: learning-teaching strategies, learning styles, and learning resources. This was a refinement of the original one created by the earlier pedagogical group (Norlia et al., 2009). Figure 2 shows the three pedagogical considerations that formed the basis for planning and designing of the mobile learning content. Each of these considerations and how they influenced the decisions are further described below.

Learning-Teaching Strategies

Basically, the learning outcomes of a course determine the type of learning and teaching strategy to be used. The instructional objectives of the lesson should indicate "what a learner is expected to do after completing the lesson". However, a well planned course will consider the learner as the main focus in the learning environment. The knowledge that learners gained and the process in acquiring are the most important of all. The different types of learning styles and the different views of learning theories from behaviourism to constructivism need to be looked at holistically before proper learning can take place. The process of determining suitable learning and teaching strategies require thorough planning in order to meet the learning objectives and learning outcomes. Appropriate learning and teaching strategies should be selected to motivate learners; to cater to their differences; to promote meaningful learning; to enhance students' ability in thinking; and to encourage interaction in class. The learning content in the podcast was designed and structured to provide a variety of learning activities that incorporate both passive and active learning. For the passive learning, the podcast may require learners to listen, view, read, think and reflect upon a certain topic and for the active learning, learners are required to have a discussion among their peers and tutors in MyVLE discussion forum.

Learning Styles

Undeniably in any mode of learning, the learners should be the focus of instruction. Awareness of their different learning styles will contribute to effective pedagogical design. Learning style refers to how a learner perceives, interacts and responds to the learning environment. Each individual has a distinct difference in their approach to learning. Kolb's Learning Style Inventory (Kolb, 1985) looks at how learners perceive and process information. Learners perceive information through sensing and absorbing information where they will learn from concrete experience to reflective observation. The second component in Kolb's Learning Style is the processing of information. Further, learners will be categorized as abstract-conceptualisation learners or active-experimentation learners according to how they process the perceived information. Each of these four categories describes how learners assimilate knowledge differently. Due to this factor, adequate support system should be provided to motivate learners in the learning process. Learners need to be motivated, they will not learn if they are not motivated. To ensure learners are kept engaged in mobile learning, the process of developing mobile learning materials has taken the following into consideration: maintain learners' attention, motivate learners, identifying the difficult or abstract topics or lessons and increase their satisfaction. Learning should be made meaningful and interactive and learners should have control of their learning. Therefore to accommodate the different learning styles, the approach taken is to present podcast content in a variety of ways i.e. visual, audio, simulation and interactive thinking. The objective of the approach is to provide opportunities for learners with different learning styles to be able to select the most desired form of content and learn accordingly at their own pace and time as distance learners.

Learning Resources

The aim of any instructional system is to promote learning. Educators should be able to determine how their learners learn before developing any learning materials. In OUM, the podcast files were developed in various formats; audio in MP3 format, and video in MP4 format and 3GP format. The team felt that these formats can be accessed by a wide range of mobile devices, which include netbooks (laptops specially designed for wireless communication and Internet access), notebooks, mobile phones and mobile players. The learning resources in the podcast include textual content, video, narration of facts, exercises or stimulating learning activities, tips, learning objects, and reminders. It was anticipated that learners should not face any difficulty in identifying and downloading the type format supported by their mobile devices. The file sizes of these podcastings were intentionally kept at approximately one minute as this would minimise time taken when downloading.

While each of these three elements played an important role in the design and development of the podcast, the team also explored the technological aspect such as accessibility, screen size, mobile platform or mobile operating system.

Technological Considerations for Mobile Learning

There are several important criteria in considering appropriate technology for mobile learning (Thirumeni and Yuzery Yusoff, 2009). The criteria are accessibility, mobile devices, and platform.

Accessibility

In Malaysia, the penetration rate of mobile phones was reported to have exceeded the penetration rate of the fixed phone lines and almost every Malaysian owns at least one mobile phone (MCMC, 2008; Karimuddin, 2008). Mobile phone users are able to access the Internet through subscription of various mobile services such as the General Packet Radio Service (GPRS), Enhanced Data for GSM Evolution (EDGE) and the Third generation (3G). These services are generally only available in most urban areas in Malaysia. A study conducted by Pastore (2004) suggested that wireless networks based on WiFi standards are a good solution for implementing an infrastructure for mobile learning application. This solution can cater to problems arising in the maximum limited bandwidth on simultaneous real-time data or video streaming. As such, the Malaysian government has begun popularising wireless Internet (WiFi) access by providing free WiFi access at over 1,500 hotspots nationwide (NST, 2008). In OUM, WiFi access is available at all the learning centres. For learners with WiFi enabled mobile phones, it is hoped that they will utilise the free wireless Internet connection at the learning centres to access the mobile learning materials. It is hoped that learners could transfer the available mobile learning materials from their peers' mobile phones, from their desktop computers or laptops into their mobile phones by using the external memory card, which is considered as a common feature in almost all mobile phones. The transferring of the mobile learning materials can also be achieved by the Bluetooth technology, Infrared file transfer or through the data-link transfer cable.

Mobile Devices

There is a wide range of mobile devices available in the Malaysian market that is able to support mobile learning. The mobile learning materials developed should also be executable in some of these mobile devices. As the readiness survey revealed that the majority of learners have at least a mobile phone, the discussion here is focussed on mobile phones. The multimedia capabilities of a wide range of mobile phone models were considered in this project. In determining the type of multimedia file type suitable to host the mobile learning materials, elements such as mobile operating system, screen size, display resolution, sound quality, file size and the presentation length were considered. The developed mobile learning materials should be displayed in a range of mobile devices of a reasonable resolution and audio quality, and with appropriate text sizes. The memory storage capacity of mobile phones was also being considered as learners may download and store several mobile learning materials, music files and pictures on their mobile phones. As for the file size, usually small file size is preferred as this will reduce the download time and indirectly lower the accessing cost for the learners. Besides the mobile phones, the mobile learning materials should also be accessible by other mobile devices as well such as mobile multimedia players, notebooks and desktop computers. From the readiness survey, the team assumed that most learners have multimedia playback capabilities and Internet access enabled mobile phones based on the year of purchase.

Platform

Ideally, a smart mobile learning portal should be able to adjust itself automatically to suit the different mobile phones used by learners (such as the screen size). The font size and type used should be able to be displayed correctly on various mobile phones and mobile Internet browsers. The use of graphics, audio and video formats as well as animations must be justified and should meet their learning objectives (Gardner, 2000). The mobile learning materials should also be made available to learners who wanted to download them into their personal computers and mobile devices. Several media type such as MP3, AAC, MP4, MOV, RM and 3GP were considered in this project.

The podcasting materials were developed based on the first three topics of the course and were presented in the audio and video formats. The audio files were developed in the MP3 format and the video file was in the MP4 and 3GP format. These file types were selected as the mobile learning team felt that most mobile phones and portable multimedia players are able to execute them. These files were limited to approximately one minute and small byte size. The size of the

MP3 ranged between 1-1.5MB, MP4 ranged between 5-6MB and 3GP ranged between 2-2.5MB. The primary reason was to ensure minimal time spent in downloading these materials.

IMPLEMENTATION

The Learning Skills for the Open and Distance Learners (OUMH1103) course was selected as this course equips learners with basic skills such as time management, stress management, note taking and basic ICT skills that are necessary for OUM learners to apply during the span of their studies. It was also the type of course favoured by the majority of respondents in the readiness survey. This project was implemented in the January 2009 semester.

A web portal, Mobile Learning@OUM web portal was made available at http://mlearn.oum.edu.my (see Figure 3) to introduce mobile learning. A few scenarios showing the benefits of mobile learning were listed on the portal. This web portal was hosted by the OUM"s server and the technical team were able to notify learners undertaking this course by linking it with the MyVLE.

The mobile learning team anticipated that learners may face difficulty in viewing this site due to the different mobile Internet browsers and the screen sizes of their mobile phones, and hence, a mobile version was prepared. An icon was placed in the portal to enable learners to change the view of the portal.

To reduce the number of 'clicks' before learners can access the mobile learning materials, and to enable immediate feedbacks and notification of any updates, a public blog called Learning Skills

Figure 3. A snapshot of the mobile learning portal

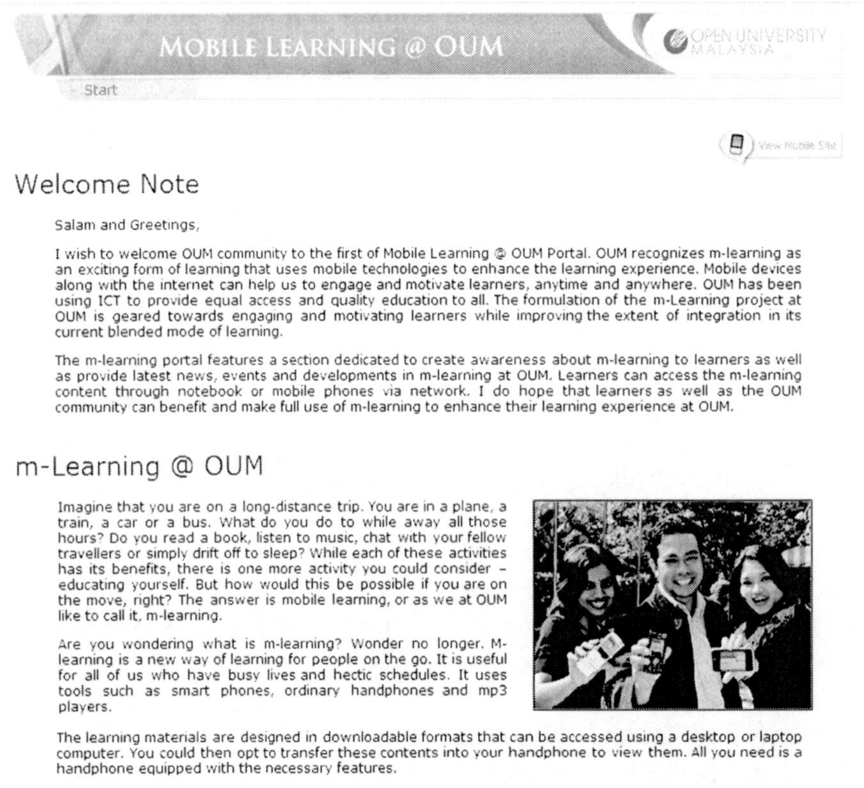

for Open and Distance Learners at http://mlearn-oumh1103.blogspot.com (Figure 4) was used. The features available in the blog, the process of uploading the mobile learning materials and the blog's layout design were simple compared to creating such features by the technical team. In addition, the blog could adjust itself automatically to suit the various mobile phones. Learners could access the blog by either entering the web address or by clicking a link available on the mobile learning portal. Instructions on how to download the mobile learning materials were available in the blog.

Figure 4. A snapshot of the mobile learning blog

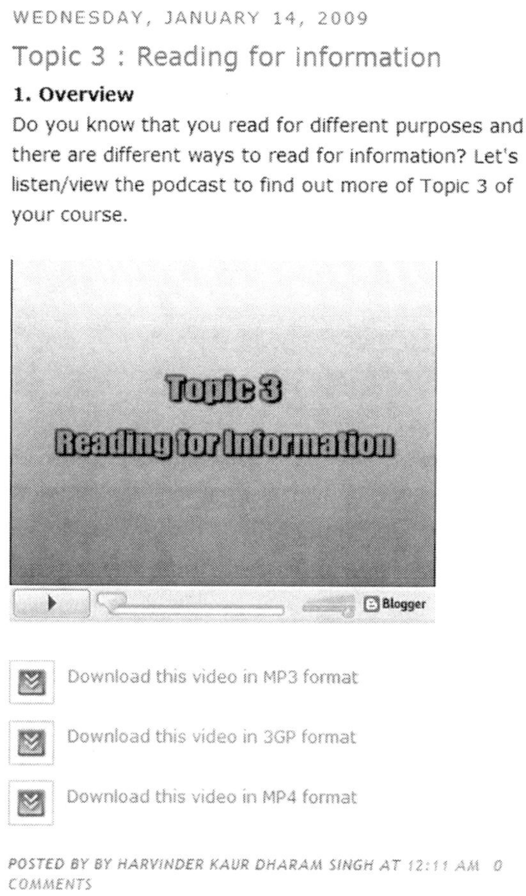

An announcement of this project was placed on the course community section in MyVLE discussion forum to notify the learners and tutors. Besides this, learners and tutors of this course, and staff at the learning centres were also informed of this project via SMSes and emails.

In regards to the development of the mobile learning materials, the course instructor identified certain topics which were important to the learners and discussed these with the graphic designers and animations and instructional designers. Instructional design and strategies were incorporated into the mobile learning materials. The process of developing one mobile learning module took about 2-3 weeks and more time was invested to develop the mobile video presentation slides. The course instructor had to narrate the content professionally.

Feedback on the Podcast

A survey based on the technological acceptance model (Davis, 1986) was developed to gauge the acceptance of podcast learning materials among the learners. Ninety-five learners who registered for this course participated in this study. Majority of the respondents were from the Shah Alam and Negeri Sembilan learning centre and about 19% from the Sarawak learning centre. The gender and age distribution ratio were similar to the readiness survey. Female respondents (52.6%) were more than the male respondents (47.4%), and the majority of the respondents were between the age of 21-25 (33.7%) and 26-30 years old (31.6%). The mobile phones owned by the respondents were considered as fairly recent as more than 60% bought in years 2007-2009.

The findings of the survey indicated that only a handful of learners were aware of the podcast learning materials as only 10 respondents (10.5%) have accessed the mobile learning materials developed for this course. The low awareness and use of the podcast learning materials was primarily due to the low awareness of such facility. Therefore, insights on the use of the Mobile@OUM web portal and the blog were not available.

Among the possible reasons for low usage of podcasting and high ignorance among the learners could be due to:

- insufficient awareness campaign to promote the availability and benefits of the podcast learning materials,
- learners were still in the midst of getting acquainted to the OUM learning environment, hence were not familiar with how to go about learning with podcast mobile learning,
- learners did not or did not know how to subscribe to the podcast services,
- learners were not techno-savvy as some learners reported that they do not know how to access the Internet from their mobile phones,
- possibility that the mobile phones owned by learners did not support the mobile learning materials, and
- learners' concern about the cost incurred in accessing the Internet via their mobile phones.

Challenges

The key challenges faced during the implementation of the mobile learning were mainly in the design and development of the content. Though, theoretically, the mobile learning development model became a good reference point for the team, it lacked the hands on instructions needed for the real time development.

The actual development was very daunting for the team as it was new to the development team, consisting of the course instructor, the audio, video and multimedia specialists and the web designer. The development team needed to learn how to create video and audio for mobile phones. Among the challenges faced was looking for methods of delivering the mobile learning materials that suit various mobile operating systems and the different screen sizes of mobile phones. The course instructor also needed to learn appropriate narration technique for broadcasting purposes.

Though the team has made many efforts to create awareness to promote mobile learning, the message did not get across to the learners. This was probably because no print-based flyers were used, which the team felt that learners were more accustomed to.

The other challenge that the team faced was the weak infrastructure which led to slow download time and availability of mobile Internet access in rural areas, and the high cost of accessing the Internet via mobile phones. The team felt that the podcasting trend among the learners, generally, was very low compared to the SMS. In Malaysia, the SMS is used by many governmental agencies, banks, universities and companies to enable their users to check on various information and transactions. As a result of the above findings, the mobile learning team ventured into delivering learning content via SMS (Abas, Lim and Woo, 2009). The rationale was that SMS would probably be a better means to mediate mobile learning as compared to podcasting.

CONCLUDING REMARKS AND RECOMMENDATIONS

The success of mobile learning is determined by the benefits which learners gained from using the mobile learning materials. This obviously was not demonstrated in the case of the podcasting of mobile learning materials at OUM. There is no doubt, the initial project had many challenges, but that in fact, motivated the mobile learning team to further experiment with other mobile technology for learning, i.e using SMS. Lessons learned from this project will be useful for the implementation of the mobile learning via SMS. Perhaps, the podcasting of mobile learning materials could be revived in the near future.

To accelerate mobile learning, strategic involvement of the learning institution, mobile service providers and manufacturers of mobile phones are important. The mobile service providers together with the mobile phone makers can provide the technical knowledge in terms of the designing and downloading of the mobile learning materials, recommendation of suitable mobile phone models and even offer learners of the institution a special subscription service and mobile phone. The learning institution can provide the learning content and apply appropriate pedagogies and instructional strategies in the development of mobile learning materials. This kind of collaboration would give learners access and experience to a new type of learning without having to worry much about the cost involved.

ACKNOWLEDGMENT

The authors would like to acknowledge the members of the mobile learning research and development team, the Institute of Quality, Research and Innovation, OUM and Teo Hui Thian, Jimmy for language editing.

REFERENCES

Abas, Z. W., Chng, L. P., & Mansor, N. (2009). A study on learner readiness for mobile learning at Open University Malaysia. In *Proceedings of IADIS International Conference Mobile Learning*, (pp. 151-157).

Abas, Z. W., Lim, T., & Woo, T. K. (2009). Mobile learning initiative through SMS: A formative evaluation. *ASEAN Journal of Open Distance Learning, 1*(1).

Abfalter, D., Mirski, P. J., & Hitz, M. (2004). Mobile learning - Knowledge enhancement and vocational training on the move. In *Proceedings of the 5th European Conference on Organizational Knowledge, Learning and Capabilities*, Innsbruck, Austria. Retrieved June 25, 2009 from http://delphi.jura.uni-sb.de/texts/mirski2.pdf

Andronache, A., Brust, M. R., & Rothkugel, S. (2007). Hycast- podcast discovery in mobile networks. In *Proceedings of the 3rd ACM Workshop on Wireless Multimedia Networking and Performance Modeling,* Crete Island, Greece, WMuNeP '07, (pp. 27-34).

Ardito, C., Costabile, M. F., & Lanzilotti, R. (2009). Enhancing user experience while gaming in archaeological parks with cellular phones. In *Proceedings of the 8th International Conference on Interaction Design and Children*, Como, Italy, pp. 270-271.

Bradley, C., Haynes, R., & Boyle, T. (2005). Adult multimedia learning with PDAs - The user experience. In *Proceedings of mLearn 2005, 4th World conference on m-Learning*, Cape Town, South Africa.

Brown, R., Ryu, H., & Parsons, D. (2006). Mobile helper for university students: a design for a mobile learning environment. In Kjeldskov, J. & Paay, J. (Eds.), *Proceedings of the 18th Australia Conference on Computer-Human interaction: Design: Activities, Artifacts and Environments,* Sydney, Australia.

Bruner, J. (1986). *Actual Minds, Possible Worlds.* Cambridge, MA: Harvard University Press.

Bruner, J. (1990). *Acts of Meaning*. Cambridge, MA: Harvard University Press.

Bruner, J. (1996). *The Culture of Education.* Cambridge, MA: Harvard University Press.

Canalys.com. (2008). *Global smart phone shipment rise 28% - Nokia retains lead, but Apple moves into number two position.* Retrieved December 14, 2008 from http://www.canalys.com/pr/2008/r2008112.htm

Chen, Y. S., Kao, T. C., & Yu, G. J. (2004). A mobile butterfly-watching learning system for supporting independent learning. In *Proceedings of the 2nd International Workshop on Wireless and Mobile Technologies in Education*, pp. 11-18.

Colella, V. (2000). Participatory simulations: Building collaborative understanding through immersive dynamic modeling. *Journal of the Learning Sciences*, *4*(9), 471–500. doi:10.1207/S15327809JLS0904_4

Craik, F. I. M., & Lockhart, R. S. (1972). Levels of processing: A framework for memory research. *Journal of Verbal Learning and Verbal Behavior*, *11*, 671–684. doi:10.1016/S0022-5371(72)80001-X

Cui, Y., & Bull, S. (2005). Context and learner modeling for the mobile foreign language learner. *System*, *33*, 353–367. doi:10.1016/j.system.2004.12.008

Davis, F. D. (1986). A technology acceptance model for empirically testing new end-user information systems: Theory and results. Cambridge, M.A: MIT Sloan School of Management.

Engestrom, Y. (1994). *Training for change: New approach to instruction and learning in working life*. Geneva: International Labour Office.

Facer, K., Joiner, R., Stanton, D., Reid, J., Hull, R., & Kirk, D. (2004). Savannah: Mobile gaming and learning? *Journal of Computer Assisted Learning*, *20*(6), 399–409. doi:10.1111/j.1365-2729.2004.00105.x

Fortier, P. (2008). Improving student nurses clinical care experience through the use of a computerized mobile handheld decision support system. In *Proceedings of IADIS International Conference Mobile Learning*, (pp. 58-72).

Gardner, H. (2000). Can technology exploit our many ways of knowing? In Gordon, D. T. (Ed.), *The Digital Classroom: How Technology is Changing the Way We Teach and Learn* (pp. 32–45). Cambridge, MA: Harvard Education Letter.

Geddes, S. (2004). Mobile learning in the 21st century: Benefit for learners. *The Knowledge Tree: An e-Journal of Learning Innovation* 6. Retrieved June 25, 2009 from http://knowledgetree.flexiblelearning.net.au/edition06/download/geddes.pdf

Goolamally, N. Abas, Z. W., Yusof, S., & Singh, H. K. D. (2009). Implementation of m-learning: Pedagogical Perspectives. In *Proceedings of IADIS International Conference Mobile Learning*, pp. 311-315.

Gwebu, K. L., & Wang, J. (2007). The role of organizational, environmental and human factors in e-learning diffusion. *International Journal of Web-Based Learning and Teaching Technologies*, *2*(2), 59–78.

Karimuddin, A. (2008). *Research and markets: Malaysia and Philippines mobile report*. Retrieved July 16, 2008 from http://3gweek.net/2008/07/16/research-and-markets-malaysia-and-philippines-mobile-market/

Kimura, M. (2006). English Language Learning by Multimedia Mobile phones. In E. Pearson, E. & Bohman, P. (Eds.), In *Proceedings of World Conference on Educational Multimedia, Hypermedia and Telecommunications 2006*. Chesapeake, VA: AACE. (pp. 2575-2578).

Kolb, D. A. (1984). *Experiential learning: Experience as the source of learning and development*. Englewood Cliffs, NJ: Prentice-Hall.

Kolb, D. A. (1985). *LSI Learning-Style Inventory*. McBer & Company, Training Resources Group, 116 Huntington Avenue, Boston, MA.

Lee, M. J. W., & Chan, A. (2007). Pervasive, lifestyle-integrated mobile learning for distance learners: An analysis and unexpected results from a podcasting study. *Open Learning*, *22*(3), 201–218.

Lehner, F., & Nosekabel, H. (2002). The Role of Mobile Devices in E-learning – First Experience with a E-learning Environment. In Milrad, M., Hoppe, H.U. & Kinshuk (Eds.), *IEEE International Workshop on Wireless and Mobile Technologies in Education*, (pp. 103-106).

Lin, N., Kajita, S., & Mase, K. (2007). A multimodal mobile device for learning Japanese kanji characters through mnemonic stories. In *Proceedings of the 9th International Conference on Multimodal interfaces*, pp. 335-338.

Malaysian Communications and Multimedia Commission. MCMC *(2008). Cellular Phone in Malaysia*. Retrieved October 23, 2008 from http://www.skmm.gov.my/facts_figures/stats/ViewStatistic.asp?cc=77141977&srid=9247989

Markett, C., Sanchez, I. A., Weber, S., & Tangney, B. (2006). Using short message service to encourage interactivity in the classroom. *Computers & Education*, *46*(3), 280–293. doi:10.1016/j.compedu.2005.11.014

McMahon, M. (1997). Social Constructivism and the World Wide Web - A Paradigm for Learning. *Paper presented at the ASCILITE conference*, Perth, Australia. Retrieved October 15, 2008 from http://www.ascilite.org.au/conference/perth97/paper/Mcmahon/Mcmahon.html

Naismith, L., Lonsdale, P., Vavoula, G., & Sharples, M. (2005). Literature review in mobile technologies and learning. *Futurelab Series*, Bristol, UK: FutureLab. Retrieved January 31, 2008 from http://futurelab.org.uk/research/reviews/reviews_11_&12/11_01.htm

News Straits Times. NST (2008). Times are changing. *The News Straits Times Publication*. Nov 1, pp. 4.

Nyiri, K. (2002). Towards a philosophy of m-learning. In *Proceedings of the 1st IEEE International Workshop on Wireless and Mobile Technologies in Education*, Växjö, Sweden, pp. 121-124.

O'Malley, C. O., Vavoula, G., Glew, J. P., Taylor, J., Sharples, M., & Lefrere, P. (2003) *Guidelines for learning/teaching/tutoring in a mobile environment*. Retrieved June 25, 2009 from http://www.mobilearn.org/download/results/public_deliverables/MOBIlearn_D4.1_Final.pdf

Oblinger, D. (2003). Understanding the new students. *Educause*. Retrieved October 15, 2008 from http://www.educause.edu/ir/library/pdf/ERM0032.pdf

Ohashi, Y., Ogawa, H., & Arisawa, M. (2008). Making new learning environment in zoo by adopting mobile devices. In *Proceedings of the 10th International Conference on Human Computer Interaction with Mobile Devices and Services*, (pp. 489-490).

Pastore, S. (2004). Experiences of mobile learning in science: technological solutions for wireless network and content delivery. Paper presented in *CUC 2005: Internet Users' Conference*, Dubrovnik, Croatia. Retrieved December 15, 2008 from http://cuc.carnet.hr/cuc2005/program/papers/abs/h1_pastore_abs.pdf

Proctor, N., & Burton, J. (2003). Tate modern multimedia tour pilots 2002-2003. In Attewell, J., Da Bormida, G., Sharples, M., & Savill-Smith, C. (Eds.), *M Learn 2003: Learning with mobile devices* (pp. 54–55). London: Learning and Skills Development Agency.

Roschelle, J. (2003). Unlocking the learning value of wireless mobile devices. *Journal of Computer Assisted Learning*, *19*(3), 260–272. doi:10.1046/j.0266-4909.2003.00028.x

Sánchez, J. Flores, H., & Sáenz, M. (2008). Mobile science learning for the blind. *CHI '08 extended abstracts on Human factors in computing systems,* Florence, Italy.

Sánchez, J., & Salinas, A. (2008). Science problem solving learning through mobile gaming. In *Proceedings of the 12th international Conference on Entertainment and Media in the Ubiquitous Era.* Tampere, Finland, (pp. 49-53).

Sariola, J. (2001). What are the Limits of Academic Teaching? In *Search of the Opportunities of Mobile Learning 2001.* Vancouver, Canada: Telelearning.

Savill-Smith, C., & Kent, P. (2003). *The Use of Palmtop Computers for Learning: A Review of the Literature.* Learning and Skills Development Agency. Retrieved June 25, 2009 from http://www.m-earning.org/docs/the_use_of_palmtop_computers_for_learning_sept03.pdf

Schroyen, J., Gabriëls, K., Luyten, K., Teunkens, D., Robert, K., Coninx, K., et al. (2008). Training social learning skills by collaborative mobile gaming in museums. In *Proceedings of the 2008 international Conference on Advances in Computer Entertainment Technology,* pp. 46-49.

Sharples, M. (2000). The design of personal mobile technologies for lifelong learning. *Computers & Education, 34,* 177–193. doi:10.1016/S0360-1315(99)00044-5

Smith, R. (2008). *GoMobile! Maximising the potential of mobile technologies for learners with disabilities.* Learning and Skills Network.

Stanton, D., & Neale, H. (2002). Designing mobile technologies to support collaboration. *Equator Annual Conference.*

Tétard, F., Patokorpi, E., & Carlsson, J. (2008). A conceptual framework for mobile learning. *Research Report 3/2008.* Retrieved October 15, 2008 from http://iamsr.abo.fi/publications/openFile.php?pub_id=464

Thirumeni, S., & Yusoff, Y. (2009). A conceptual framework for OUM m-learning portal. In *Proceedings of IADIS International Conference Mobile Learning* (pp. 250-254).

Thornton, P., & Houser, C. (2004). Using Mobile Phones in Education. In *Proceedings of the 2nd IEEE International Workshop on Wireless and Mobile Technologies in Education, pp. 3-10.*

Traxler, J. (2005). Strategic aspects of wireless and mobile learning. *Landscape Study in Wireless and Mobile Learning in the Post-16 Sector.* Retrieved August 19, 2008 from http://www.jisc.ac.uk/eli_outcomes.html

Traxler, J. (2009). Current State of Mobile Learning. In Ally, M. (Ed.), *Mobile Learning: Transforming the Delivery of Education and Training* (pp. 9–24). AU Press, Athabasca University.

Yatani, K., Onuma, M., Sugimoto, M., & Kusunoki, F. (2004). Musex: A system for supporting children's collaborative learning in a museum with PDAs. *Systems and Computers in Japan, 35*(14), 54–63. doi:10.1002/scj.10696

Chapter 11
Educational Podcasts at University Campus Suffolk

Tim Goodchild
University Campus Suffolk, UK

Sam Chenery-Morris
University Campus Suffolk, UK

ABSTRACT

This chapter will explore the introduction and development of podcasts at University Campus Suffolk (UCS). The podcasts discussed in this chapter have all been developed in relation to pre-registration health and social care courses within the Faculty of Health at UCS. UCS is a relatively new university, and has a wide range of professional courses including nursing, midwifery, radiography, operating department practice and social work. The chapter will begin with a discussion of where podcasts sit in the paradigm of mobile learning and then a brief history of podcasting. The introduction of podcasts at UCS has been ad-hoc and mostly in response to ideas for developing the wider student learning experience. This ad-hoc approach has led to the development of a model for their educational use. Three case studies will be outlined, followed by presentation of the model. These case studies will show how podcasts came to be utilised, and the progression of our thoughts and experiences which have informed their current and future development at UCS. Small scale evaluations throughout the developmental period, and informal student feedback have helped inform the progression of podcasting at UCS. These evaluations have driven the increased use of podcasts at UCS, with students enjoying the experience of using podcasts, and also the ability to digest the podcasts at a time of their choosing. However, it should be noted that because of the nature of the developmental process, full scale evaluative research is only now being undertaken.

DOI: 10.4018/978-1-61692-849-0.ch011

INTRODUCTION

Podcasts are audio or video files that can be 'subscribed to' and then delivered to a computer or compatible handheld device such as an iPod or mobile phone. They have been used at UCS since 2007 as an educational tool for a variety of courses. The development of podcasts at UCS was not planned as part of a large scale implementation of podcast development, or as a pedagogical solution for mobile learning, but was rather as an idea to meet a perceived need for students wanting to get access to learning material when and where they needed to.

Podcasts are considered part of mobile learning, although there seems to be no widely agreed learning theory for 'mobile learning', and consequently, how podcasts are viewed within this theory. Instinctively, mobile learning may mean 'learning whilst mobile' or 'learning via mobile or handheld devices'. However, it is also apparent via even a cursory glance through educational literature, that there are many divergent views as to what mobile learning truly represents.

The term "mobile learning" can simply be viewed primarily as an extension of e-learning (Shuler, 2009). Indeed, early theories put forward focus on this idea of a further dimension to e-learning (Traxler 2005). This approach has also been coined as a "technocentric" approach – where the focus resides with the technology itself, such as a mobile phone handset or an iPod, rather than with the learner or the learning taking place. It could be argued that this use of the term 'mobile learning' is just an alternative version of the term 'open and distance learning'. Perhaps "handheld learning" is more cogent with this technocentric view of mobile learning, where handheld technology is central.

The technocentric view of mobile learning as an extension of e-learning, or a focus on handheld technology does not take into account the unique nature of mobility and learning, and rather absorbs it within the existing e-learning or distance learning spectrum. It also posits mobile learning as attached to a current iteration of technology, and consideration should be placed on the underlying educational experience which takes place, perhaps via the technology. The rapid rate of technological advancement, especially in the mobile arena, deters a concrete association between any iteration of technology and educational theory.

Podcasting fits in with the technocentric view of mobile learning, as podcasts can be delivered and played on mobile devices. However, they are not solely for handheld devices and are easily accessible on computers as well, and as the longevity of podcasts and their ability to evolve have demonstrated, they are not tied to a particular iteration of technology.

There seems to be a widening acceptance for the view that mobile learning should refer to learning whilst mobile, or learning that takes place away from the traditional sites of education such as a classroom, library or in a study/office (Pachler 2007). In this view the focus is not on the devices, but rather on the mobility of the learner. O'Malley et al (2007) states that mobile learning is:

"any sort of learning that happens when the learner is not at a fixed, predetermined location, or learning that happens when the learner takes advantage of learning opportunities offered by mobile technologies."

Podcasts are 'mobile learning' in this respect, as they are delivered to the learner's devices, and therefore, to wherever the learner may be. It may be that the podcasts are listened to on a computer at home, but they may also be listened to in the car, the bus, whilst undertaking leisure activities, or even at work. Podcasts promote the idea of mobility of learning because they can become intrinsic to a handheld device, and free the learner to participate in learning wherever and whenever they choose.

In September 2007, prior to commencing podcast use in education a small survey was undertaken to determine technological usage of nursing students at UCS in the Faculty of Health. Sixty-seven pre-registration nursing students completed the questionnaire (see Appendix A) out of a possible 92, with the results showing that 36% of students knew what a podcast was. This questionnaire was repeated early in 2009 to see if there had been any changes, with 78 pre-registration nursing students responding out of a possible 160. The number of students who knew what a podcast was in 2009 had doubled to 72%. This demonstrated the rapid rise in the awareness of students of the phenomenon of podcasts. The term podcast has become commonplace in society (BBC News, 2005), perhaps surprisingly as it was only coined in 2004 (OUP, 2009).

The use of podcasts in education, however, is not a well known or utilised phenomenon, with studies only appearing in recent years. For example, as recently as 2009, many lecturing staff at UCS were unaware of podcasting as a tool during staff development sessions, even though, podcasts have been employed successfully at UCS for some time. For example they have been used in the teaching of anatomy and physiology, with podcasts delivered at least once a week throughout the duration of a full time module. The material was 'extra' to the timetabled curriculum, and feedback has shown the popularity of using podcasts in this way.

Brief History of Podcasts

Many people think of podcasts as downloading audio or video content from the internet. This in essence is partly correct as a podcast is a digital media file, usually audio or video, however, it is not just downloaded from the internet, but rather published or delivered to the person wanting to listen to or watch the podcast. It is this automatic 'delivery' of content which separates podcasts from a media file which is hosted, and accessible. Podcasts can be subscribed to using a technology known as RSS (Really Simple Syndication), and then can be listened to or viewed on a computer or portable media player or phone. The term podcast is a portmanteau from iPod and broadcast, although you do not need an Apple iPod to listen to or watch a podcast.

The history of podcasts can be dated back to the release of a program called Napster (Businessweek, 2000) which was an online music file sharing program. In essence, Napster allowed its millions of users to copy and share MP3 files (mostly of music) without financial cost, and also to organise digital music on a personal computer. The music industry responded aggressively to the appearance of easily traded 'free music', and Napster was eventually closed down after 2 years in 2001. Whilst Napster allowed digital sharing of media, it did not offer a subscription to any content.

Between 2001 and 2004 there were several small scale developments around the concept of sharing audio content, and subscribing by RSS, a 'web feed', where digital media is fed to users who subscribe to the central resource, and also contains information about the author and the content.

The first podcasting client was released in September 2004 called iPodderX, but had to terminate after a cease and desist letter from Apple Inc., over trademark issues. It was later released as 'Juice', which is still available in 2009. Also in September 2004, Google began to receive search queries for the term podcast, where it elicited 24 "hits" (Nataatmadja & Dyson 2008). Interestingly, in 2009 the number of hits for the search query 'podcast' is 139 million (Google Trends 2009).

Apple introduced their digital music player called 'iPod' in 2001, and by 2004 these had been steadily gaining in popularity. Apple included the iTunes software client alongside the iPod, which was primarily intended to organise digital music on an iPod, but by 2003, Apple had launched the 'iTunes Store', which allowed users to purchase and legally download digital music. In 2005, Apple introduced a "podcast library" (LeClaire, 2005), which featured over 3,000 podcasts available to its users (Steve Jobs, Apple Inc, CEO claimed there were 8,000 available, CNET News, 2005).

Apple had made the RSS technology simple to use, and with a simple click of the mouse, iTunes users could subscribe to a podcast of their choice, or listen to individual episodes. In 2005, Apple released its 5th generation iPod, which incorporated video playback and allowed video podcasting (also termed vodcasting). In 2009, the podcast section of iTunes contains a vast array of podcasts covering a diverse selection of topics, and podcasts also form part of the content for the ITunes University (iTunesU).

UCS Case Studies

The first development of podcasts at UCS arose from an idea to support student nurses with learning clinical skills. A method to enable students to access clinical skills learning material whilst in practice was explored, and podcasting seemed a good fit. Financial support was gained from the institutions 'Teaching Quality Enhancement Fund' (TQEF) to support the purchase of recording and editing equipment. As part of the bid for the TQEF money, students were surveyed to ensure enough would have access to the required technology to use the podcasts. This took place in September 2007. First year pre-registration nursing students were surveyed to see what technologies they were aware of and engaging with (see questions/responses in Appendix A and B). As part of ongoing podcast development, this survey was repeated in 2009.

The results demonstrate that the age range of the students has increased with more students over the age of 36. This may have led to an negative impact on the number of students who accessed the educational podcasts within UCS, since it is often quoted that the younger students often called the YouTube generation (Johnson 2008) or millennial students (Siemens, 2008) might use podcasts more frequently. However, the age of the students did not make any noticeable difference to the results.

Results suggest that around 60% of students own an MP3 player (63% in 2007, 58% in 2009), although this is probably higher, as mobile phones often now integrate MP3 players, and students may have thought this just applied to specific MP3 players. The results also suggest that around 70% of students own a device capable of playing video (68% in 2007, 73% in 2009). These results would probably have included mobile phones, as specific "video players" have not penetrated the market in a substantial way. A more probable answer is that students had devices such as an iPod, iPhone or PlayStation Portable which is capable of playing MP3 audio files and video files, as well as other functions.

The greatest shift in three years was the rise in the number of students who knew what a podcast was and how to download and listen to one (36% in 2007 to 72% in 2009). What we did not ask the students was whether they prefer to download the podcasts to their handheld devices or whether they listened or watched the podcasts on a home or university PC. This would have helped answer the question of whether the podcasts are consumed in mobility or when at a fixed place. However, feedback has demonstrated that the podcasts are consumed in a variety of spaces, including students' homes, whilst travelling and on work placement.

Inspiration for Making Podcasts

Inspiration for using podcasts educationally came from the idea of letting student nurses have access to videos of clinical skills whilst they were working on clinical practice placements. The main driver behind the project (a senior lecturer at UCS) was considering the use of videos, and was looking into the use of videos on students' mobile phones. After purchasing an iPod for personal use, the idea of utilising podcasts followed.

At first, the idea was purely for the use of videos with clinical skills, but as student feedback was positive to this approach, the idea grew to use podcasts in other areas of teaching, and to include simpler types of podcasts such as audio. The area of anatomy and physiology was considered as suitable, as it has a reputation for being a 'tough' subject to learn. It was hoped that the podcasts would help to inspire the students to learn in their own time, and perhaps most importantly of all, would be a 'fun' way for students to learn difficult concepts. Blended learning in this context is not just a blend of "face-to-face teaching" and "online teaching", but rather a blend of methods to give the students a varied and enjoyable experience. The third case study is the most recent development of podcasts, which is podcasts of interviews with service users and clinical staff and clinical tours. These were released at the end of 2008 as part of a wider 'virtual children's nursing practice placement' for pre-registration nursing students, and has led to the development of a 'midwifery virtual practice placement' in 2009, and just recently, the development of a 'virtual polyclinic'.

Case Study 1: UCS Clinical Skills Podcast (T. Goodchild)

The benefits of podcasts, their ease of use and applicability for learning was not really understood at the onset. The first video podcasts recorded in Autumn 2007 were "adult basic life support", "taking a temperature", "taking a pulse" and "hand-washing technique". They took around two hours to plan and film and around an hour to edit. The second batch of podcasts commenced filming in January 2008 included "paediatric basic life support" "taking a manual blood pressure" and "female catheterisation". The third batch of podcasts was filmed from March 2008 onwards and included aseptic technique with more podcasts of advanced clinical skills being developed in 2009.

The podcasts were released in QuickTime (MP4) format, with instructions for students on where to obtain QuickTime for free should they require it. A separate folder on the Virtual Learning Environment (VLE - *sometimes termed Learning Management System or LMS*) was also set up with the videos in 3GP, a format suitable for mobile phones – however the use of the 3GP format has been discontinued due to low demand from students. The skills laboratory lecturer is responsible for ensuring that the clinical skills podcasts remain up to date and reflect best current available practice.

The clinical skills podcasts are not part of a specific course, but are part of the "clinical skills" VLE area. Therefore they are available to all students enrolled in the faculty. The podcasts average around 500 downloads per month, although this is higher during term times, and very low during holidays. Some of the podcasts are applicable to all students such as 'basic life support', whilst others are only relevant to nursing students.

Feedback & Evaluation

Clinical Skills Movies were made available from January 2008, and audited for the first two months. The clinical skills VLE area averaged 1,000 individual visits per month during January

Figure 1. Basic life support clinical skills podcast playing on an iPod

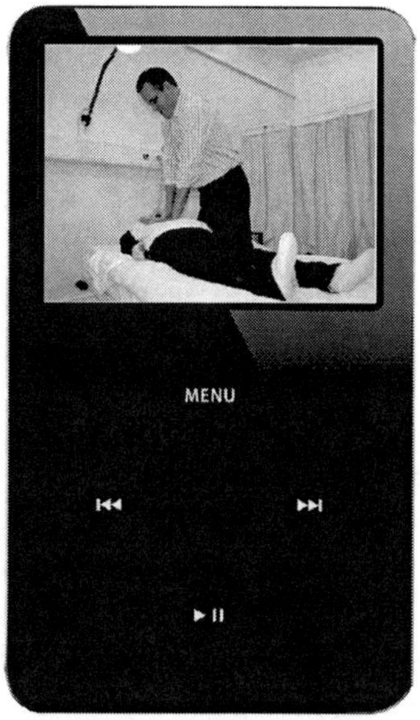

and February 2008, with 5% of these at weekends, and 15% outside work hours. There were around 1,000 downloads of the clinical skills podcasts each month in January and February 2008.

A feedback tool was available to students from the clinical skills area on the VLE, which also asked for feedback on the use of podcasts. In February 2008, the results showed 63% found the podcasts useful, 5% did not find them useful, and 30% said they were not applicable. The results may include some experienced students who are already knew how to undertake these skills or they felt that they did not need to perform these clinical skills.

When asked whether the students would like more podcasts developed, interestingly 74% said yes, 21% said not applicable and 5% did not answer. This may be explained by the proportion of students who may not understand the terminology podcast or those students who do not have the technical knowledge to download and play the podcast, or perhaps those who thought the podcasts were of no clinical relevance to their profession.

The students were asked if they would like to see anatomy and physiology podcasts developed, 90% of the students agreed or strongly agreed with this question, only 10% answered 'no', again this might be attributed to students who do not study anatomy and physiology. The success of the clinical skills podcast, led to consideration of how else podcasts may be used in education. Having experimented with video and found it time consuming; a decision was made to try using audio podcasts, which was thought would be simpler to produce.

Case Study 2: USC 5 Minute Anatomy & Physiology Podcast (T.Goodchild)

These podcasts were made from February 2008 and are still being produced. They include general discussions about certain topics, such as around the brain and brain stem, as well as more specific subjects, such as particular tissues or nerve conduction. This series of audio podcasts are accessible by health students undertaking a diploma or degree, including pre-registration nursing students and foundation degree students who undertake a 70% online Human Anatomy and Physiology module. The podcasts are not course specific, but rather serve as a general learning aid for anatomy and physiology, they are embedded within modules. Audio was chosen as this was thought adequate for conveying the learning in short time frame, and also as it would be simpler to produce. Audio has a chequered history in its use in education (Chan & Lee. 2005), with its most recent use being audio cassettes and compact discs to deliver the experience, often to distance learners.

All the podcasts are under 5 minutes long (hence their title of *UCS 5 minute A&P podcasts*). The majority of the podcasts are monologues, although a "talking heads" approach has also been taken where one lecturer asks questions to another lecturer. These questions were formulated in response to students' questions, such as in the

Figure 2. Accessing UCS A&P podcast via an iPhone

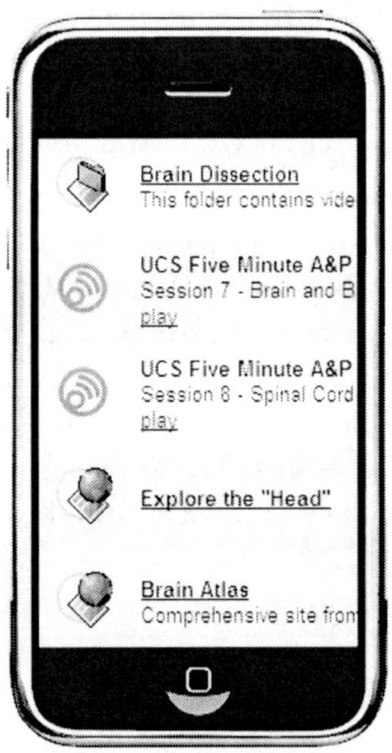

Figure 3. Accessing the A&P podcast through the UCS VLE

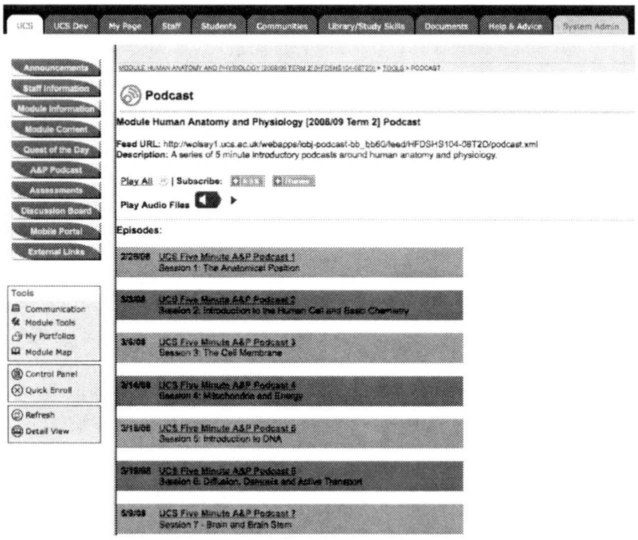

immunology podcast; questions include whether 'the lymphatic system is the immune system', and 'what is the difference between T and B cells'.

The A&P podcasts have been successful in terms of student's feedback. The students have reported that they like the audio-only podcasts; they have downloaded and subscribed to the series to listen to when travelling, and whilst at home doing other tasks. The ability to listen at any time and in any place is prominent, and the audio only format of the podcast promotes this, as the listener can be 'hands-free'.

The planning and construction of the A&P podcasts was less time intensive than the video podcasts, amounting to between 20 minutes for the question and answer podcasts, and around 1 hour for each monologue (monologues needing a rough 'script' to ensure the area is covered correctly). The two lecturers responsible for the talking heads podcasts are experienced lecturers who have taught anatomy and physiology for a number of years and readily know both the subject area and the commonly asked student questions. Initially, there were 10 podcasts delivered over a 6 week anatomy and physiology module. These have also been re-used in other modules, and new podcasts are being produced each term by the original and a new team of lecturers.

Feedback and Evaluation

The UCS Anatomy and Physiology (A&P) podcasts were utilised during a pre-registration first year nursing module. At the end of the module students take part in an A&P multiple choice examination. Following the introduction of the A&P podcasts (along with some online support material for A&P teaching) the average exam results were improved. In April and July 2007 the exam results were (mean) 42% and 43% respectively. After the introduction of the podcasts and online material, the mean exam mark was 61% in April 2008, and was 58% in July 2008. Referral rates also dropped by an average of 1/3, although the higher marks did not appear to improve significantly. Unfortunately the curriculum (and assessment type) changed in 2009, and there are no further comparable results.

In the nursing modules the proportion of students using the podcasts ranges from 50-75% for each module (cohort sizes range from 80-120). The foundation degree healthcare students are in smaller cohorts of around 10-20 students, but also have higher rates of using the podcasts with all students accessing the podcasts at some point. The Foundation Degree students are studying a module which is taught 70% online, hence their contact with the podcasts is greater.

These modules were evaluated at the end of the term. The podcasts received purely positive comments from the students who accessed them.

Qualitative comments from students included:

- *I think these are a really good idea and a good way for me to absorb and understand the subject, the immunity audio is very informative, for me personally I would like more of these in the future.*
- *Yes these were very useful. I will go back to them time and again up to the exam.*
- *I like the podcasts on A&P and the lecture recordings.*
- *I have accessed some of the audio lectures, and yes I do think it would be a good idea to expand on these so we can look back on anything we are not sure about.*
- *I found these very useful I enjoyed them*
- *I think this is an excellent way of learning and have found what I have listened to so far to be very useful. Yes I think it would be very helpful if more were made.*
- *I found these very useful and would like to see more topics covered in this area to, as they are helpful to listen to.*
- *I accessed recordings. Tim & Chris's immunity talk was excellent. Gave a concise overview of immunity. Good to have that to refer back to.*
- *I think they are really good, they are better than I thought they would be... I would like to see more next term.*

Analysis of download data also shows download activity peaking after particular sessions are delivered (for example, in the days after a formal lecture on immunity, download activity of the immunity podcast increased), and there was also a large increase in activity in the week prior to the formal examination on anatomy and physiology. This suggests students were using the podcasts as a revision tool for their learning, which also corresponds with the informal feedback received.

Case Study 3: UCS Children's Insight Nursing Placement (T.Goodchild, K. Hayward & L, Nobes)

Nursing students at UCS undertake theory based 'practice-placements' to gain insight into children's nursing (NMC, 2004). UCS has previously used a workbook which is linked to classroom time and self study to complete this placement. However, a project was initiated in 2008 to use video and audio podcasts to support a 'children's nursing insight placement' for pre-registration nursing students. The video and audio podcasts would form a basis for work completed and submitted via the VLE. The podcasts include video tours of clinical areas, interviews with clinical staff, service users and their families.

The series was designed to "bring practice to the student", as specialist clinical placements (such as children's nursing) are increasingly harder to find for the large number of nursing students. The previous workbook delivered by academics for these students to complete, was based around scenarios for the student to reflect on and work through. The podcast series is designed to compliment the theoretical component of the placement and is supported by student reflection; it gives the student a more comprehensive insight into children's services via 'real' clinicians and service users.

This series of podcasts was more time intensive to plan, coordinate and film, with planning taking around two months, including arranging service users interview time and clinical staff time, as well as gaining agreement from healthcare institutions to use staff time and facilities. The podcasts were produced in video and audio format, and range from 5-15 minutes in length.

Consent was a major issue for this series of podcasts, as previously the academic team had been used and had given consent, but this time service users including children were used, and their confidentiality needed to be protected. Consent was agreed prior to the interview taking place, and agreement was sought post-production. Service users and institutions retained the right to withdraw at anytime. The children interviewed represent a cross section of children's service users including two who experienced the acute services, two with a long term condition and a teenager's experience. Hearing the child's voice and their interpretation of their experience has been a very powerful learning tool.

The podcasts also included interviews with acute and community staff, such as staff nurses, hospice staff and specialists. One example is the role of the music therapist in life limiting care. The therapist offered a brief introduction to his role, a discussion about a particular service user who composed a song with him, then fading out to hear the song composed and sung by a child at the end stage of life. Then the therapist explored the positive impact this song has had on the child, his family and professionals following his death. The family of the child has encouraged the music therapist to use the song for educational purposes, and this has been a very effective example of using this delivery method to improve the student experience.

There are also video tours of clinical departments at acute hospitals and the children's hospice, so students' who will not have the opportunity to work in this environment can see the facilities available, and gain an understanding of the environments service users encounter.

The students who access this series are made aware that there is a strong disclaimer on protecting the public, and if any of the podcasts' were to be used inappropriately or in the public domain, the students would be disciplined for breaking confidentiality and may lose their place on the course, and be referred to their professional body (NMC 2008).

Insight Evaluation Preliminary Qualitative Feedback

Students who have undertaken the children's nursing insight placement are asked for feedback on the experience. Two nursing cohorts have thus far undertaken the placement, and there has been no negative feedback on the use of video and audio podcasts. The only negative feedback has been around the lack of an 'actual' placement, which is understandable, and also a complaint from the previous insight placements.

- *I think [podcasts] gave me the chance to look into child nursing in more detail and found it challenging to complete the various activities but yet very interesting and very motivating.*
- *I really enjoyed the [podcasts] and found that these where perhaps the best insight I could have received, as I was listening first hand to what real people with real experiences had to say about the hospital care.*
- *I feel I know more about the alternative health care environments' for children since completing this insight.*
- *I think the only way you could improve this resource is to involve more of the [podcasts].*
- *The videos were really useful because I wasn't aware of what sort of activities go on inside a children's hospice or what a typical day in a hospice would include.*
- *Watching the video of the children's nurse at the [NHS] hospital has helped me gain an insight. It was good to actually hear one of the staff nurses talk for herself rather than someone giving an overview of what roles are expected of a children's nurse.*
- *I think that the videos were an excellent learning resource and should be used in*

all the insight placements, as they really do give a good insight!

Midwifery placements are also difficult to access, so the project is being continued to create a midwifery insight placement on the VLE with the support of podcasts. This is due to be released in autumn 2009. The mental health, community and learning disability specialities are also considering the use of podcasts in this area, and recently, the development of a 'virtual polyclinic' has begun, with video and audio podcasts are core component of this.

Development of a Model for Educational Podcasts at UCS

Whilst developing podcasts for use in courses at UCS it was noted that there were similar elements to how the podcasts were used and why they were being utilised. This was also guided by informal feedback from students. The model (see Figure 4) was developed in early 2009 for staff development purposes, and then upon encountering a similar model (Edirisingha et al. 2007), was adapted further. It highlights the five elements of educational podcasts at UCS: Function, Formality, Flexibility, Freedom and Fun. These will be examined in relation to the podcasts used here at UCS.

Function

The primary question to answer when developing a podcast for educational use is 'what is the pedagogical purpose of the podcast?' It is important to consider the educational context the podcast will be used in, and also where students will be using it. There are different types of podcast (audio, video, screen capture) and as in traditional educational tools, the appropriate tool should be used for the appropriate teaching task.

The clinical skills podcasts needed to be visual and therefore it was decided that video was the appropriate method. The initial intent was for these to be used for students to review skills they had either witnessed in clinical practice or wanted to perform themselves. A visual demonstration is far more effective and quicker than audio description, especially if students wish to replicate the skill in question. Conversely, the A&P podcasts were

Figure 4. A Model for educational podcasts at UCS (adapted from Edirisingha et al. 2007)

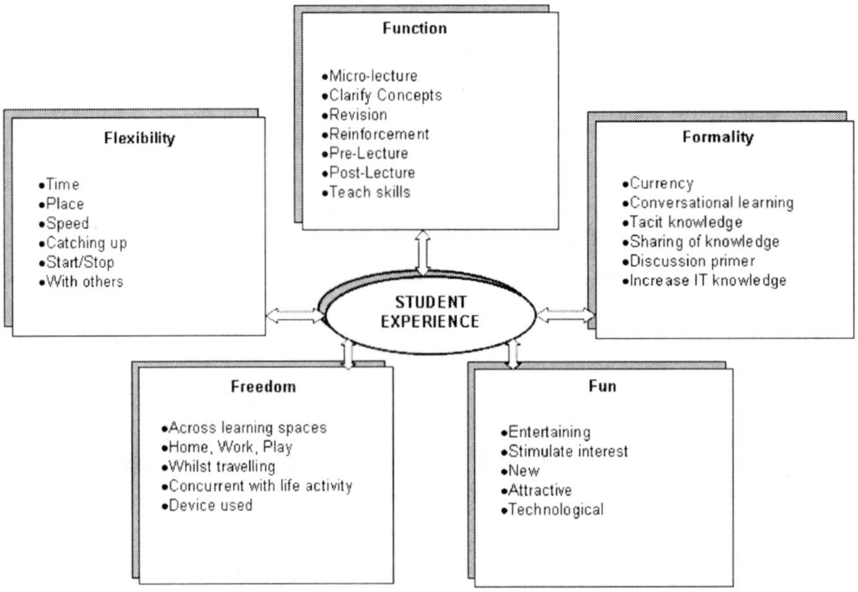

designed for reinforcement purposes and for students to revise and learn from at their own pace and audio was considered adequate. The Nursing Placement podcasts were unique, giving insight into complicated concepts.

Laurillard (2002) states that the success of learning technologies depends on embedding them in the existing learning context, and all the podcasts used at UCS have been embedded within existing educational modules or experiences. Some have been part of a 'redesign' of existing delivery, but they are seen as an essential component of the students learning, and not as a 'bolt-on'.

Formality

Podcasts can be "current" and based upon the latest information or news, such as an interview around a 'hot topic'. In this scenario the podcast cannot be pre-produced and time released, and may come across to student groups as informal. Informal podcasts at UCS have proved popular where there is no script, but lecturers discuss a particular topic or issue. It may be that one lecturer plays the role of the student in a tutorial and asks question or for clarification, whilst the other may provide knowledgeable answers – this may provide insight into more tacit knowledge. Edirisingha et al (2007) found that podcasts enriched the student experience and that podcasts let users learn tacit knowledge by listening to these conversational podcasts.

Many of the students commented on the informality of the interaction between the two lecturers in the A&P podcasts, students seem to like the discussion aspects of the short podcasts where a lecturer took student questions and posed them to another lecturer, and then would ask further probing questions to clarify the subject. Informal feedback suggests that students who struggle with the A&P topic, have found these conversational podcasts of more use than the formal monologue style of podcasts.

The Fothergill (2008) study noted that students liked the informal nature of podcasts, stating that listening to them was not like undertaking course work. Students at UCS have also mentioned this, a half way house between education and entertainment. If the podcast is used as an alternative to deeper engagement with the subject it may be detrimental to the learning process. Conversely Cake (2006) noted deeper engagement with the subject in anatomy when students were motivated and podcasts have been offered as a solution to motivate student engagement.

The children's nursing practice placement podcasts were designed to be formal as they were representing clinical partners who would expect a certain 'presentational standard', and service users. The clinical skills podcasts also needed to be current and up to date, as students may use these to refine their clinical skills.

The social presence of the lecturer, in all the UCS podcasts, must not be underestimated; this can be seen as a powerful tool that the students may actually be feeling informal support from the lecturers' presence as well as the educational benefit of the podcast content. The human voice can help students feel less isolated (Manning 2005), and with students at UCS going out into clinical practice for 50% of their course, the connection back to the university via the podcast may help increase the sense of belonging for these students. The IMPALA (Salmon & Ediringsha 2008) study also confirms that podcasts seem to improve the emotional aspect of learning by increasing the lecturers' presence and have contributed to the 'student-lecturer communication and relationships' (Salmon and Edirisingha 2008 p.5). This positive student lecturer relationship has been said to improve student motivation and encourage deeper engagement with topics such as anatomy (Cake 2006).

Flexibility

Podcasts are used to increase the flexibility of the learning experience. Podcasting entire lecturers of an hour or more have been found to be too time consuming for many students to sit and watch or listen, and also may have few benefits in comparison to the face to face experience. Shorter podcasts, such as the A&P podcasts enable the students to learn in small bite sized chunks. All podcasts can be paused by students to take notes, or reflect - this can be purposefully enabled by the lecturer with reflection points in longer podcasts.

Statistics of the times of download also demonstrate that there is a peak of downloads before the students assessment week (i.e. an examination for A&P modules) indicating that students are choosing to download some of the podcasts as a revision tool. Informal feedback has also been given by some students indicating that they like to listen to the podcasts with others, so they learn together. Again, these points reinforce the notion of flexibility, and student choice.

Students can tailor their own individual learning needs with the flexibility of podcasts. That they can stop, take notes and play back sections. Also as the exam loomed the number of hits on the podcast increased again, for revision purposes, the student was in control of their learning time. Perhaps the technology is being used as a hidden curricula activity, promoting the use of technology, or it may be in response to student demand as consumers, who want to enjoy the latest technology.

Freedom

Podcasts may be designed to be used with certain activities (such as the clinical skills podcasts, which were initially designed to be used at the workplace prior to, or after performing a skill), or in a certain place. Feedback has shown that the shorter, audio only podcasts are used whilst doing things other than 'learning', such as travelling to and from university, doing chores and even listening on the beach. Students are using them across learning spaces, from traditional areas of learning such as campus and home, to travelling or recreational spaces. Nataatmadja & Dyson (2008) found that one of the main positive aspects for students to use podcasts in education, was the "anytime, anywhere" aspect of podcasts. This greater access for students is put forward as an element of 21st century learning (Glenn 2009), along with the ideas of greater student involvement and customization in their learning.

The A&P and clinical skills podcasts were initially designed as supplementary material to the course, although for the most recent A&P modules, the podcasts are now seen as a vital element of the student learning experience. The interviews with service users and clinical placement tours were essential for a student to pass the children's insight placement, and therefore were accessed by all.

The perceived benefit of freedom may only be a perceived and not an actual reality. Although many of the UCS students said that they used their travelling time or were engaged in another activity at the time to listen to the podcasts, we did not evaluate where the students usually learned. The handheld mobile learning notion therefore needs further exploration, as to why a particular location is preferred, and how this affects the experience.

From the literature there are conflicting findings. In one study (Mount and Chambers 2008) the students still chose to listen to podcasts in an indoor environment, usually a study area. They noted that few students actually downloaded the podcast for listening to later, preferring to stream the podcasts through the university VLE. This was also the case in the France and Ribchester's study (2008) where few students actually downloaded the podcast to their MP3 player contrary to the tutors' expectations. Off campus listening exposed an anomaly within the literature with differing student experiences recorded, according to Rothwell (2008), 44% of students listened to podcasts 'always or nearly always' off campus yet she later states that 61% of students 'listened off campus'. As our professional courses are 50% off campus the numbers of students listening off campus is probably higher, and we know via informal feedback that many students do use handheld solutions to listen to the podcasts.

Fun

Fun is possibly the most important part of the model, in that podcasts are designed to increase the enjoyment of the course for the student. If the podcasts are purely educational, it has to be considered how this can be achieved.

The "fun" element of podcasts was thought about in the initial development period in 2007, as student experience is important on all parts of a course. Students have reported that they have liked learning to use the new technology, and this has stimulated their interest more. Students from cohorts where no podcasts have been used have approached lecturers involved in these case studies asking when they can have the podcasts for their courses, as they sound really useful, indicating the 'attractiveness' of using new ways of learning for students.

An important consideration is if students have the requisite knowledge to enjoy the experience of using podcasts, or will possible technological problems induce a negative perception of podcasts. We knew at UCS that a percentage of students did not know what podcasts were (circa. 30% in 2009) or their potential use. To counter this, and ensure that a negative perception was not created due to technical inability, the teams took time to offer instruction to students who required it. Also the VLE support team at UCS were told of the podcast initiative so they knew where they were and how to support students should they make enquiries. Finally, the lecturer involved added contact details to the VLE podcast area, and the video podcasts, so students would know where to ask for help should they require it.

Students are consumers in the 21st century and have an expectation of receiving 'value'. Podcasts can add enjoyment to a course, and therefore contribute to perceived value. Cox et al (2008) found students felt they were getting a better quality of contact with the lecturers, and felt a positive attitude towards the lecturers who had put in the effort to create the podcasts. Fothergill (2008) also used humour deliberately in his podcasts; feedback from previous students encouraged the use of humour stating it humanised the podcasts, kept the students interested and encouraged the students to listen to the end where the joke was.

The recent student comments below (March 2009) as feedback to podcasts (comments made as part of nursing module evaluation via the VLE), give an overall picture of student's feelings towards podcast use in their courses.:

- *I am a in my third week of starting on the Adult Nursing course and I only really wanted to say I had read the information in the book first, but I have just listened to your first A&P podcast and it has really helped me understand a bit more easily.*
- *I have downloaded and have been listening to your podcasts from VLE. I particularly like the ones where you are discussing the subjects together.*
- *I must admit I am finding A & P a little bit tricky and I definitely think the podcasts will help.*
- *I think they are very good, I have only recently discovered them, but very useful,*
- *I have now played a few of the podcasts and they are really good as you can listen to them whilst doing something else and it just reinforces stuff you already know a bit about.*
- *I find the podcasts excellent, I have been listening to them whilst in the car, doing the housework, ironing etc.....*

- *Have just listened to a couple of the podcasts I think they are great, really helpful to refresh my mind on past lectures!*
- *These are great, keep them coming please*
- *Really helpful as can listen to a few 5 min ones on the bus in and then home again. Excellent.*
- *The podcasts are great. I listen to them on the bus to and from Uni. Please keep doing them!*

Connectivism Related to Educational Podcasts

Students at UCS are encountering an increasing array of e-learning content and experiences as they make their journey through their education. Podcasts can be seen as a passive experience for students, that is then 'delivered' to them, and forms part of this varied array of 'e-learning' which will make up their overall educational experience. As the number of podcasts and e-learning content increases, students find themselves as the central component of a network of 'nodes' as put forward in the idea of connectivism (Siemens 2005, 2008). Siemens proposed that knowledge is distributed across networks, and learners would establish connections across these nodes of knowledge. As the library of podcasts increase at an institution or that are available to students, then students may become fluent in subscribing to nodes that are relevant to them for a particular time. For example, anatomy and physiology (A&P) can stay relevant for learning across various disciplines, and students not enrolled in the A&P course may connect to the 'A&P podcast node' just for a particular period during their course, as well as to other parts of their ever changing network.

This works in conjunction with the idea of informal learning as proposed in the 'Model for Educational Podcasts' above. Informal learners learn across various spaces and contexts (Cross 2006). Students on professional pre-registration courses at UCS learn within a formal context, also within work-based clinical environments. Then there are other contexts such as home, office and even travel and during tasks. Not only do students connect to various nodes of learning, but also various spaces where learning takes place. These varied learning spaces which students move through may provide additional 'nodes' of informal learning.

Part of the informality of podcasts when held on handheld devices is that the actual learning episodes for the learners are impossible to predict. As educators we cannot be sure when students will engage with the learning material, or within what context. This notion is discussed by Scanlon, Jones and Westcott (2005), when discussing the use of mobile technologies in informal science settings. Their feedback has demonstrated that students use the podcasts in very different environments (e.g. work, travelling, university, home), and the impact on their learning across these varied spaces and contexts is not yet known.

UCS Podcasts in Retrospect

Salmon & Edirisingha (2008) proposed a "10-factor design model" to follow when considering the use of podcasts in education (see Table 1). The podcasts developed at UCS can be related to this model in retrospect, as consideration of the factors did take place, although often without knowing.

In retrospect, the only developmental planning behind the first clinical skills podcasts was to give the students greater access to learning when they are in practice by carrying "mobile" content. So the only 'pedagogical rationale' noted from Salmon & Edirisingha (2008) model was the improving usefulness of teaching and learning resources. However, this was not improving the resources, as there were none there to begin with. Rather the rationale in pedagogical terms was to support students remotely more effectively (the mobile element), whilst they are in the context of clinical learning experiences.

Table 1. A 10-factor podcast development model. Salmon & Edirisingha (2008)

Factor	*Opinions*
1. Pedagogical rationale	Limitations of lectures in teaching complex and difficult topics Limitations of conventional approaches in teaching use of software tools Limitations of conventional feedback approaches Issues faced by first-time online learners Issues faced by distance learners Developing competency in: • Collaborative skills; • Active learning skills; • Presentation skills; • Essay writing skills; • Reflective skills; • Research skills; • Articulation and communication skills; Improving the usefulness and attractiveness of teaching and learning resources
2. Medium	Audio podcasts Audio and vision podcasts
3. Convergence	Integrated with other media such as VLE Stand alone
4. Authors and contributors to podcasts	Subject or module leader University teachers Other university staff Fellow students Senior students Other stakeholders (experts of local community)
5. Structure of podcasting	Single-session podcasts Multiple sessions (weekly, fortnightly or monthly) Targeted podcasts for specific reasons, such as assessments or exams
6. Reusability	Temporary or reusable podcasts
7. Length	Short or long (less or more than 5 minutes) podcasts
8. Style	Formal, informal; and style of presentation: monologue, dialogue, interview or other
9. Framework	Signposting, navigating, planning
10. Access system	Via VLE A feeder service (RSS)

The podcasts were to be hosted within the "Clinical Skills Community" on the UCS VLE, alongside other materials used to support clinical skills teaching and in practice. The authors were academics who had the initial idea for use, but also who taught clinical skills at times to pre-registration students, and who were also clinically competent, although experts were co-opted in for accuracy and authenticity (for example, the "Aseptic Technique" podcast was performed by a wound care specialist nurse, and "Paediatric Basic Life Support" was performed by a paediatric nursing lecturer).

The podcasts were planned in a series as time and funding allowed, and it was envisioned early on that the "Clinical Skills Associate Lecturer" would be involved in keeping podcasts 'current' for future use. A formal style was chosen, and it was the aim to keep all podcasts below 5 minutes in length, as if they were to be used whilst on clinical practice placement, we did not want students watching educational material for lengthy periods. Students could subscribe via an RSS option, although this was not known about at the onset, but rather discovered early on in development.

Anatomy and Physiology Podcasts in Retrospect

In contrast, by the time the A&P podcasts were to be developed, the team had a strong conceptual idea for what was wanted pedagogically and practically. A series of monologue audio podcasts was planned covering a variety of subjects the students covered. An initial batch of 10 podcasts was planned. These would be formal, short (less than 5 minutes in length), audio only podcasts hosted within specific VLE module areas, and delivered by the module teams.

The pedagogical basis for the A&P podcasts was to support student learning, by offering complex concepts in a short, simple format. This would enable students to learn in their own time and space, but also to pause the audio, take notes or reuse for revision/learning purposes. Student feedback during the module was positive, and the team chose to keep developing the podcasts past the initial planned period and is now ongoing.

A development idea was to have a "conversation" between 2 lecturers, where one lecturer would ask questions which students passed on to the team, and also probed for clarification on certain areas. Students who were struggling with the content reported that the conversational podcasts were particularly useful. The A&P podcasts are permanent, and are also re-used with other modules where A&P is studied. Further developments are underway to film dissection of organs as podcasts to develop physical examination podcasts, and also to deliver screencasts (PowerPoint slides with a voice over) as podcasts on particular subjects.

Children's Insight Nursing Placement Podcasts in Retrospect

The podcasts for the children's nursing insight placement were very specific. There were to be video tours, and video interviews with service users and clinicians. This was a difficult and complex teaching area, and also an area where the lecturers involved believe the approach would enhance the student experience. The video tours were split into different areas, and were therefore short, whilst the interviews were released according to length of the interview (between 10-15 minutes).

The team knew that a closed VLE area would host the podcasts due to confidentiality, and also students were instructed face-to-face about the confidential nature of the podcasts. They were ambitious as external specialists were filmed in their own residences/workplaces – which took a great deal of organisation. Again, these are to be re-used for further cohorts, although they have led to a development of "life-stories of service user's" which are to be made available to more students in the future.

FUTURE RESEARCH DIRECTIONS

Questions arising from the UCS experience are varied. In particular the longevity of the learning from podcasts has not been explored. Advocates of mobile learning may argue that computer technologies have created a culture for increasingly active learning. The learning from podcasts and other handheld learning devices may be more passive than commonly believed, unless students use them to engage in discussion, or actively create podcasts themselves. The advent of the iPod Touch with WiFi built in, and the popular iPhone and other internet 3G phones may herald a greater deal of interaction for students who utilise podcasts as part of their course. Certainly being able to receive new podcasts without the need for a computer is a huge benefit for the internet enabled handsets, and one that may change the way podcasting is utilised.

The popularity of the shorter 5 minute podcasts may be problematic, as it is questionable whether in-depth learning can be 'condensed' into small, easily accessible chunks. But, as feedback shows, students seem to like this style of short, informal learning. As autonomous learners they can connect to the short podcasts that interest them and then find new nodes to meet their personalised learning requirements. Research could explore the validity of learning from these bite-sized chunks.

Podcasts, and handheld devices, mean that students may be learning across or within varying learning spaces and contexts. Academic staff have little or no control over where a student will choose to interact with the learning material. There is little evidence over the effect of learning through a variety of spaces has on the learning experience, and the efficacy of the learning.

Podcasts may be seen as a more economically viable alternative to traditional sources of learning/information by institutions. However, this concern has long been standing for e-learning in general, and for UCS, podcasts are seen as complimentary to traditional educational content, and are part of a blended approach to learning.

If the increasing use of technologies is consumer led, and students as consumers are demanding greater successful use of technology in their courses, is there a real generational gap in the use of technological solutions such as podcasts? There could be a potential generational gap in lecturing staff as well as in students. Anecdotal experiences at UCS suggest that there is no gap, with older generations keen and able to utilise podcasts.

CONCLUSION

Many forms of teaching use some kind of technology, whether it is paper and pen, CD-ROMs or podcasts. At UCS, students have driven the adoption of e-learning and technology in their courses. As they experience a particular successful use of a technology and understand its benefits, they become more vocal in demanding increased use of these approaches.

Handheld technologies have provided flexibility in meeting individual student needs. They provide the student with immediate access to rich resources beyond the classroom, and in locations of the students choosing. However, further work needs to be undertaken to establish where students prefer to learn, and if they are using podcasts in a true 'mobile' sense, just how effective is the learning in comparison with focussed study in a traditional learning environment.

Podcasts can present information in new ways to enhance the student experience. Of particular note is the enjoyment students find in using podcasts as part of their courses. It increases the variety of methods used, is not intimidating, and provides an opportunity for students to perhaps experience a new way of learning. Students have expressed clearly that podcasts are an enjoyable way to learn, and in consumer led education, this can only be seen as positive.

The issues of teacher competence still need to be addressed. At UCS there are staff development and support sessions to design and deliver podcasts, and a recently developed online course in how to develop and use podcasts in education, although these are not compulsory, and usually is only taken up by academic staff that already have an interest in the uses of technology in their teaching. As a result of podcast use as described in this chapter, teaching enhancement funds have been used in the summer of 2009 to buy equipment to give to academic staff to promote the use of audio podcasts in their courses.

There is no clear evidence to demonstrate podcasts' long term learning potential and there may be issues of equity among students. However, the experience at UCS demonstrates that students, once they experience the use of podcasts, are keen to use them throughout their courses, and that age or experience does not preclude their use.

The popularity of podcasts seems to rest on the ability to use them for effective learning and revision, in a variety of environments, and in an enjoyable way. Ultimately, student feedback has driven the continued and growing use of podcasts at UCS, and as feedback remains positive, and the voice of students demanding podcasts increases, the use of podcasts at UCS will continue to develop and be utilised in new ways across the student experience.

REFERENCES

Businessweek (2000). *The highs and lows of Napster*. Retrieved April 21, 2009 from: http://www.businessweek.com/2000/00_33/b3694003.htm

Cake, M. (2006). Deep dissection: motivation students beyond rote learning in veterinary anatomy. *Journal of Veterinary Medical Education, 33*(2), 266–271. doi:10.3138/jvme.33.2.266

Chan, A., & Lee, M. J. W. (2005). *An MP3 a day keeps the worries away*. Student Experience Conference 2005. Charles Stuart University. Retrieved June 20, 2009 from: http://www.csu.edu.au/division/studserv/sec/papers/chan.pdf

CNETNews. (2005). *Apple WWDC Keynote Steve Jobs*. Retrieved April 2009 from: http://news.cnet.com/2100-1047_3-5748045.html

Cox, B., Macharia, R., Short, N., & Whittlestone, K. (2008) in Salmon, G., Ediringsha, P.,(Eds), Podcasting for Learning in Universities. Glasgow: McGraw Hill.

Cross, J. (2006). *Informal Learning: Rediscovering the Natural Pathways That Inspire Innovation and Performance* (Essential Knowledge Resource), Jossey Bass, San Fancisco: Pfeiffer.

Edirisingha, P., Rizzi, C., Nie, M., & Rothwell, L. (2007). Podcasting to provide teaching and learning support for an undergraduate module on English language and communication. Turkish online journal of distance education. *8*(3). Article 6.

Fothergill, J. (2007, January). *Profcasts: supporting online learning of campus-based students.* Paper presented at the Learning Futures Conference, University of Leicester, UK.

France, D., & Ribchester, C. (2004). Producing websites for assessment: a case study from a Level 1 fieldwork module. *Journal of Geography in Higher Education, 28*, 49–62. doi:10.1080/0309826042000198639

Glenn, M. (2009). *What does it mean to educated in the 21st Century*. Open EdTech 2008. Retrieved December 10, 2009 from http://www.slideshare.net/timbuckteeth/what-does-it-mean-to-be-educated-in-the-21st-century

Google Trends. (2009). *Podcast Google Trends*. Retrieved December 10, 2009, from http://www.google.com/trends?q=podcast&ctab=0&geo=all&date=2004

HEFCE. (2009). *Enhancing learning and teaching through the use of technology: A revised approach to HEFCE's strategy for e-learning*. Retrieved July 17, 2009 from http://www.hefce.ac.uk/pubs/hefce/2009/09_12/

Johnson, B. (2008). The rise and rise of the YouTube generation, and how adults can help. *Guardian* retrieved July 17, 2009 from http://www.guardian.co.uk/technology/2008/oct/06/youtube.youngpeople

Laurilliard, D. (2002). *Rethinking University Teaching: A conversational Framework for the Effective Use of Learning Technologies* (2nd ed.). London: RoutledgeFalmer. doi:10.4324/9780203304846

LeClaire, J. (2005). *Apple including podcast library in iTunes 4.9.* Retrieved December 10, 2009 from: http://www.technewsworld.com/story/EIGWrJ6Jx65C9M/Apple-Including-Podcast-Library-in-iTunes-49.xhtml

Manning, S. (2005). The promise of podcasting. *Pointers and Clickers, 6*(2). Retrieved December 10, 2009 from http://www.ion.illinois.edu/resources/pointersclickers/2005_03/Podcasting2005.pdf

Mount, N., & Chambers, C. (2008). Podcasts and Practicals. In Salmon, G., & Ediringsha, P. (Eds.), *Podcasting for Learning in Universities*. Glasgow: McGraw Hill.

Nataatmadja, I. & Dyson, L.E. (2008) The role of podcasts in student's learning. *iJIM, 2*(3), 17-21

News, B. B. C. (2005). *Wordsmiths hail podcast success*. Retrieved May 20, 2009 from: http://news.bbc.co.uk/1/hi/technology/4504256.stm

NMC. (2004). *Standards of Proficiency for Pre-registration Midwifery Education*. London: NMC Publishing.

NMC. (2008). *The Code Standards of conduct, performance and ethics for nurses and midwives*. London NMC Publishing.

O'Malley, C., Vavoula, G., Glew, J., Taylor, J., Sharples, M., & Lefrere, P. (2003) *Guidelines for learning/teaching/tutoring in a mobile environment. Mobilearn project deliverable*. Retrieved April 21, 2009 from http://www.mobilearn.org/download/results/guidelines.pdf

Pachler, N. (2007). *Mobile learning towards a research agenda*, WLE Centre, Institute of Education. Retrieved April 21, 2009 from http://www.wlecentre.ac.uk/cms/files/occasionalpapers/mobilelearning_pachler2007.pdfb

Rothwell, L. (2008) in Salmon, G., Ediringsha, P.,(Eds), *Podcasting for Learning in Universities*. Glasgow: McGraw Hill

Salmon, G., & Ediringsha, P. (Eds.). (2008). *Podcasting for Learning in Universities*. Glasgow: McGraw Hill.

Scanlon, E., Jones, A., & Waycott, J. (2005). Mobile technologies: prospects for their use in learning in informal science settings. *Journal of Interactive Media in Education*, 21(5), Retrieved December 10, 2009 from: http://jime.open.ac.uk/2005/25/scanlon-2005-25.html

Shuler, C. (2009) Pockets of Potential: Using mobile technologies to promote children's learning. Retrieved December 10, 2009 from: http://joanganzcooneycenter.org/pdf/pockets_of_potential.pdf

Siemens, G. (2005). Connectivism: A learning theory for a digital age. *International Journal of Instructional Technology and Distance Learning, 2* (1). Retrieved December 10, 2009 from http://www.itdl.org/journal/jan_05/article01.htm

Siemens, G. (2008) Learning and Knowing in Networks: Changing roles for Educators and Designers. Retrieved December 10, 2009 from http://it.coe.uga.edu/itforum/Paper105/Siemens.pdf

Traxler, J. (2005). Mobile learning: it's here but what is it? *Interactions*, 9,1. University of Warwick. Retrieved December 10, 2009 from http://www2.warwick.ac.uk/services/ldc/resource/interactions/archive/issue25/traxler

APPENDIX A

Questions asked in UCS technological survey of nursing students in 2006.

What is you age?
Do you own an MP3 player?
Do you own a mobile device capable of playing video?
Do you have access to a PC at home?
Do you have the internet at home?
Do you know how to download video from the internet?
Do you know what a podcast is?
In 2009 in the repeat survey, the additional questions were asked:
Would you like to access more podcasts as part of your courses?
Do you have Broadband internet at home?

APPENDIX B

Results from UCS technological survey of nursing students 2006 and 2009

Year		2006 n=67 out of 92	2009 n= 78 out of 160	Change
Age range	*17-25*	65%	57%	- 8%
	26-35	23%	24%	+1%
	36-45	12%	19%	+7%
MP3 player?	*Yes*	63%	58%	-5%
	No	37%	42%	+5%
Video Device?	*Yes*	68%	73%	+5%
	No	32%	27%	-5%
PC at home?	*Yes*	95%	NA	-
	No	5%	NA	-
Internet at home?	*Yes*	85%	95%	+10%
	No	15%	5%	-10%
Can you Download Video?	*Yes*	71%	70%	-1%
	No	29%	30%	+1%

continues on following page

continued from previous page

Year		2006 n=67 out of 92	2009 n= 78 out of 160	Change
Do you know what a Podcast is?	*Yes*	36%	72%	**+36%**
	No	64%	28%	**-36%**
Would you like more?	*Yes*	NA	54%	-
	No	NA	12%	-
	Unsure	NA	34%	-
Do you have BB?	*Yes*	NA	92%	-
	No	NA	8%	-

Section 3
Research and Innovative Pedagogy

Chapter 12
Imagine Mobile Learning in your Pocket

Cecilie Murray
Delphian eLearning, Australia

ABSTRACT

Students and teachers are embracing mobile technologies in their social lives. How is this reflected in K-12 schools? This chapter examines the experiences of students and teachers in a range of mobile learning projects in the K-12 environment. Four research projects highlight the experiences of students and teachers as they grapple with mobile technologies and the wireless environment, with implementation and technical issues, with learning approaches and pedagogical innovations. The projects focused on Literacy, Mathematics and cross-curricular learning with Australian primary and secondary students as well as students in international collaborative projects. In each project, students demonstrated improved attitudes to school, greater engagement and participation in learning and enhanced performance. Teachers learned a diversity of approaches to classroom management and curriculum planning, and demonstrated significant pedagogical change. The benefits of mobile learning were also reflected in the community, with parents taking greater responsibility and interest in their children's learning opportunities.

INTRODUCTION

Many young people today use mobile technologies as a matter of course in their lives outside school. They are engaged and motivated by emerging technologies, such as mobile phones, media players (MP3 and MP4), gaming devices, digital cameras, and with web based software that enables social networking, communication and collaboration within their peer groups As these behaviours and technologies converge, teachers and students are attempting to adapt a range of commercially available mobile devices to the educational environment.

These technologies and software have converged with the consolidation of the mini-computer and wireless access, to provide a rich environment for mobile learning – the use of

DOI: 10.4018/978-1-61692-849-0.ch012

small handheld devices anywhere, anytime. The smartbook, a hybrid mini-laptop and smartphone device provides continuous connectivity, increased battery life, 3G connectivity and GPS in a slightly larger device with a full keyboard. Innovative technologies such as these have the potential to fundamentally change learning as we know it in schools.

While the gap between technology devices used in everyday life and those used in schools continues to widen, 'public agencies [Departments of Education] and schools are beginning to explore multimobile services – convenient, flexible tools that enable users to access information in a range of formats using mobile devices – as a way to support…staff and students' (The New Media Consortium, 2009, p.16). Innovative teachers are excited by the opportunities to work at the leading edge and are keen to make a difference to student engagement and learning, but many are intimidated by the fast pace of change and the increasing range of technologies available for learning.

Mobile devices can have a positive impact on learning experiences for both educators and students (McFarlane, Triggs & Yee, 2008; Ng & Nicholas, 2009; Perry, 2003; Twining et. al, 2005; Vahey & Crawford, 2002). This principle has been trialed in these four projects in an Australian context, focusing on how mobile learning might be used to increase engagement, motivation, information and communication technology (ICT) curriculum integration and effective learning in K-12 schools.

This chapter examines the experiences of K-12 students and teachers in four Australian mobile learning research projects, as they grapple with mobile technologies and the wireless environment, with implementation and technical issues, with learning approaches and pedagogical innovations.

These studies, focused on the use of iPod, iPod Touch and Nintendo DS, indicated that the engagement value of these mobile devices motivated students to learn more creatively and independently. In each project, students demonstrated improved attitudes to school, greater engagement and participation in learning and enhanced performance. Teachers developed a diversity of approaches to classroom management and curriculum planning, and demonstrated significant pedagogical change. The benefits of mobile learning were also reflected in the community, with parents taking greater responsibility and interest in their children's learning opportunities.

BACKGROUND

While research on mobile learning, that is the use of small, handheld devices for learning anywhere and anytime is emergent in nature, given the recent convergence of innovations in mobile technology and social software, a generally positive impact on student learning has been reported.

Stead (2006) found that the spread of handheld portable devices has meant that schools can embrace a 1:1 policy, that is, one student to one computer or handheld device. However, this has highlighted two distinct aspects of mobile learning– 'safe learning' and 'disruptive learning'. "The first extends what we are already doing into new places, and the second helps us think differently about learning: learning in a more personalised way, handing over more control to the learners themselves" (Stead, 2006, p.11).

American research (Willard R-12 Schools, 2004) found that the vast majority of teachers believed the devices can have a positive impact on students' learning, were easier to integrate into classroom activities than desktop computers, support an increase in homework completion, and are an effective instructional tool.

Educators have long accepted that students can learn from viewing and interacting with visual media on screen, such as video and television (Edgar, 2008, p.3). This of course includes interactive digital learning objects linked to the curriculum, for example, the simulations and animations developed for Australian schools by

The Le@rning Federation (2001–current). Multimedia has the power to engage students visually, cognitively, emotionally and socially if the content is of high-quality and is teamed with sound pedagogical application. Studies on 1:1 devices indicate that a device in the hands of each learner increases student engagement with learning and 'promotes a shift in teaching practice to more collaborative, small-group work that is student centred and problem based, demanding more higher-order thinking' (Metiri Group, 2007, p.13).

As with any emerging technology in education, there are detractors and supporters of mobile learning devices. Carvin (2007) talks about K-12 schools banning students from carrying MP3 players, fearing they may use them for cheating, while Universities are embracing them, giving each new student an iPod for learning. This highlights the need for policies for mobile devices, in just the same way that schools have policies and guidelines for monitoring the use of computers.

The Horizon Report for Australia and New Zealand (Horizon.AU) placed mobiles on the far horizon in 2008 because of slow adoption rate and low availability of bandwidth. However, the adoption rate is now considered to be increasing more quickly, with reduced costs for bandwidth. (Horizon.AU Report Shortlist, 2009, p.2)

Mobiles can be an effective way to take work home, and 'the use of iPods in the classroom holds and heightens student interest, engages students in learning, and provides yet another means for expressive and receptive literacy. Thus, professors reported that students seemed more engaged in classes where they could use the iPods' (Dogeby, 2007, p.2)

Jobbings' (2005:3) framework for podcast activity aligns closely with the Victorian Essential Learning Standards (curriculum framework for Preparatory to Year 10 school levels in Victoria) for Information Communication Technology (ICT): ICT for visualising thinking, for creating and for communicating, as students plan and design their video podcast, draft and review their audio scripts, record and edit their video material, publish and review their video podcasts. The process also develops literacy and numeracy as they timeline and synchronise their video podcast elements.

Becta (2007) concludes that "the use of ICT seems to engage pupils, resulting in more on-task behaviour, greater persistence and deeper understanding…and it appears to support and encourage greater collaborative activity, inquiry or problem-based learning and independent study" (Becta, 2007, p.29).

The research in this chapter in the Australian context bears out these mainly positive understandings and benefits, and furthermore attempts to enrich the literature by considering the following questions:

How can we enable students at school to use the technologies with which they are most comfortable?

How do teachers learn about new technologies and extend their repertoire when professional learning opportunities are limited?

What are the enablers for successful school-based innovative mobile technology projects that bring about fundamental change?

The four research projects described in the chapter investigated student use of mobile technologies both at home and at school and attempted to examine the changes in teacher pedagogy. Student and teacher perspectives on the future of handheld, mobile technologies were revealed.

Methodology

The projects used a number of research methodologies, both quantitative and qualitative, including action research, case studies and other aspects of research activities such as evaluations and interpretive research. Data was collected through interviews, online questionnaires, attitude surveys, teacher reflective journals, blogs and video podcasts. Students and teachers in all four projects were surveyed online before the projects commenced to establish quantitative baseline data on their knowledge, skills, attitudes and use of emerging technologies. Exit surveys were conducted at the close of the projects to provide a basis for analysing their responses to emerging technologies for learning. All teachers and students reflected on their feelings, observations and attitudes at various times and were interviewed as a final reflective piece for each project.

Action research formed the basis of teacher reflection based on their work, generally in teams. It gave the teachers new opportunities to reflect on and assess their teaching practices; to explore and test new ideas, methods and materials; to assess how effective the new approaches were and to share feedback with fellow team members. Their reflections took the form of online reflective posts, blogs, video podcasts and interviews. Research on student learning was based on student engagement and learning, and relevant areas such as the development of literacy and numeracy skills, participation in class and with homework, and willingness to come to school.

Video interviews were conducted before and after the research period in the case of the Nintendo DS project, along with pre and post Mathematics performance testing for the research classes, compared with the control group using paper and pens only to complete the program.

MAIN FOCUS OF THE CHAPTER

Successes and Challenges

While each of the four projects had a different focus, generic objectives were around improvements in student learning, changes in teacher pedagogy, planning for and implementing innovative curriculum and classroom practices, and managing the technical problems. Issues highlighted during the implementation were teacher confidence and training, wireless access, device synchronisation, copyright and intellectual property.

The following sections briefly describe each project: its focus, context, technology, participants and location and outlines what happened in each case and what was found.

iPodagogy

Background

The iPodagogy research project used the video iPod Classic and video podcasting to investigate the potential of the video iPod to a) engage and stimulate students in a cross curriculum approach, embedded within the Victorian Essential Learning Standards (VELS), and to b) improve teachers' use of ICT in the classroom through the concept of iPodagogy, that is, innovative ways of teaching using iPods and video podcasting. It teamed the iPod with an online learning environment, Studywiz, for all students in a Year 8 class at Heathmont College in Victoria, Australia.

It was essentially a learner-centred project, where the views and ideas of students about how emerging technologies can enhance engagement were investigated and tested in a practical environment. Teachers were also encouraged to look creatively at their teaching practice and to work on strategies that could be articulated for others on using emerging technologies.

Twenty-five Year 8 students (13-14 years old), selected as representative of typical students across the state of Victoria, each with a video iPod Classic, were given six Apple MacBook laptops to share with six of their teachers, to integrate them across the core curriculum areas. The specific objectives of the research project were to investigate the potential of the resources and approach to support learning and teaching, focusing on:

- the use of ICT outside the classroom by young people and the impact of these activities on their learning
- the capacity of iPods that synchronise audiovisuals with an integrated online

learning environment to enhance student engagement
- the capacity of iPodagogy (using iPods and video podcasting for teaching and learning) to embed innovative teaching practice.

The six participating teachers of English, Studies of Society and the Environment (SOSE), Mathematics, Science, Music, German, Health and Physical Education, were not expected to be highly skilled in using ICT across core curriculum areas, simply to be motivated and enthusiastic about making a difference. In terms of experience, the teachers represented a mix of longevity of service, senior responsibility, new recruits and industry experience.

The online learning environment, Studywiz provided teachers and students with an intuitive eLearning space that could be synchronised with iPods and accessed from home. The learning content management system, Equella was integrated with Studywiz to provide teachers with a repository to manage the digital content used in this project.

Heathmont College was selected as a typical school with a cohort of typical students. Then, it was not a well-funded high-tech school. Three corporate partners were brought on board in a unique association, in order to provide the necessary infrastructure for the school. Apple Australia provided hardware, Etech Group provided the online learning environment and The Learning Edge International provided a learning content management system. Technical support and skill-based training for teachers was also provided by the companies. Research on an agreed framework was conducted by Delphian eLearning for the Department of Education and Training, Victoria.

What Happened

In the teacher pre-entry survey, teachers reported that they were 'expert' or 'confident' in functions such as word processing, spreadsheets, presentation software and 'reasonably confident' with multimedia, email and Internet research. Most indicated that they had little experience with Web 2.0 applications or handheld mobile technologies.

The students, all 14 years old, indicated that they developed their ICT skills by themselves, with some support of family and friends, but less support from specialist or general classes at school. Predominantly, the greatest use of technology by students at home on a daily basis was using a mobile phone, MP3 player, desktop computer, X-Box or Playstation Portable. Most of the students in Year 8 had confidence in their abilities with technologies and considered themselves expert in many online areas. Wherever they had access, either from school or home, the students' main interests were in listening to music or keeping in touch with friends. Most of the students were confident technology users, with many considering themselves expert in using online services.

"Education professionals recognise that the students generally know more than their teachers about 'doing' digital. So even if they wanted to… they cannot be the classroom authority figures when their digital literacy is not on a par with their students, and when their information resources do not compare with those of the internet" (Spender, 2007, p.1). This is essentially the landscape that confronted the six teachers in the iPodagogy project. In this research project, pedagogy was at the heart of the teacher conversations during the planning and early implementation stage when the technologies were being installed and introduced to students and teachers.

In English, the teacher was really enthusiastic that "for once I had access to iPods and Macs – humanities misses out in ICT" (English teacher, 2007). The students were engaged in producing a reflective family piece that entailed drafting and proofing a script, videoing and reviewing the output. The students' reflections involved making a podcast for an audience to view, in which they revealed their family life to others. As they observed each other and saw what some families' lives were like, many were brought to tears and "...realised that life is about others, not just yourself" (ibid). The technologies enabled all the students to appreciate each others' lives and to reflect with constructive support for each other. In the teacher's words, "I'd do it again... loved it, challenged me a lot but really loved it" (ibid). Being able to take home their personal video podcasts on their iPods and share with family members gave them an added incentive that pen and paper did not provide.

Students can derive great benefit from using video podcasts to explain mathematical operations. A series of video podcasts were made by the Mathematics teacher on Directed Number and then uploaded to students' iPods for viewing. Students reported that they used them for revision at home and before tests, resulting in improved test scores for a number of students. One student commented that the podcast breaks down the mathematical operation into small segments, making it easier to understand what is happening. The teacher was able to develop individual learning plans for each student and used the video podcasts to supplement self-paced learning of Maths concepts. What made the difference for the students was being able to view the podcasts on mobiles anywhere and anytime for Maths revision, even when their peers thought they were listening to music.

Figure 1. Directed number video podcasts developed by the Maths teacher

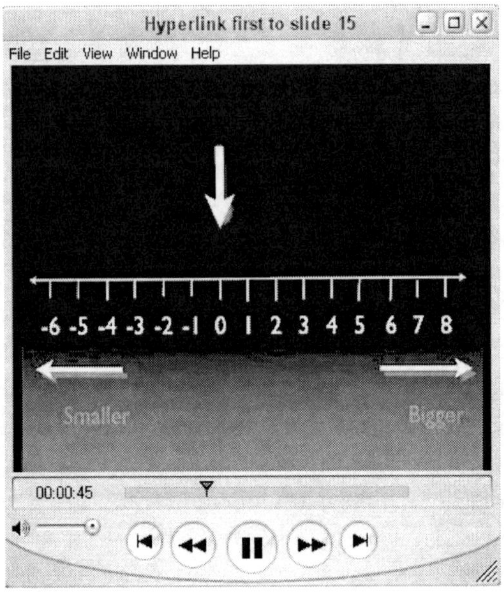

In Science the teacher customised her pedagogy to include group work to enable students to record their practical experimental reports as video podcasts. iPods stimulated creativity and engaged students in sharing information and reflecting on their roles as group members and on their interaction with the scientific experimental process. Students hypothesized the effect of adding salt to water on its boiling point and recorded temperatures at specified time intervals to compare results. They filmed and took still photos during the process and prepared video podcasts which were uploaded to Studywiz for others to see. Students were able to assess their group's performance and to self-assess. In the words of the Science teacher, "I became more flexible in my teaching practice and reflected on it more. It was motivating" (Science teacher, 2007).

Health and Physical Education may be areas not associated as easily fitting with technology. Nevertheless, the Physical Education teacher developed very professional video podcasts of cricket techniques including batting and bowling. The students were asked to learn visually from the video podcast, to take photos of their own technique and to upload their photos to their iPod for comparison. "In PE it's hard to include ICT – we need something that's portable and contains content" (Physical Education teacher, 2007).

Studies of Society and the Environment had ready applications for student video podcasts through interpretations of events, in this case the weather. Students developed video podcasts using software to create climographs and water cycles, interpret maps, present research and class discussions, while embedding teamwork as a behavioural strategy. According to the teacher, "The boys were engaged and empowered by the technologies, particularly boys who were challenged. The girls who were sound in written work and books found it quite confronting. Working in groups, sharing the equipment equitably was a challenge…the students found it empowering and helped me which in turn helped our relationship as a whole class. The power dynamics have changed" (SOSE teacher, 2007). Engagement with the technology enthused them to work together, thereby increasing their ability to learn content in greater depth.

In Languages Other Than English (LOTE), downloading podcasts and video podcasts to an iPod had immediate benefits, as student could listen over and over again to the German vocabulary or conversations and learn as and where they wanted. Similarly, they developed their own presentations on their families in German and as the teacher commented, there is "huge potential in LOTE" based on the demonstrated benefits to students (LOTE teacher, 2007).

Music and iPods have a strong connection. The Music teacher commented that "students don't need a knowledge base; they can make music by clicking and dragging." Using Acid Music students were able to create and synchronise loops in time and key, take their work away on the iPod to reflect and project changes, then rework them at school with teacher assistance. The students rendered their files to MP3 and uploaded to the iPod which motivated them to reflect and make refinements at home. "Using iPods made my life easier as a teacher, storing a whole library, including all the pieces of [exam] repertoire including sheet music as well" (Music teacher, 2007). Being able to store a comprehensive aural library on a small device gave students and teachers examples of style and technique for aural comprehension previously unavailable in schools.

Findings

At interview, students were asked to reflect on whether they had discovered anything new about themselves and the way they learned using the emerging technologies. Roughly 50 per cent reported that they learned better by doing things and liked to be active learners "I like to learn actively, don't like to sit there writing…it made me plan and think more about my work" (Student A, Year 8, 2007). In a traditional class environment where students generally sat in rows and were taught content, the shift to teamwork for using technologies in an active, more practical learning environment was stimulating for student learning.

Some students learned that they could perform well as part of a group and these comments were particularly interesting from those who are on the social periphery: "[I] learned that I can take control of the group a bit to get the job done as soon as possible" (Student B, Year 8, 2007). Student reflections were also quite revealing about the way they learned: "Reflecting on video made me think more"; "we matured a bit and learned to work together" (Student C, Year 8, 2007). Some students who coped well with print literacy and the traditional curriculum, struggled with technologies: "[Using technologies] the learning was more complicated, normal work is easier" (Student D, Year 8, 2007).

Student perceptions were that they were not performing as well as other Year 8 classes, based on uniform progression in curriculum; however, their teachers claimed that they were ahead of all the other Year 8s in all curriculum areas, simply having covered the work in alternate ways, a perception that was not always clear to the students. The teachers reported that this Year 8 class had progressed further than the other Year 8s in all curriculum areas and also in behaviour, motivation and responsibility by the end of the project. As a result of students viewing a series of Maths video podcasts developed by the teacher, and used as revision for understanding operations, test scores improved for a number of the students: "the podcast breaks down the mathematical operation into small segments, making it easier to understand what was happening" (Maths teacher, 2007)

Teachers also discussed the impact of the technologies on the increased level of higher-order thinking required by the students, which added to their perception that the work was harder. They were asked to comment on whether an iPod for each student promotes greater independence and engagement in learning and they reported that work output increased and improved in quality; students were "more independent and focused"; "engagement increased for boys particularly, they would do the work and listen to music"; for one teacher "none of them ever came without their iPod – responsibility is an expectation". (iPodagogy teachers, 2007).What the teachers discussed was the degree to which the technologies had prompted improved teamwork, greater focus and engagement, resulting in improved learning.

Teachers reported a significant change in their pedagogy, with a sharp movement to a more consultative style of facilitation, based around the introduction of teamwork. What was not clear was the impact of new and emerging technologies on higher-order thinking and the resulting stress this placed on students, particularly those used to a traditional model of schooling. In this project it was clear that both teachers and students felt that there was pressure on the students to work in a different way, to learn new skills quickly and to engage in deep thinking about what they were doing. In the words of one teacher, using the iPods and video podcasting "...made the kids think more and they are not ready for that – they've never done it before and we're reinventing the wheel for them" (English teacher, 2007). On the other hand, in terms of outcomes, it was also revealed that "the kids engaged with the podcasts and still remembered everything after four weeks which doesn't happen normally" (Maths teacher, 2007). Most of the students themselves admitted that the podcasts had improved their performance and test scores. "I used the teacher podcasts – did good on the Maths test – if I didn't understand I went over and over it and then understood" (Student E, Year 8, 2007).

Implications for Mobile Learning

What we have seen in this research project is that personal, portable, handheld devices that provide flexibility, accessibility and frequency of use, despite their small screen size and limited battery life, can increase motivation, organisational skills, active learning and self-directed learning. They provide 1:1 access in schools leading to more engaged students and their portability extends learning into the home. If students are using mobile technologies in their lives outside school to communicate, socialise and learn, schools should have an opportunity to leverage mobile learning to engage students in the Middle Years in authentic, deeper learning. With the rapid development, ownership and cost of mobile technologies, particularly handheld devices, schools will increase the availability of mobile technologies, and the need for greater professional learning for teachers.

The iPod Touch in Three Primary Schools

Background

The iPod Touch Project enabled teachers in three primary schools to use eight iPod Touch devices, with classes of up to 30 students. This project investigated students' attitudes toward emerging handheld technologies, and examined the use of the iPod Touch with emphasis on the impact on student learning, teacher pedagogy, curriculum and assessment and on external technical issues involved in implementing emerging technologies. Students shared their perceptions of mobile technologies for learning and provided a snapshot of what they expected from their schooling.

Three schools with different cohorts of students were involved: a small, regional school; a large school in a significant population growth corridor; and a school with a significant English as a Second Language (ESL) cohort. All three schools had senior management who were committed to ICT as an enabler of learning.

The project investigated the potential of the iPod Touch to engage and stimulate students and teachers in a cross curriculum approach embedded within the VELS, and to improve teacher practice through the use of mobile devices to stimulate teacher reflection and transform their teaching.

The specific objectives of the research project were:

- to gain an insight into the pedagogy, technical issues and professional learning needs of teachers as they integrate a new technology into their classrooms
- to develop teaching, learning and assessment resources which can be shared with the broader educational community and which provide leverage for deeper understanding and investigation.

The overarching research question was: How can the use of handheld technology (iPod Touch) contribute to changes in pedagogy used in teaching, learning and assessment?

Each of the three teachers developed their own action research approach to the use of the iPod Touch devices and developed their own essential research questions, based around literacy. They developed a rationale and approach consistent with their individual student cohort needs and backgrounds.

What Happened

During the first term with the iPod Touch devices, the teachers familiarised themselves with the functionality of the iPod as a personal device, introducing them slowly to the students, and gathering ideas from the students as to how they might be used for learning.

Corio South Primary School is in the outer suburbs of the Geelong area located near a large industrial area south-west of Melbourne, and caters for a large cohort of ESL and lower socio-economic students. A significant proportion of the students in the Year 6 classes were recent arrivals in Australia, non-English speaking students performing below their capability. The school also had significant numbers of disengaged students, some with extreme behaviour and chronic absenteeism. Corio South Primary School focused on the 'need to engage all students in Grade Six in purposeful and powerful learning' (Delphian eLearning, 2008, p.6) around the essential question to guide inquiry: How can the iPod Touch improve our student outcomes in Literacy?

Using the iPod Touch, students engaged in a constructivist approach to learning and worked on an open-ended project that collaboratively explored the task through a range of approaches. Whilst the objective was to improve student literacy using the iPod Touch, the class had a focus on healthy eating, nutrition and fitness, in preparation for the school sports being held in Term Three. Students used the devices and suggested ways it could be applied in their learning.

This approach became the hallmark of the implementation at Corio South Primary where the students used the devices as one tool among a whole array of ICT – interactive whiteboards, desktop computers, digital still and movie cameras, Nintendo Wii and so on. Students were engaged in assessing and developing learning uses and applications for the devices from the very outset, including researching the impact of mobile technologies, copyright, privacy and ethical use.

The teachers trialed the iPod Touch in a wide variety of classroom activities, focusing on improving students' reading, writing, speaking and listening. After careful exploration and discovery of its functions and possibilities, the devices were carefully integrated as tools in students' everyday learning. The teachers developed their term planner and purposefully integrated the iPod Touch where appropriate, as a regular tool for learning, with the result that the iPod Touch functions had a much greater role and impact on the final term's curriculum objectives. The many refugee children in the classes were eager to use the iPod Touch, and lessons were carefully planned to help with their understanding of English using the iPod Touch, to test its effectiveness in the classroom setting. The school Annual Implementation Plan identifies ICT as an enabler of learning and this is exemplified in the teacher's curriculum planner where ICT and mobile learning are embedded (Delphian eLearning, 2008, p.63).

Online books, iTunes, apps (applications for use with Apple's iPod Touch or iPhone downloaded through iTunes from Apple's App store), podcasts and video podcasts were used to enhance reading. Map reading, search and direction activities on the iPod Touch also provided successful reading activities. Personal learning blogs were developed and students were encouraged to read and comment on each others' blogs, and to visit other children's blogs around the world. Students also published their work using Keynote which was uploaded to the iPod Touch for others to read and share. Students researched the Nintendo Wii to assess its benefit for literacy programs. The Wii has the ability to keep records of activity and this information was used collaboratively with literacy work. The Wii Sports, Mario and Sonic at the Olympics and Wii Fit, stimulated discussion, reading and writing by providing authentic contexts for collaboration between students. While non-English speaking students were catching hoops on their arms using the Wii Fit, at the same time learning directions left and right, some students were supporting them in their language acquisition, while others were noting down the points gained in the game and analysing the class performance, average and mean scores (Delphian eLearning, 2008, p.10).

Students also used the following iTunes apps to assist with their writing:

- Check Word for editing their written work
- Blanks for extending vocabulary of high achievers

Figure 2. Using the Nintendo Wii to support Literacy programs

- Brain Lantern for discussion, language and writing
- iPhrase for ESL students and those who have difficulty with sentence construction
- Jumble Solver for solving anagrams
- Word Warp as an anagram game with a time limit
- Spell Number for improving spelling of numbers
- Dictionaire for word comprehension.

Critical, creative discussion and reflections were enabled by the use of illustrations, movies and animations on the iPod Touch devices. Student engagement was demonstrated when they interacted with each other in a lively manner when being taught the basics of essential applications for the iPod Touch such as Kahootz, Comic Life, Keynote, Rubrics, Garage Band, iTunes, iPhoto and I Can Animate. Issues such as language, self-esteem and working together, formerly areas of difficulty in this diverse class, became catalysts for learning. Podcasts for the iPod Touch such as Harry Potter and Hannah Montana created animated discussions on copyright, where students' literacy skills were assessed, such as identifying opinions of others, proposing relevant viewpoints and extending their ideas in a constructive manner.

Internal data from Corio South indicated that significant improvements were made in the targeted areas of reading, writing, speaking and listening. Whether this improvement can be attributed to mobile learning, however, is not conclusive. As the teacher commented positively, "My classroom observations would suggest that the iPod Touch kept the children focused on their work tasks without distractions or disturbances" that would usually occur (Corio South teacher, 2008).

Courtney Gardens Primary School, set within a population growth corridor south-east of Melbourne, is a leading ICT school, having been involved in a range of innovative ICT projects, particularly in the area of student-developed television broadcasts. Courtney Gardens Primary School was confident in their students' ability to quickly adapt to using any new technology-driven device and their fearless approach to mastering the iPod Touch. Roughly one third of the classroom teachers immediately volunteered to participate in the project, despite the fact that very few of these teachers had ever held an iPod before the trial. With an ever increasing trend towards information-on-demand, and smaller, one-to-one portable devices, they wanted to investigate whether these devices would have a noticeable impact on student learning, and whether they could be used on a wider scale across the school. Their essential question to guide inquiry was: How can handheld devices improve student outcomes in writing, particularly non-fiction writing?

Courtney Gardens Primary was keen to see if handheld mobile devices (like the iPod Touch) could improve student outcomes in a school-wide focus on writing, particularly in non-fiction writing. Ten teachers from various grade levels volunteered to try out the devices in their classrooms. The plan was to roster the iPod Touch devices throughout the ten classrooms to see how different teachers would use the devices in innovative ways to assist the delivery of instruction in their regular lessons.

The iPod Touch devices required the schools to install an Apple Airport for wireless access which conflicted with the security parameters governed by the Department of Education's EduPass Security regime. With the possibility of the school's secure networks being compromised, the project at Courtney Gardens reached a stalemate for quite some months. By the end of the project, their iPod Touch devices had still not been connected to the network, although a suitable patch and security certificate had become available. Thus, the teachers had to develop creative ways to use the iPods, other than on the Internet. With a fresh approach, they uploaded to the iPod Touch their current podcasts on Spelling Rules, Indonesian language snapshots and student-made movies. These were used for media literacy activities by the ICT teacher as students learned to review other students' work and evaluate film and video content. The students learned to evaluate characters, settings and camera angles, concepts that were easily transferable to their own creations. A range of Mathematical operations were also developed as 'Maths Minutes' in Term 4. These minute-long video podcasts were developed by the students under guidance and screened across the school's television station each morning.

Portable learning devices like the iPod Touch certainly appear to meet an unfulfilled niche in schools. Despite the difficulties with accessing the network, Courtney Gardens Primary School staff indicated that "...the iPod Touch is simple to use, relatively durable and has the ability to access the Internet anytime and anywhere [which] certainly gives them a high potential for widespread use in our school"(Courtney Gardens teacher, 2008). They also indicated that it would be of great use as an assessment device for all teachers, as it would enable teachers to assess immediately in the classroom and to instantaneously upload the data to a central assessment database.

Epsom Primary School, a small primary school within the greater Bendigo area north of Melbourne, is surrounded by rural views. Two people at the school were involved in the project, the Principal and the ICT Teacher. Epsom Primary School was considering the idea that students are digital learners and their intention was to find out 'if the iPod Touch would assist children to learn through digital delivery of content.' (Delphian eLearning, 2008, p.7) Teaching practices in the upper primary school had been traditional and teacher-centred. They were looking for an approach that would move teachers to a student-centred learning environment via the use of the iPod Touch. Their essential question to guide inquiry was:

How can the iPod touch be used to enhance teaching and learning and support a student-centred classroom?

Epsom Primary School focused on the potential of the iPod Touch to enhance teaching and learning, and create change in teaching strategies or pedagogies. The teachers at Epsom used their previous background in mobile devices at the school to develop a range of curriculum activities for students across the year levels. The Principal developed Maths podcasts on division for use in his Year 5/6 class. The Year 6 students used the podcasts on the iPod Touch along with classroom materials such as MAB blocks (Multi-base Arithmetic Blocks) and worksheets to teach the mathematical operations to Year 5 students. Most students said they preferred to learn this way and the teacher also described that their learning was more active and engaged. Students also used materials and number lines to develop understanding of multiplication and division of decimals (to two decimal places) and simple common fractions. At interview, the Principal commented that the students were engaging in deeper learning when they were peer coaching younger students in basic Maths concepts (Delphian eLearning, 2008, p.19).

Figure 3. Peer tutoring Maths with podcasts on the iPod Touch

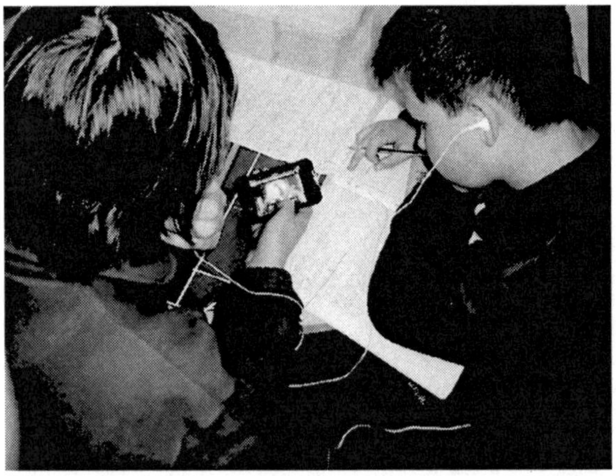

Preparation for a zoo excursion meant that students researched endangered animals on the iPod Touch and produced podcasts of their research findings. Younger students were able to take the iPod Touch with them on the excursion. Other applications on the iPod Touch such as the Weather was used to download information which was graphed for comparison and interpretation. The Maps icon also gave access to a range of local and national maps, which the students used in geographical, spatial and mathematical activities.

The students used Etch a Sketch app to write on the screen and practise letter formation and words. Using this app, a shake of the iPod Touch means it is ready to write on again. Engagement was very high for this activity that enables English Level 1 students to learn to form letters correctly and to use a range of writing implements. Apps that have become student favourites include Wurdle, Comic Touch, Etch a Sketch and Writing Pad.

Students used Google Maps and Google Earth on the iPod Touch for Geography to develop the ability to identify and collect evidence from primary and secondary sources and ICT based resources. They then recorded, represented and interpreted data in different types of maps, graphs, tables, sketches, diagrams and photographs.

In the Geospatial skills dimension, students interpreted maps of different kinds and at different scales, including street directories, atlas maps, ordnance survey maps and topographic maps. Students identified and collected information from maps, plans, photographs, satellite images, statistical data, and ICT resources; and recorded and represented data in different types of maps, graphs, tables, sketches, diagrams and photographs.

The teachers identified three conditions for the iPod Touch or any other ICT tool to be integrated effectively into the classroom:

- teacher familiarity with the device and its capabilities
- enthusiasm and a positive approach to innovation, and
- time to rethink their teaching approach.

Findings

The introduction of the iPod Touch devices into primary school classrooms has been received by students as exciting, motivating and, in their word, 'awesome'. From the teachers' perspective, it has required a significant amount of time, curriculum innovation and pedagogical change. It has further required a great deal of technical expertise to solve the connectivity issues. Teachers tended to use the eight iPod Touch devices in their classrooms of thirty students as one part of their educational technologies repertoire.

What has become clear is that mobile devices like the iPod Touch have real learning value in teaching and learning contexts, and that it is the imagination of teachers to integrate such devices in innovative ways that is the most valuable resource. In each of the primary schools, the iPod Touch has been integrated into authentic classroom learning activities in meaningful ways. They were used in tandem with Interactive Whiteboards, laptop and desktop computers and Nintendo Wii to develop student skills and understandings across all curriculum areas. While they can stand alone as useful wireless devices for learning, most schools employ a range of technologies in the classroom to cater for different learning styles.

Of particular note is the confidence and independence that most children displayed when using the devices for learning activities, regardless of year level and age. ESL new arrival students, who had no English language skills, were asked to develop a podcast about the basic hygiene of hand washing. The podcast was developed in their language and they added English sub-titles as an English literacy activity. They took quickly to the new technologies and produced quality podcasts, mentored by other students in the class. The podcasts were then taken home to share with parents and community members. Similarly, Year 6 students developed an elaborate video podcast on how to use the iPod Touch for all students in their school, while Year 3 students were filmed giving sophisticated advice on the uses of mobile devices for learning. Students developed their own web pages and discussed their thoughts in blogs on the iPod Touch.

The VELS ICT dimensions suggest students use ICT for Creating, Communicating and Visualising Thinking. It was evident in classrooms that students were using the iPod Touch as one ICT tool for those purposes but also for research, identification, observation, description, analysis, explanation, synthesis and evaluation. Students used a range of apps which provided an array of educational tools that teachers and students found engaging and beneficial for literacy and numeracy, particularly for 'at risk' students, those who are disengaged from schooling, as well as ESL and refugee students.

Interview and in-school survey data suggests that there is evidence of critical evaluation of the devices for learning by the students. When asked about the potential of the devices for use by other students and schools, the feedback was remarkably similar across the schools. Having become accustomed to the size, portability and functionality of the iPod Touch, students generally indicated a preference for the iPod Touch for more 'personal' tasks such as email and viewing podcasts, but preferred the computer for tasks where you needed to use more fingers, such as writing. They did suggest that this was because the iPod Touch only has a small note-taking application, and it was a clear recommendation that if the iPod Touch had word processing and spreadsheet functions, then that would suffice as a key device for primary students. For some students one finger typing was slow and the general recommendation was that a stylus or pen would be an improvement. They commented favourably on the portability of the devices and the fact that "you can carry them around in your pocket", and they enjoyed the visual nature of the functionality, the fact that "you can look and learn" (Corio South students, 2008).

Student-student mentoring while using the iPod Touch was widespread. Students were observed working together, teaching each other concepts from video podcasts on the iPod Touch while using tangible learning objects in the classroom such as blocks, worksheets, computers, Nintendo Wiis, data projectors, video editing suite, and other ICT devices. A limited number of devices appeared to promote cooperative learning and has stimulated a change in teacher pedagogy.

Students took to the apps with enthusiasm. "They liked the mobile learning concept and will do anything to participate in the activities, even doing homework and writing for the first time, when other strategies have not worked for them" (Corio South teacher, 2008). "In the life of the project, the launch of the iPhone has prompted a huge array of apps to be developed…and it is these applications that have opened up the use of the iPod Touch as a tool for learning, particularly for ESL, students at risk and those with difficult behavior or learning disabilities' (Delphian eLearning, 2008, p.27).

In a couple of the schools, students were able to take the devices home to show their parents and families examples of their work. One school indicated that parents were accustomed to technologies coming home with students, and that the students were able to show their parents current approaches to Maths problems and digital books for literacy development. Essentially the school uses the devices to help educate parents about contemporary schooling. Refugee parent-teacher interview sessions were conducted with the student as facilitator, using the iPod Touch to demonstrate their learning to their parents.

All of the teachers involved in the project have indicated that learning with the iPod Touch was beneficial, as it increased student motivation and encouraged collaborative work. It supported ICT skill development and interpersonal learning through engaging and persuasive discussions amongst the students.

"As a result of this engagement with purposeful learning, student self esteem has increased. From using iPod Touch and the Nintendo Wii, students have improved their memory, they think logically, use problem solving strategies, have acquired mathematical skills, become more resilient with learning challenges. Each student has developed skills and strategies that promote learning, and has developed a positive attitude to learning within and beyond the classroom" (Corio South teacher, 2008).

In conversations with the teachers, some commented that using the devices prompted them to re-think their practices, to find new ways of using the iPod Touches. At Epsom, one teacher described his approach to student mentoring where children learned to teach each other, utilising the visual, tactile nature of the iPod Touch, which appealed to a wider range of learning styles. "I'm now thinking of different ways to make it easier to learn…engage the kids and understand what it's like for them" (Epsom teacher, 2007).

In terms of the functionality of the iPod Touch for curriculum purposes, teachers and students were asked to consider what works and what doesn't. Most cited the video clarity as great, as it has been used for activities such as media literacy, accessing quality video from the Australian Broadcasting Commission (ABC), research and data collection and for classroom topics such as paper recycling, endangered animals, healthy eating, fitness, literacy and numeracy. Teachers commented that it was engaging for students, motivating and prompted a desire to learn, especially for reluctant learners or those with behavioural difficulties or ESL background. It was felt to be particularly useful for Literacy and Numeracy work and its potential was viewed as having increased significantly with the launch of the iPhone and the array of apps that have been developed.

Student feedback was honest and knowledgeable. Generally, students loved the device with its portability, clear, visual interface and immediacy of content delivery. They reported enjoying the responsive touch screen and the quality of the screen real estate and sound. They criticised the lack of camera and audio/voice recorder, which are present in the updated iPod Touch and the iPhone. In terms of usability, most student feedback indicated that all year levels could use it with ease, and the majority preferred watching movies on the iPod Touch as opposed to a computer. The keyboard size posed some difficulties in the early stages, but after considerable use, students reported that it was not an issue.

The project website featured blogs where teachers interacted with each other to share what worked and what didn't; uploaded video podcasts and uses of the devices in various curriculum activities; solutions to technical issues and so on. Generally the teachers acknowledged that they were able to embed the iPod Touch in their pedagogy with sufficient time to learn about the functionality and to plan creatively. They also indicated that the device caters for multiple intelligences and preferred learning styles, so embedding the iPod Touch in an inquiry learning model would facilitate change to pedagogy.

The iPod Touch is essentially a personal or individual device and as such there have been issues that teachers have had to manage carefully in this project. Sharing the devices has compromised their ability to be synchronised with the correct content, but has, however, provided an opportunity for teachers to educate students on the safe and appropriate use of mobile devices and the Internet in general. Student agreements on Acceptable Use were in operation in most schools, and the intellectual property issue with student created material, was raised. Copyright was a significant issue in terms of music and books. Instead of breaching copyright, students created their own music in Garage Band and uploaded it on to the iPod Touch. Generally, in all schools, the iPod Touch prompted teachers to discuss issues of copyright and intellectual property with students, and they clearly saw this as beneficial and relevant to class work with web pages, blogs, instant messaging and email.

Technical issues required significant time from the teachers and from technicians, and generally speaking, impacted on the implementation of the project. Teachers reported that the iPod functionality within the classroom and publishing the children's work on the iPod, were tricky issues to solve, let alone the worry of wireless access and school network security issues.

The synchronisation of the iPod Touch has been an issue in all schools, with issues such as lost content and lack of ability to synchronise on some of the devices, cited. Use of personal mobile devices in a shared context also presents issues with synchronisation to the one computer. Minimal battery life for classroom use has prompted the teachers to develop novel ways to charge all the devices at once, such as developing a 'charging toolbox'.

In terms of curriculum use, all schools indicated that the biggest disadvantage of the iPod Touch was that it did not read Flash files, so the digital learning objects from national initiatives such as The Le@rning Federation cannot be accessed on the device. Many of the websites schools had hoped to integrate into their planning were not accessible through the Safari browser on the iPod Touch. Many of the sites were reliant on Flash Plug-ins that were not supported by iPod Safari. Since a lot of educational digital content is developed in Flash, it is a major drawback with the device.

Implications for Mobile Learning

In general, the iPod Touch devices have been integrated in imaginative ways to suit the learning areas, topics and concepts being taught. The devices have also had a viral effect on other teachers in the schools, who became involved in using the devices with their classes.

Mobile learning devices have stimulated both teachers and students to work creatively together to improve literacy and numeracy. Student performance data revealed that students had improved in both numeracy and literacy, along with increased skills in teamwork, interpersonal skills and technology skills. The use of the iPod Touch has encouraged student interaction in blogs, podcasts and web pages, and it has been particularly important in encouraging reluctant and ESL learners. All three schools have plans to purchase handheld devices to extend their integration as learning tools for today's students.

Key enablers of effective implementation of emerging technologies include wireless connectivity, technical support and collegiality. With these in place, teachers can focus on their core task of using the technologies to improve learning outcomes. This project has made discoveries that are constructive and can be applied to other 1:1 initiatives.

Global Mobile Learning

Background

The global mobile learning project investigated the potential of mobile handheld devices, such as the iPod Touch, to support learning and teaching in the context of a collaborative project where students interacted with cultural and literacy activities via a common online learning environment, Studywiz.

Three schools from Australia, Singapore and the US participated in the project over a two year period, collaborating and interacting via the Studywiz Global Community and building on the lessons learned across the schools. Students varied in age (8-18) and in grade (primary, middle and senior school) and provided data across key age groups for comparison and analysis.

It was a learner-centred project developed in two phases, where the views and ideas of students about how mobile technologies can enhance learning were investigated and tested in a practical context. Teachers were also encouraged to look creatively at their teaching practice and to work on strategies that may be articulated for others around using mobile technologies for learning.

An Action Research approach was employed to assist teachers to develop the research project in the context of their individual classrooms and schools. Research questions were developed based on three elements – learning, teaching and consumer behaviour. The project had two phases, the initial one being an introductory collaborative project between two schools, followed by a second, deeper collaboration based on curriculum activities.

Phase One

The United World College of South East Asia in Singapore and Shepparton High School in Victoria, Australia collaborated in Phase One to share their understandings about global citizenship and cultural identity.

United World College of South-East Asia (UWCSEA) is a Kindergarten to Grade 12 international school for approximately 3000 expatriate students in Singapore. The students are in Year 11 (aged 16-17) and are studying for their first year of the International Baccalaureate Diploma. The school has plans to expand by 2012 to 5000 students in world-class facilities where ICT is an everyday tool for learning.

Shepparton High School is a Government 7-12 Secondary College in the major commercial centre of the rich dairying, fruit growing and agricultural region of the Goulburn Valley, 200 km North-East of Melbourne, Australia. It was established in 1909 as the Shepparton Agricultural High School and has grown over the past 100 years to a school of about 900 students from Years 7 to 12.The Year 8 students (14 years old) include Afghani, Iraqi, Polish and Koorie students. The school has new learning spaces configured to enhance student learning with high levels of ICT. There is a history of strong student- teacher relationships and personalised learning programs for each student.

The essential question researched by the two schools was: How can a mobile device (such as the iPod Touch) improve our students' understanding of themselves in a global context?

What Happened

The two schools in Phase One participated in the introductory collaborative project. In the early phase of implementation, the teachers spent a great deal of personal time getting to know the iPod Touch, using it for their own purposes, developing learning activities for students, and most importantly, finding out how to integrate it with their school IT network.

The classes involved in the collaboration were Year 8 students from Shepparton, who were involved in a personalised learning curriculum, and Year 11 students from UWCSEA studying the Theory of Knowledge, which explores the ways in which we know and deal with knowledge (reason, perception, emotion and language). The two teachers developed scaffolded curriculum questions to suit their own context and curriculum needs, and embedded thinking questions to extend students' responses that displayed deeper and more meaningful interaction.

Shepparton High School commenced the project with vigour and enthusiasm, lead by an innovative teacher who was keen to have her Year 8 students integrate the use of the iPod Touch in their personalised learning program. The underlying concept is based upon student self management and choice, along with an analysis of student learning styles, study skills and emotional intelligence. Year 8 students are engaged in theme-based units, each covering skills and concepts consistent with the Victorian Essential Learning Standards. The teacher identified and created a wide range of student activities using the iPod Touch across curriculum areas including research, discussions, photo and video activities, blogs, events, quizzes, games and a QR Code activity that integrated with their personalised learning program. These activities were designed to engage and stretch each student. The teacher utilised the whole range of tools inside the Global Studywiz to develop their skills and to spark their imaginations.

"One of the elements of our 'personalising learning' teaching program is to provide a choice of activities for students which suit their learning styles. In practice in the classroom, we provide a folder of activities, with each learning style assigned a certain colour. Students access this folder to select their task then prepare a learning contract and task plan…[as] each member of my class has an iPod Touch with the Studywiz mobile interface, these learning styles are easily set up for them in a series of galleries, to be viewed on their iPods whenever required" (Shepparton teacher, 2008). Students used a variety of iTunes apps that they downloaded to the iPod Touch and found that Air Share and Scribble were particularly useful.

As part of her Action Research, the teacher developed a series of reflective podcasts that were uploaded to the Global Studywiz. As time passed, she was able to log developments in student learning, and record issues and challenges with using the devices in a classroom context. Her podcasts were initially audio podcasts, which progressed to video podcasts of high quality that demonstrated what students were learning. The teacher noted in her reflections that the devices encouraged full student attendance, and that other teachers in the school were keen to use them in Music and in English.

At UWCSEA the students initially used the iPod touch for online research in Science and History, while they familiarised themselves with the Touch interface instead of the traditional physical keyboard and mouse, an experience described by one student as 'quite scintillating' and 'an eye opening experience: to be online anytime, anywhere'. It was intended that students would study art and use their devices to capture different forms of art and create multimedia galleries that would be podcast. The iPod Touches were also used to share calendars and to access the online storage space (Studywiz eLockers) where students kept their files.

Even in the early stages of the research, UWCSEA collected a substantial amount of student feedback, probably best summarised by one student comment that the iPod Touch proved to be "[a] lightweight, portable way to access the internet, extremely useful for stressed out, busy IB students who have too much to carry around anyway" (UWCSEA student, 2008).

The curriculum activity based on cultural identity and difference was successfully implemented through the discussion facility in Studywiz. Students were required to outline their views on living in Australia or Singapore and to discuss the similarities and differences. 'What's it like to live in Singapore/Australia' was set up, with suggestions about what to write, and guidance on how to write it. Students interacted in the global Studywiz environment in discussions, posting photos in the gallery and uploading their activities. Students responded to each other on what it meant to be an Australian or a Singaporean, what their school and country was like and what they valued.

Phase Two

Shepparton High School collaborated with Chormann Elementary, a Southgate Community school in Michigan, USA on improving student literacy in reading and writing, speaking and listening, media, digital and visual literacies.

Shepparton High School, as described above, is a Government 7-12 Secondary College in the major commercial centre and agricultural region of central Victoria, Australia. It is a significant multicultural school.

Chormann Elementary is a primary school in the Southgate Community of schools. Southgate is in Wayne county, eleven miles south-west of Detroit, Michigan. The city is located inside the Detroit metropolitan area. There are an estimated 30,136 people in Southgate. It is an urban area with a middle class intake in the 'car capital' of the world.

Their essential question was: What impact does mobile learning and multicultural text have on student achievement and cultural appreciation?

Project objectives for Phase Two built on Phase One and aimed to identify factors that promote:

- improved learning when students interact via Studywiz using the iPod Touch
- improved and innovative teaching practice
- international curriculum collaboration by students
- the production of student demonstrations of learning such as video podcasts, digital resources, group discussions.

In Phase One, differences in age of the students and curriculum content impacted on the amount of interaction between the schools, with the result that in Phase Two, students of similar ages and common learning areas were linked up to facilitate closer communication and collaboration.

The Year 5 class (aged 10-11) at Chormann Elementary had a set of iPod Touches to be shared amongst class mates and with other classes. It was not a 1:1 initiative and the iPod Touches were used to complement the range of ICT technologies used within the class. The teacher had access to an interactive whiteboard, to computers and to the iPod Touches for the Literacy project. Shepparton High extended their personalised learning program with each student having their own iPod Touch.

The two schools negotiated a focus on Literacy in all its forms - reading and writing, speaking and listening, media, digital and visual. The wide ranging approach to literacy in all its forms was based on the shape of the new Australian Curriculum and on the 21st Century Skills Framework being implemented in many US states. Each class focused on a novel and the teachers developed activities around the characters and events, relating back to the students' own lives. Shepparton High School chose Roald Dahl's 'Boy' and Chormann Elementary school chose Neil Gaimon's 'Coraline' as novels to explore. Students had access to audio books as well as hard copies.

The students at Shepparton joined with the students at Chormann Elementary to discuss common topics raised in the novel reading. The teacher's comments on the impact of connecting with learners across the world are revealing: "I really believe having the wider audience and the ability to read the comments left by other students creates a more engaging way for students to share their thoughts about the book and its themes" (Shepparton high teacher, 2009).

The two teachers involved utilised a range of approximately twenty literacy apps and resources for the iPod Touch that encouraged the students to learn to use them, to experiment, to create their own work and to become independent learners as part of their Literacy project. In a multicultural class, with new arrivals from China during the term, the teacher found it useful to introduce them to the App Dictionary.com. "Our English teacher has been using Dictionaire for the word wizards in Literature Circles classes. Our geography teacher has been firing general knowledge questions that can be answered using 'All Countries' and our Maths teacher, a confessed technophobe, is also being thrown into the deep end...and loving it" (Shepparton high teacher, 2009).

Students were encouraged to create their own podcasts using a podcast script writing template. They were routinely accessing their learning from the Studywiz Mobile interface on their iPod Touch. The only other means of delivering learning resources on the iPod Touch to students is through syncing with iTunes, which the teacher felt was a barrier to using these devices effectively. Studywiz Mobile bypasses this barrier. Students at Shepparton High were using discussions, galleries, links to websites and bulletin messages. They regularly used Studywiz Mobile to access Word documents and PowerPoint presentations that were available as resources.

Some activities were more suited to the mobile interface than others, for example, the Studywiz Mobile Calendar enabled students to access the school's shared calendar. As the iPod Touch does not have texting or phone capabilities, the students would have preferred to make full use of Studywiz Mobile Mail. However, scrolling through the list of contacts can take quite some time to locate the recipient of a message.

Considerable time was spent in planning how to introduce the iPod Touch, and as in most emerging technology projects, committed teachers spent a great deal of their own time in preparation and learning.

"I have spent many an hour playing with my own personal iPod Touch and reading about learning with mobile devices. Needless to say I have been doing tons of research" (Chormann elementary teacher, 2009). Significant reflection was evident in the pedagogical implications of using mobile learning and the fundamental change to pedagogy required for a successful outcome, as indicated in the following reflection. Using mobile learning prompted teachers to rethink what they were doing in the classroom:

"The week back has been quite a challenge with the Touches. I did not appreciate the WHOLE new classroom management style that has to be developed in order to accommodate the students and help me keep my sanity!!!" (Chormann elementary teacher, 2009).

As the teachers' confidence developed, so did the group interaction and sharing among the students. At one point, one teacher described the emotional impact when she realised that things were going well and that the students were working in groups in a calm atmosphere. "As a class we have been so calm with the Touches this week. Students were careful and generous with the sharing of the Touches. It can bring a tear to a teacher's eye!" (Chromann elementary teacher, 2009).

Findings

Mobile learning in these school classrooms has been exciting and motivating for the students. It has required a significant amount of time, rethinking curriculum innovation and fundamental pedagogical change. The teachers were also required to source a range of technical expertise to solve the connectivity, synchronisation and battery life issues.

The degree of innovation has required that the teachers extend their imagination, their creativity and their traditional pedagogies in the classroom to realise the full potential of the devices. Despite a significant degree of risk-taking by the teachers, all reported that they had only just begun to understand the possibilities of the iPod Touch for learning when teamed with the Studywiz learning environment.

Mobile devices that synchronise with an online learning environment are valuable learning tools if their potential can be harnessed by creative, imaginative teachers. In each of the schools, mobile learning has been integrated into authentic classroom learning activities in meaningful ways in literacy, numeracy, geography, civics and philosophy.

The students in both Australia and the US displayed confidence and independence when using the devices for learning at both the primary and the middle school levels. In general, the devices have been integrated in imaginative ways to suit the Literacy learning area. Both groups of teachers were challenged in terms of changing their pedagogical practices to a student-centred model which is essential for students in a personalised learning environment. The teachers were pushed to rethink what and how they were teaching 21st century learners.

Mobile learning forced a shift in pedagogy for most of the teachers, who indicate that they had to rethink what they are teaching and how they are teaching it, what is important and what their aims are. The teachers in this project indicated that using the mobile devices promoted independence and engagement, particularly for less able students, by keeping them on task. As the students were able to have individual input, the teachers commented that this highlighted their desire to learn.

"Many parents of students who participated in the project have reported increased student ICT use and new ICT skills in their homes and at school" (Delphian eLearning, 2009, p.30). Students find the iPod Touch functionality intuitive and easy to navigate, and tend to prefer the touch sense screen. Students have been complimentary about the interface with Studywiz Mobile and the ability of the iPod Touch to synchronise with Studywiz Learning Environment.

Any internet-enabled mobile device brings the inevitable issue of students using it in class for purposes other than learning, such as listening to music or browsing on social networking sites. Protocols need to be developed for their use in class that complements learning. If they are not embedded properly in curriculum plans, students will tire of the responsibility of looking after the devices, while not using them.

In schools, as it is a personal device, the main concerns relate to synching to a school iTunes for downloading apps. The synching facility available through Studywiz Mobile, accessible from school and home, has provided a simple tool to alleviate these issues. Regular updating of applications on the iPod Touches rejuvenates student interest in remembering to bring them, keeping them charged and using them for learning.

From a teacher's perspective, access to the Studywiz Mobile interface for day to day teaching in the palm of their hands has had huge implications for the schools. It enabled immediate access to their classes to have students commence work quickly, saving both time and frustration with regular computer set–up time. One thing they indicated they would use every lesson, if it were a standard inclusion of the mobile interface, is a 'grading' system, similar to a recently released App called GradePad.

Major technical issues highlighted include the need to have an improved charging process for the iPod Touches, as batteries regularly went flat each day. Students charged them early in the evening and listened to them during the evening and on way to school in the morning, with the result that when they were required to be used in class, they had little battery life left. Their popularity meant that students used them incessantly. Students took their USB cables home, with the result that additional USB cables for synching were required at school. Not having to carry them home from school limited the loss of the student owned USB cables.

Time consuming and technically difficult jobs place significant pressure on the teachers. Without iTunes installed on the school network, downloading movies onto the iPod Touches proved difficult – students all needed individual access and extra teacher time was required to facilitate this process.

Implications for Mobile Learning

Main issues raised with the iPod Touch include that in the model used, there was no video or audio recording or phone features. The iPod Touch does not provide communication tools like mobile phones or the iPhone. Since students indicated that almost 100% of them have a mobile phone and see this as their main communication mechanism, the original iPod Touch required additional functionality, since updated in the second generation iPod Touch.

Mobile learning and mobile devices like the iPod Touch engaged and motivated students and improved their engagement. This was demonstrated by improved attendance, more active participation in class, better preparation and organisation for class, more regular completion of school work and homework and enjoying learning as a 'fun' activity. It also has benefits for teachers and schools in terms of enabling attendance logs to be sent instantaneously to school administration, and as immediate communication available from one teacher to another. Parents reported increased student ICT skills at home and at school.

The Nintendo DS

Background

The Nintendo DS Project investigated the potential of the Nintendo DS, to improve student learning in Mathematics. A controlled experiment was established at Xavier College and Trinity Catholic Primary School in Victoria, Australia. Students in four Year 5/6 classes (aged 10-12) used the Nintendo DS to complete the Professor Kageyama Maths Training Program, while a control group of students completed the same Maths program with traditional pen and paper. The aim was to compare student outcomes and the impact of the mobile technology on student performance by identifying the influential factors.

The students used the Nintendo DS to work through the Maths training program that focuses on addition, subtraction, multiplication and division. For twenty minutes each day, for a period of ten weeks, students used the Nintendo DS to consolidate their basic Maths skills.

Two rounds of video interviews were conducted with a random sample of students from the three classes at Xavier College and Trinity Catholic Primary School. The two schools have different cohorts of students, one from a high to middle socio-economic community and the other from a lower socio-economic community with a high proportion of recently arrived, non-English speaking migrants. In total, thirty students were interviewed before the commencement of the classroom research activity and again at the end of the classroom activity. The factors affecting uptake of the technology at home and at school were examined, along with the students' performance in Mathematics.

In the pre-research interviews, students were asked to comment on their ownership and use of technology at school and at home. The majority indicated that they already had mobile devices at home, in particular, the Nintendo DS, which they used for playing games. Their perceptions of Maths and homework were canvassed along with their understanding of their parents' attitudes to using the Nintendo DS at school for learning. In the post research interviews, students were asked for their responses to the Nintendo DS, their perceptions of it as a tool for learning and for their impressions about its impact on their Maths skills and understandings.

The essential research question was: How does a common handheld gaming device, the Nintendo DS, contribute to improved student performance and engagement in Mathematics? Over a period of one term, students were engaged daily for twenty minutes on the same Mathematics program, using either paper based materials or using the Nintendo DS. The Nintendo DS project provided comparative data on the acquisition of Mathematics skills and understandings using paper based materials and the Nintendo DS devices with Year 5/6 students in two schools.

What Happened

Early indications were that the students found the mobile learning engaging and motivating. According to online survey and interview data, the vast majority of the students already owned, or had used a Nintendo DS at home, and were keen to have the opportunity to use the devices at school for learning. Their interests outside of school fell into three main areas – reading, computer and sport. These activities ranked significantly higher than any of the other interests, such as friends, family, animals, homework, music, art etc. Student personal ownership of technologies was spread across computers and handheld devices, with the computer being the most commonly owned, followed by Playstation and Nintendo DS, Nintendo Wii and mobile phones.

When asked at interview about their parents' attitudes to using technologies at school, the majority indicated that their parents liked it or thought it was great, that computers are the 'normal thing', 'good for helping us learn' and that they 'need to know how to use them for the future'. Some indicated that their parents did not like them playing games, as they do not help with learning, while one student indicated that his parents like him to create rather than play games.

Findings

Students were asked to give their general impressions of the Nintendo DS, with the majority commenting positively about its use for learning. All students thought it was fun to use and better than worksheets; it helped in learning to do basic calculations more quickly and it was exciting using it in groups to compete against each other. Some students expressed surprise that such a gaming device could be used for learning in schools:

"Didn't expect to learn with a DS...thought it was meant for games, so it's a bit of a surprise" (Student A, Xavier College, 2009). Similarly, quite a number of students commented on the fact that it "trains your brain reflexes and [you] calculate faster" (Student A, Xavier College, 2009).

When asked whether using the Nintendo DS had improved their Maths, students were in definite agreement that it had, with comments such as: "I'd recommend it as it's made a difference to my Maths...I'm faster and my understanding of division, high level subtraction and addition has improved" (Student B, Xavier College, 2009). "It's more intriguing; it makes people want to learn more...gives Maths learning fun, excitement and inquiry to want to learn more" (Student C, Xavier College, 2009).

In terms of assisting their learning in other learning areas, most students noted a transfer of knowledge to other subjects. As one student commented, "It's a different way of learning...we'll be using it in English next term." (Student D, Xavier College, 2009). Another identified memory and recall as useful in other subjects: "Good for most subjects – memory and recall" (Student E, Trinity Catholic primary, 2009), while another indicated that it would assist writing: "Makes your writing clearer as it won't read messy writing" (Student F, Xavier College, 2009).

Significantly, two students identified their perception of the clear link between games and learning by commenting that "If you think you are playing a game, you learn more" (Student G, Trinity Catholic primary, 2009). Another student saw a bigger picture when he indicated that "It helped us settle into school" (Student H, Xavier College, 2009).

Their perceptions of their parents' attitudes to use of the Nintendo DS at school were varied. Some students indicated that they and their parents were 'shocked' that a gaming device was going to be used as follows: "Parents were pretty shocked at first – they like brain training" (Student I, Xavier College, 2009). However, the majority indicated that their parents were supportive: "My parents enjoy it and say it will improve my Maths" (Student J, Trinity Catholic primary, 2009). The final comment from one student sums it up well, "Parents [were] really surprised...normally it would rot your brains" (Student I, Xavier College, 2009).

Student performance in Maths was compared before and after the project period. Within the groups of students using the Nintendo DS, two-thirds of the students showed an improvement in test scores on general arithmetic operations, with one third of the total students showing greater than 10% improvement. Conversely, one quarter of the students showed a decline in performance. Generally speaking, the Nintendo DS may have some impact on enhancing arithmetic operations skills.

The teachers were as enthusiastic about the use of the Nintendo DS for Maths Training as the students were. They commented upon the positive impact of the devices in a number of areas including:

- student engagement in Maths learning
- increase in the pace of mental calculations
- improvement in number skills and basic calculations
- growth in student confidence regarding Maths competitive challenges
- focus on learning independently without distractions
- ability to transfer skills of concentration and independence to another learning area
- development of responsible classroom behaviour.

As reported by the teachers, most students improved their performance in basic calculations. The less able mathematicians gained confidence, as the Nintendo DS provided the necessary scaffolding for them to improve. They noticed that those students usually struggling with Maths, improved to the extent that they challenged students who formerly ranked higher than them within the class. As the Xavier College Year 6 teacher commented,

"The ranking by ability of students in the class has noticeably changed with the weaker boys now flying along with their Maths and competing with more able boys in the networking activities". He also remarked on the development of ICT skills promoted by the Nintendo DS, which he backed up with discussions in the Studywiz online learning environment, where he posted questions or discussion starters and the boys responded. This helped them to concentrate and also improved their communication and ICT skills.

For the students at Trinity Catholic Primary School, the Nintendo DS was a real opportunity to use an emerging technology in their learning. The Year 5/6 teacher indicated that he'd seen "...a huge difference in the students in terms of engagement. We use it as a warm up activity in number and the kids see it as an opportunity. They love the competitive aspect and they absolutely love becoming quicker with their calculations".

One important outcome has been the development of a focused and independent learning culture within the classroom as a direct result of the concentrated effort each day. One of the Year 5 teachers commented, "It's enabled me to establish a classroom and behaviour routine where the boys quietly give out the devices and start working. It's promoted a good classroom organisation and routine".

The Year 6 teacher was also impressed with student focus and concentration when he said, "I know that I can confidently ask them for 20 minutes of quality time and quality work to focus on a particular thing, such as handwriting, and I'll get it". It appears that the approach to structured learning exercise has had spin-offs in other learning areas and in the students' general work habits.

One teacher summarised the impact on student learning when he spoke about the effects of the brain training approach that is integral to the Nintendo DS programs. "The kids love the 20 minutes on the Nintendos and it affects them for the rest of the day as it puts them in the zone to learn."(Teacher A, Trinity Catholic primary, 2009).

In terms of pedagogy, all three teachers interviewed indicated that using the Nintendo DS with students had definitely improved their teaching practice and changed their thinking. "I know that I can confidently ask them for 20 minutes of quality time and quality work to focus on a particular thing, such as handwriting, and I'll get it. They love the routine and they organise themselves" (Teacher A, Xavier College, 2009). While some of the teachers had high-tech classrooms and could use a variety of technologies as and when they need it, others were less confident.

On one hand, a confident teacher remarked, "I'd definitely use handhelds in the future as they are quicker to open and load than laptops and they provide immediate use when the boys need it" (Teacher A, Xavier College, 2009). Another commented, "I never saw them as useful for learning until a few weeks ago. It's changed my approach to the use of technology as it promotes real engagement" (Teacher, Trinity Catholic primary, 2009). He has developed an interest in multi-literacies using technologies and will be a catalyst for a rethink of the use of technology across the whole school.

Summary

These four mobile learning projects were clearly diverse in the nature of their implementation and application in teaching and learning. However, some common outcomes may be identified relating to the research questions. The four projects investigated mobile technologies and the wireless environment, implementation and technical issues, learning approaches and pedagogical innovations.

In order to enable students to use technologies at school, schools need to plan to systematically trial and integrate emerging social technologies in projects such as these. Learnings can be shared and transferred across school systems in an effort to spread innovations. Teachers need to be given time, support and professional learning opportunities to extend and broaden their repertoire of pedagogies. With the Internet, pervasive tools such as social networking software and online collaboration tools need to be employed as vehicles for refreshing pedagogy and retraining teachers. The enablers of technology innovation that bring about fundamental change in schools remain consistent: curriculum based innovation, robust IT infrastructure, tailored professional learning and sound technical support.

RECOMMENDATIONS AND IMPLICATIONS

Across the range of K-12 mobile learning projects described above, some common characteristics emerged that have implications for future projects in terms of planning and implementation, learning and teaching, professional learning and technical support. The following recommendations may assist the effective integration of mobile learning in K-12 settings:

Plan

- Well defined learning goals provide for skill development and independent learning.
- Adequate time is required for teachers to become familiar with the mobile technology functionality, prior to implementation.

Implement

- Create curriculum activities that integrate mobile learning consistent with student usage.
- Focus on the power of visual media and social networking to significantly influence literacy, numeracy and life chances for disadvantaged students.

Learning

- Use mobile devices to stimulate learners to *want* to learn and provide a means to pursue knowledge independently.
- Promote personalised learning with handheld mobile devices (especially for Literacy, Numeracy, ESL, 'at risk' and reluctant learners).

Teaching

- Stimulate teachers to rethink their pedagogy by considering *how* to integrate mobile devices in curriculum activities.
- Enhance learner autonomy and independence by using mobile learning devices in the classroom and environs.
- Integrate mobile learning devices with other technologies (such as interactive whiteboards, laptops, Nintendo Wii) for purposeful learning.

Professional Learning

- Employ mobile learning devices to support teacher ICT capacity and confidence.
- Promote online collaboration about mobile learning as an effective mode of teacher networking for sharing knowledge and the transfer of good practice.

Technical Support

- Prior to project start-up, conduct a thorough technical needs analysis of issues that may impact on successful implementation.
- Provide adequate technical support to manage major technical disruptions.
- Ensure there is one conduit or process for resolution of technical issues.

FUTURE RESEARCH DIRECTIONS

Mobile devices such as the iPhones and iPads have huge potential for the administration of teaching and learning, providing an excellent range of productivity tools to alleviate the 'time poor' nature of K-12 teachers. Classroom management, attendance tracking, homework, parent communication, location tracking using GPS features are becoming available on mobile devices.

Similarly, functionality for curriculum design and development, lesson planning and learning assessments are emerging. Studywiz already integrates the iPod Touch with the online learning environment. Emerging iTunes apps enable schools to integrate the iPhone, iPad or iPod Touch with a learning management system or student assessment system. Teachers can plan lessons, monitor student attendance and progress and engage students through a range of thinking and eLearning tools.

While education sectors consider how to use mobile learning in K-12, for example in policies, funding, device management, assessment rubrics and professional learning, research must lead the way with pilot trials and studies on what mobile learning has to offer schools in terms of productivity gains and teaching and learning tools. The end point is to engage students in K-12, to stimulate them to want to learn and to enjoy learning.

CONCLUSION

These projects have made constructive discoveries that can be applied to the use of mobile technologies and 1:1 initiatives in the classroom. Handheld devices have stimulated both teachers and students to work creatively together to improve skills for the 21st century. The use of the iPod Touch has encouraged student interaction in blogs, podcasts and websites, and it has been particularly important in encouraging reluctant and ESL learners in Literacy and Numeracy.

Mobile learning supports personalised learning by stimulating learners to *want* to learn, and causes teachers to rethink their pedagogy. Students show improved attendance, more active participation in class, better preparation and organisation for class, more regular completion of schoolwork and homework, and enjoy learning as a fun activity.

It is essential that clear, precise learning goals that support a culture of learning and skill development in the classroom are identified, for effective implementation of mobile technologies. It is recommended that schools focus on the power of visual media and social networking, to significantly influence completion of school work, homework, participation, organisation and attendance. For example, the range of educational iTunes apps available today has the capacity to enhance the use of mobile learning.

What the research further reveals is that mobile technologies have the potential to change teacher pedagogy to a learner-centred approach that alters the teacher-student relationship dynamics and promotes greater levels of higher-order thinking by students in classrooms and at home.

Key enablers of effective implementation of mobile learning technologies include wireless connectivity, technical support and collegiality. With these in place, teachers can focus on their core task of using the technologies to improve learning outcomes.

In summary, students are keen to investigate and use mobile learning, as the use of new mobile technologies engages and motivates them, engendering an interest in attending school, participating in class activities and learning from and with their friends.

REFERENCES

Carvin, A. (2007). The iPod of the Beholder: Can MP3 Players Enhance Learning? *PBS Teachers*. Retrieved July 20, 2009 from http://www.pbs.org/teachers/learning.now/2007/05/the_ipod_of_the_beholder_can_m.html

Condie, R., Munro, B., et al. (2007). The Impact of ICT in Schools – a landscape review, *BECTA ICT Research*. Retrieved July 20, 2009 from http://partners.becta.org.uk/page_documents/research/impact_ict_schools.pdf

Delphian eLearning. (2007). iPodagogy: Using iPods and Podcasting for Learning, *Department of Education and Training,* Victoria. Retrieved July 20 2009 from http://www.delphian.com.au/images/documents/ipodagogy%20research%20report_final1.pdf

Delphian eLearning. (2008). iPod Touch Research Report, *Department of Education and Early Childhood Development*, Victoria. Retrieved July 20 2009 from http://www.eduweb.vic.gov.au/edulibrary/public/ict/ipodtouchresearch2009.doc

Delphian eLearning. (2009). Global Mobile Learning Research Report 2009, *The Etech Group*, Australia. Retrieved July 20 2009 from http://www.delphian.com.au/images/documents/mobile%20learning%20research%20report_etech2009_final.pdf

Dogbey, J. (2007). Using iPods for Instruction, *The Principals Partnership*. Retrieved July 20, 2009 from http://www.principalspartnership.com/iPods.pdf

Edgar, D., & P. (2008).Television, Digital Media and Children's Learning, *Victorian Curriculum and Assessment Authority*. Retrieved October 30, 2009 from http://www.vcaa.vic.edu.au/earlyyears/Edgar_Paper_20081223.pdf

Jobbings, D. (2005). *Exploiting the educational potential of Podcasting.* Retrieved July 20, 2009 from http://www.recap.ltd.uk/articles/podguide.html

Metiri Group. (2006). Technology in Schools, What the Research Says, *Cisco Systems*. Retrieved July 20, 2009 from http://www.cisco.com/web/strategy/docs/education/TechnologyinSchoolsReport.pdf

Ng, W., & Nicholas, H. (2009a). Introduction of pocket PC in schools: attitudes and beliefs in the first year. *Computers & Education*, *52*, 470–480. doi:10.1016/j.compedu.2008.10.001

Perry, D. (2003). *Handheld computers (PDAs) in schools*. Coventry, UK: Becta.

Spender, D. (2007). Digi-kids and a new way of learning, in *The Age Newspaper*. Retrieved July 20, 2009 from http://www.smh.com.au/news/technology/digikids-and-a-new-way-of-learning/2007/05/18/1178995417590.html

Stead, G. (2006). Mobile technologies: transforming the future of learning, in *Emerging Technologies for Learning*, BECTA. Retrieved July 20, 2009 from http://partners.becta.org.uk/upload-dir/downloads/page_documents/research/emerging_technologies.pdf

The New Media Consortium. (2009). 2009 Horizon Report ANZ Edition. *COSN.* Retrieved October 30, 2009, from http://www.nmc.org/pdf/2009-Horizon-Report-ANZ-Edition.pdf

The New Media Consortium. (2009). Horizon Shortlist, *Horizon.AU*. Retrieved July 20, 2009 from http://horizon.nmc.org/australia/2009_Shortlist

The New Media Consortium. (2009). 2009 Horizon Report K-12 Edition. *COSN.* Retrieved October 30, 2009, from http://www.nmc.org/pdf/2009-Horizon-Report-K12.pdf

Twining, P., & Evans, D. with Cook, D., Ralston, J., Selwood, I., Jones, A., et al. (2005). Should there be a future for tablet PCs in schools? *Journal of Interactive Media in Education*, 20, 1–13. Retrieved July 7, 2009, from http://jime.open.ac.uk/2005/20/twining-2005-20-paper.html

Vahey, P., & Crawford, V. (2002). *Palm Education Pioneers Program Final Evaluation Report.* Retrieved on July 7, 2009, from http://www.palmgrants.sri.com/PEP_Final_Report.pdf.

ADDITIONAL READING

Ally, M. (Ed.). (2009). *Mobile Learning: Transforming the delivery of education and training*, AU Press, Athabasca University. Retrieved July 20, 2009, from http://www.aupress.ca/books/120155/ebook/99Z_Mohamed_Ally_2009-MobileLearning.pdf

Consortium of Networking Schools. (2004) A Guide to Handheld Computing in K-12 Schools. *COSN*. Retrieved July 20, 2009, from http://www.cosn.org/Resources/tabid/4307/Default.aspx

Herrington, J., Herrington, A., Mantei, J., Olney, I., & Ferry, B. (Eds.). (2009). *New Technologies, New Pedagogies: Mobile Learning in Higher Education*, Wollongong: University of Wollongong. Retrieved July 20, 2009 from http://ro.uow.edu.au/cgi/viewcontent.cgi?article=1092&context=edupapers

McFarlane, A., Roche, N., & Triggs, P. (2007). Mobile Learning: Research Findings, *BECTA*. Retrieved July 20, 2009, from http://partners.becta.org.uk/upload-dir/downloads/page_documents/research/mobile_learning_july07.pdf

McFarlane, A., Triggs, P., & Yee, W. (2008). Researching mobile learning - Interim report to Becta, *BECTA*. Retrieved July 20, 2009, from http://partners.becta.org.uk/upload-dir/downloads/page_documents/research/mobile_learning.pdf

The New Media Consortium. (2009). Horizon Report 2009. *COSN*. Retrieved July 20, 2009, from http://www.nmc.org/pdf/2009-Horizon-Report.pdf

Traxler, J. (2008). Learners: Should We Leave Them To Their Own Devices? *BECTA*. Retrieved July 20 2009, from http://emergingtechnologies.becta.org.uk/upload-dir/downloads/page_documents/research/emerging_technologies/learners_johntraxler.pdf

Chapter 13
Using Mobile Phones for Teaching, Learning and Assessing Irish in Ireland:
Processes, Benefits and Challenges

Katrina A. Keogh
National Council for Curriculum and Assessment, Ireland

ABSTRACT

This chapter outlines the processes, benefits and challenges of two pilot projects which investigated the integration of mobile phones into the teaching, learning and assessment of Irish in post-primary schools in Ireland. Existing literature examining the status of Irish in Ireland and previous research into the use of mobile phones for teaching and learning languages are described. Findings from the two pilot projects indicate that mobile phones can help to promote the increased use of oral language skills, can increase student motivation and enjoyment for the task at hand and increase students' competency in the language being taught. Mobile phones also proved successful in providing opportunities for students' oral language practice and self-assessment and teachers' formative and summative assessment of students' language skills.

1. INTRODUCTION

"I think using mobile phones to teach teenagers Irish is a good idea because most teenagers are always stuck to their phone and they couldn't live without it..."

(School student aged 14)

DOI: 10.4018/978-1-61692-849-0.ch013

Most schools in Ireland have a policy that states that students' mobile phones must be powered off during lessons, study periods and any learning-centred activity such as field trips. Some schools do not allow students to carry mobile phones in school. For students, school is not a phone-friendly environment.

Yet is it clear how engaged students become when they are using their mobile phones to send a text message to a friend, check whether they are busy on Saturday or to check their email; how at ease they are with this technology and how it is seamlessly integrated into their day to day lives; resulting in the perception that *they couldn't live without it*. Also consider the number of students who own a mobile phone, versus the information and communications technology (ICT) facilities available in schools. If we consider mobile phones to have the potential of a mini-computer (Prensky, 2004), and that the majority of students carry a mobile phone in their pocket, the possibilities for seamlessly integrating ICT into education become more attainable.

In 2007, the National Council for Curriculum and Assessment (NCCA) started to investigate whether mobile phones had a place in school and attempted to harness students' interest in and use of mobile phones by initiating a pilot project investigating the use of mobile phones for teaching, learning and assessing Irish. The integration of mobile phones had the additional benefit of enabling Irish to be spoken outside of the classroom. The project was grounded in a request from the Minister for Education and Science, to investigate ways to integrate ICT into the teaching and learning of Irish, with the particular focus of increasing the communicative use of Irish. Irish is a compulsory subject for students in post-primary schools, but one, which evidence suggests, they have little enthusiasm for (Smyth, Dunne, McCoy & Darmody, 2006). The premise of the project then was to match the most desirable technology with the most undesirable subject.

The initial pilot project examined the use of mobile phones for vocabulary SMS delivery and oral practice and assessment through a phone-based question and answer system. Sixty nine students and their three teachers participated in this phase of work, named the MALL project – Mobile-Assisted Language Learning. The NCCA worked in partnership with the National Centre for Technology in Education[1] for this phase of work.

The MALL pilot project was extended the following year to include a third use of mobile phones – student voice to voice chat, a mobile phone feature rarely used in Mobile-Assisted Language Learning (Kukulska-Hulme & Shield, 2008). The participating cohort also increased to 368 students, 16 teachers and six schools - three in Northern Ireland and three in the Republic of Ireland. This second phase of work was referred to as the FÓN project, which stands for *Foghlaim Ón Nuatheicneolaíocht*, or *learning through new technology*.[2] Both phases of work included an associated online interface, where teachers administered the mobile interface and content, accessed student recordings and provided feedback to students on their recordings. The second phase of work added an online interface for students, where they could access their recordings and any associated teacher feedback. A further project partner worked with NCCA and NCTE on the FÓN project – Foras na Gaeilge.[3]

This chapter examines the background, design, implementation and evaluation findings from both pilot projects, with particular emphasis on the impact of the integration of the technology from the students' and teachers' perspectives. Evaluation questions focussed on whether student competency and motivation increased through using mobile phones for learning and practising Irish and the effectiveness of this technology for teachers in facilitating oral language practice, and formative and summative assessment.

2. BACKGROUND

2.0 Introduction

Irish in Ireland has been reported to have "an overwhelming dependence ... on the public education system" (Mac Gréil & Rhatigan, 2009, p. 3). With falling numbers of people speaking Irish as a first language (Ó Giollagáin, Mac Donnacha,

Ní Chualáin, Ní Shéaghdha & O'Brien, 2007) and less favourable student attitude towards Irish as they progress through school (Smyth, Dunne, McCoy & Darmody, 2006), the increase in interest in Irish-medium or immersion schooling and competent teaching at all levels of education (Mac Gréil & Rhatigan, 2009) are important in keeping the spoken language to the fore. Further action is also required to improve student motivation for and attitude towards learning the language, especially in English-medium or non-immersion settings.

Mobile devices, more specifically, mobile phones, have been used for language learning during the course of research projects, pilot projects or university-level (third-level) integrated scenarios for a range of target languages (for example, Kiernan and Aizawa, 2004; Kukulska-Hulme & Shield, 2008; Kennedy & Levy, 2008; Chinnery, 2006; Stockwell, 2008; Thornton & Houser, 2005). Projects from post-primary level have rarely been documented or indeed used within the context of teaching and learning Irish.[4]

The integration of mobile phones into the teaching, learning and assessment of Irish has potential for increasing the use of Irish communicatively, increasing student motivation and positively impacting student learning and competency.

2.1 Irish in Ireland

Irish is taught through primary and post-primary (K-12) education in all schools in the Republic of Ireland. It is a compulsory subject for all except those who may be given an exemption by virtue of being born or educated outside the state, or of a particular learning disability. School-going children aged 4 to 5 years, commence Irish during their initial year at school. The majority of schools teach through the medium of English, with approximately 9% of schools across primary and post-primary level in the Republic of Ireland, teaching through the medium of Irish or in an immersion setting. The schools in both phases of the project were English-medium.

In 2007, in an attempt to revive Irish through increasing its use communicatively, the Minister for Education and Science increased the proportion of marks allocated to the oral assessments for both state examinations – the Junior Certificate (taken at age 15) and the Leaving Certificate (taken at age 17 or 18). The increase in weighting for the oral component of the examinations was substantial, resulting in almost a doubling of marks for each examination. The changed weighting took effect for the Junior Certificate examination in 2010 and will take effect for the Leaving Certificate examination in 2012. Of note is that the oral examination in the first examination is optional and taken only by a handful of students. All students take the oral assessment in the school leaving examination.

In the case of Northern Ireland, Irish is taught as an optional ab-initio modern language at post-primary level. The GCSE (taken at age 15) and A-Level (taken at age 17 or 18) state examinations both include an oral component. When students are selecting their subject choices for examination years, Irish is offered as an option among a range of other subject types – Arts, Sport or Sciences. Teachers try to promote Irish within their schools and raise its profile among students. Trends in recent years have seen the same numbers of students selecting Irish as an examination subject, with between 1.3% and 1.4% of candidates selecting Irish during the period 2004-2008 (Joint Council for Qualifications, 2008).

At primary level in the Republic of Ireland, the revised Primary School Curriculum (Government of Ireland, 1999) shifted the teaching of Irish to a communicative approach. This shift has led to improved student attitudes and increased enjoyment for children learning Irish (NCCA, 2008; Department of Education and Science, 2008). However, the increased focus on communicative tasks and perhaps the lack of teacher understanding of the three stages of the communicative approach,[5] have led to reduced levels of children's overall attainment in Irish (NCCA, 2008; Department of

Education and Science, 2008), especially in children's accuracy in Irish. Harris provides further evidence of children's decreasing levels of attainment in Irish over time (Harris, 2006). Some of this reduction can be attributed to changing language demographics in Ireland. Current national trends indicate decreasing numbers of people speaking Irish as a native language. If these trends continue, it is estimated that Irish will cease to be spoken as a native language within 15 to 20 years (Ó Giollagáin et al., 2007).

At post-primary level in the Republic of Ireland, students have been shown to have negative attitudes to learning Irish, which tend to further deteriorate as their years of schooling progress (Smyth et al., 2006). Irish was reported within the top six subjects that students wished they had not taken and the most frequently cited subject to be least liked, with 32% of students studying Irish reporting that it was their least favourite subject (Smyth et al., 2006). Smyth, Dunne, Darmody and McCoy (2007) described how these same students categorised Irish one year later in the longitudinal study as one of the most difficult, least interesting and least useful subjects to learn.

In a second language learning environment, traditionally, observation of students' oral language skills has been restricted to class time, where the teacher observes or interacts with students individually or within pairs or groups. For students, it can be difficult to mimic the communicative use of oral language outside of school, unless the student benefits from access to target language speaker(s). While Ireland does have an Irish language television station and radio stations, these media do not provide opportunities for communication, such as conversation, through the target language.

The MALL and FÓN projects attempted to transfer students' enjoyment of using mobile phones to their perspectives of learning Irish, in an attempt to improve student attitude towards Irish. Through providing increased and alternative opportunities to access and practise Irish, it was hoped that improvements in student competency would also result. A further potential advantage offered by the technologies, was the opportunity for students to practise their oral skills and for teachers to listen back to these and provide feedback, outside of the time restrictions of the classroom.

2.2 Mobile Phones in Language Learning

Mobile phones have been shown to be the most commonly used mobile device reported for teaching and learning languages (Pęcherzewska & Knot, 2007 reported in Kukulska-Hulme & Shield, 2008). As time moves on, more and more papers are being published which examine their use as language learning tools. Mobile phones offer ever increasing functionality, with e-mail, internet access, text, camera, MP3 and chat facilities available. The terms m-learning, mobilearning and mobile-assisted language learning (MALL) have all been used to describe the use of mobile devices, including mobile phones, for learning, with MALL more specifically referring to language learning. Moblogging refers to the use of mobile phones for web-logging (blogging).

Projects have enlisted different features of mobile phones, for a variety of language learning purposes - to provide learners with feedback and assistance via voice (Twarog and Pereszlenyi-Pinter, 1988), vocabulary quizzes, word and phrase translations via voice and email, (Brown, 2001), vocabulary instruction via email (Thornton and Houser, 2002; 2003; 2005; Stockwell, 2007; 2008; Dias, 2002a), vocabulary instruction via SMS (Levy and Kennedy, 2005; Kennedy & Levy, 2008; Kiernan & Aizawa, 2004; Lu, 2008), activities and learning websites accessed via web browsing (Nah, White & Sussex, 2008; Taylor & Gitsaki, 2003; Hsu, Wang & Cormac, 2008) and English lessons from BBC World Service's Learning English section via SMS (Godwin-Jones, 2005).

Few studies report on the use of mobile phones for speaking or listening activities.[6] The focus instead has been on delivering materials to or gathering materials from students. Kukulska-Hulme and Shield (2008) confirmed this trend in their meta-study on the use of mobile phones for language learning purposes. They also pointed to the frequent use of asynchronous modes of communication, which do not occur in real time (for example email), rather than synchronous ones, where people communicate with each other there and then (for example, a telephone conversation). Mobile phones provide an ideal medium to facilitate synchronous communication. Some research points to the need for scheduling in order to facilitate synchronous communication among learners, and how this can often hinder the success of a project if students are unavailable at specific times, as well as violating the "any time, anywhere principle of m-learning" (Kukulska-Hulme & Shield, 2008, p. 281).

Mobile phones have proven popular among students, with high rates of penetration among them. Most studies employing mobile phones report high levels of existing mobile phone use among their students. For example, Levy and Kennedy (2005) reported that only one of their 91 students did not use a mobile phone, with 96% of those who do, having their mobile phones with them "constantly" or "most of the day." The high proportion of students who carried their phones with them provided the impetus for mobile phone use to deliver vocabulary SMS over PC-based email, as students would receive SMS messages regularly rather than just when they checked their email. Dias (2002) also reported a high level of mobile phone penetration of 96% among his students. While mobile phones were provided to students for use during the MALL and FÓN projects, it is interesting to note that 99% of students participating in FÓN possessed their own mobile phones. Indeed, across Ireland, the mobile phone penetration rate reached 121% at the end of 2008 (ComReg, 2008) with some people owning more than one mobile phone.

Benefits and challenges have been attributed to using mobile phones to facilitate a variety of language learning activities and purposes. The small screen on a mobile phone has proved limiting to the amount of text which can be sent and viewed on a mobile phone screen (Kiernan & Aizawa, 2004; Stockwell, 2008; Shudong & Higgins, 2006). This challenge has been overcome in certain instances, by materials being designed as short and 'bite size', lending themselves to the small screen-size (Kennedy and Levy, 2008; Lu, 2008; Dias, 2002a) or through the use of WAP[7] websites, which can be viewed on a mobile phone screen (Nah, White & Sussex, 2008). Prensky (2004) argues that the small amount of material which can be viewed on a mobile phone screen forces publishers to "rethink their design and logic for maximum effectiveness, rather than just adding pages" (p. 6). Kiernan and Aizawa (2004) also maintain that the short messages which can fit on a mobile phone screen encourage beginner language learners to use only the language being learnt when composing messages, rather than straying into their first language, as can be the case in face to face communication.

The small key, one-finger input format on a mobile phone keypad has been cited as a challenge in some studies (Stockwell, 2008; Shudong & Higgins, 2006), especially where the language being learnt may use a different script to the learner's native language. Elsewhere, this input format has been cited as a benefit, as being better for learners who have not yet learned to type efficiently (Kiernan & Aizawa, 2004). Evidence from the number of emails and SMS messages sent via mobile phones in personal communication by students,[8] indicates that they have mastered the art of 'thumb-typing.' Students' familiarity with mobile phones, also implies either no or a gentle learning curve for them in adapting to use their mobile phones for the language learning purpose their teacher puts in place.

Depending on the mobile phone features used for language learning, cost can become a factor (Dias 2002a; Kiernan and Aizawa 2004, Stockwell, 2008). One part of a study due to include voice-to-voice communication between students was discontinued due to the prohibitive costs associated with it (Kiernan & Aizawa, 2004). Certain networks charge per WAP webpage accessed. The cost associated with sending SMS or email via mobile phone is generally quite low, but where a requirement to pay any cost is present, the question of equality of access can come into play. Some studies indicated that mobile phones were supplied or call/SMS costs were paid for by those conducting the research (Kukulska-Hulme & Shield, 2008). This approach would have implications for the long-term feasibility or scalability of any project.

The portability, any time/place access and immediacy of communication (Wishart, Ramsden & McFarlane, 2007) afforded by mobile phones, and mobile devices more generally, have often been cited as the main reasons and benefits for the adoption of these technologies. This type of access to information and ability to communicate readily, fits with learners' increasingly mobile and connected lifestyles (Shield & Kukulska-Hulme, 2008; Prensky, 2004). Stockwell (2008) reported his students' use of mobile phones instead of PC because of their anytime/anywhere access which meant they could check emails between classes, while waiting for friends, on the train and in other classes. They also found mobile phones quicker to use, as they were always switched on and didn't require time to boot up as a PC would. Conversely, some students reported requiring a quiet environment to digest the content of emails. The need for a quiet environment is also discussed by Shudong and Higgins (2006), who ponder whether people can learn on the move or whether the physical environment is too distracting or inappropriate. The mobility afforded by mobile phones, coupled with their high penetration rates, provides opportunities to take language learning beyond the classroom and transform language learning into a more autonomous experience for students.

The successful integration of mobile phones is closely linked to the way in which they have been employed and the reason they have been employed. Mobile phones, when used in a setting that creates a genuine physical distance between learners or learners and their teacher in which communication via the device is the only way to interact with others seems to create a more realistic setting for communication. In contrast, young people in a community who were asked to send their views to a pilot scheme - Pleasurable Cities - exploring options for young people to have a voice in their local community, were not inclined to do so as they were already able to discuss the issues face to face among different groups they were members of (Lee, 2006 reported in Kukulska-Hulme & Shield, 2008). This authentic 'gap', be it physical or in knowledge, provides the impetus for successful language learning via mobile phones (Kukulska-Hulme & Shield, 2008; Kiernan & Aizawa, 2004).

Similarly, the integration of mobile phones has also been successful, where they have been used for legitimate purposes led by the pedagogy being employed, and not simply for the sake of using technology. Mobile phones in these instances have only been used to perform a task which cannot be accomplished through other means. Kennedy and Levy's study (2008) provides one example of this – SMS was used to deliver vocabulary and Italian language-related information in place of email or more traditional paper-postings as they wanted to "seek the students' attention at regular intervals, rather than leaving it up to them to decide when to check for messages" (p. 317).

The next section examines how the MALL and FÓN projects integrated SMS, voice calls to a recorded system and student voice to voice calls, and whether these features facilitated language learning in a way that more traditional methods could not.

3. MOBILE PHONES FOR TEACHING AND LEARNING IRISH: THE MALL AND FÓN PROJECTS

3.0 Introduction

This section outlines how mobile phones and associated online interfaces were facilitated for teachers and integrated into the teaching, learning and assessment of Irish by the participating teachers and students themselves. Each of the uses of mobile phones, associated content and methods used to integrate the technology in a blended learning environment[9] are described.

The MALL and FÓN projects had the same aims, with the FÓN project focusing a little more on the potential for the technologies to facilitate formative and summative teacher assessment and student self-assessment.

The aims were, to investigate the use of ICT to:

- promote student oracy in Irish and promote the use of Irish for communicating
- increase student motivation for Irish
- help students progress their competency, across the four skills, reading, writing, speaking and listening, where possible
- assist teachers in assessing students and students in self-assessment.

3.1 MALL Project

The MALL project was the initial phase of work in integrating mobile phones for teaching and learning Irish in post-primary schools. Three teachers and their 69 male and female second year (ca. age 14) students participated from their new school in the Republic of Ireland, which was housed in temporary accommodation while the permanent school building was being constructed. The temporary accommodation meant that internet access in the school was intermittent and tended to be unreliable. This phase of work was a scoping phase, to investigate whether a more extensive project would be useful.

This phase of the project ran for a total of three months. The initial weeks were used to scope the project, design and develop the software system required,[10] and to work with the three teachers to design and record the content they wanted to make available to their students through the system. All content presented to students was led by their teachers, based on the Junior Certificate Irish syllabus and reflected the teachers' course plans for the weeks to follow. The content was developed around topics and across a series of levels, so that it could be differentiated for students studying at different levels. Two weeks were allowed for training teachers in the developed system and for them to familiarise themselves with the components they would interact with. The teachers and students actively integrated the mobile phones into their day to day teaching and learning for a period of five weeks.

SMS and call functions were the mobile phone features employed in the MALL project. SMS was used to deliver vocabulary words, phrases and quizzes daily to students. They were also sometimes used by teachers to remind students of homework or upcoming assessments for Irish. The call function was used to access a voice response system (both during and outside of class), where students were presented with a series of questions in Irish, to which they left a recorded response, similar to a voice message, for each one. During each voice response session, students were presented with a series of ten questions. They could listen back to each of their own responses in turn, and re-record their answers until they were satisfied with their production. The final recorded responses for each question were saved, for later online access by their teacher. As each student used an individual user-number and PIN to log into the system after dialling the project number, their recorded responses were labelled with their individual details.

The teachers interacted with the system through an online interface. Here they could send an SMS with vocabulary words, phrases or quizzes, or other course-related reminders to their students. They could also access student recorded responses by session. After listening back to each student's recorded responses, teachers could leave some feedback for the student on their recording. This feedback could be collated into a printable feedback sheet or downloaded by students in the form of a podcast, alongside their recorded responses.

Laptop computers were also investigated as a way to facilitate written chat among students through a messaging system. This component of the project experienced problems relating to the unreliable internet connection at the school, and failed to be integrated fully by the teachers and students involved, after a few unsuccessful attempts. This meant that mobile phone use had a greater impact, as they could be used without any requirement for internet connectivity, instead requiring only mobile phone network coverage.

Each participating student and teacher was supplied with a mobile phone for their use during the course of the project. These phones had network restrictions imposed on them, so they could only be used for the purposes of the project. They could dial out to the project line only, receive SMS but not send them, and web-browsing was disabled. All associated call and network costs were paid by the organisations facilitating the project. The school adapted its mobile phone policy, to allow the project mobile phones to be switched on and used during Irish classes, and for students to use them during breaks.

One SMS was sent per weekday, usually during Irish class or outside of school hours. Students were encouraged to use the vocabulary words and phrases received, as frequently as possible during their Irish class and within other assignments. The answers to the vocabulary quizzes were often discussed during Irish class. The voice response element was sometimes used during Irish class to reinforce materials covered, but more frequently used by students as a homework task or self-determined practice session.

3.2 Findings from the MALL Project

A mixed methods approach was used to gather qualitative and quantitative data about the MALL project.

Post-project teacher and student evaluation questionnaires were used. Teachers were asked to keep reflective diaries to record their experiences from the initial planning stages of the project through to the completion of the project. Onsite observation was also used throughout the project, especially while in action in classrooms. Some impromptu feedback was gathered from teachers and students during observation visits.

Only one teacher responded to the teacher questionnaire and one collective reflective diary was kept, which compiled the views and feedback of the three participating teachers. 61 of the 69 participating students responded to the questionnaire, providing a response rate of 88%. As the data gathered from teachers would never have been statistically significant, it was used to get a sense of their experiences to feed into the potential next phase of work. While respondents' perceptions of their increases in competency are subjective, the questionnaire was used to gauge their reactions to the system and their impressions of the impact of integrating mobile phones.

3.2.1 Student Oracy and Communication

Students had the opportunity to speak Irish as often as they wished during the project, with the average student clocking up approximately two hours talk time during the active five weeks of the project. Compare this figure to the opportunities a student would have to be heard in the classroom, when restricted to five 40 minute classes per week, among a class of 20-30 other students. 78% of the students reported that they had spoken more Irish than they usually would have, and commented favourably on the opportunities the MALL system offered them to practise whenever they wanted to. Among the benefits reported, students referred to their increase in fluency, saying that they did not have to search for words, that words came to them faster and more easily than before. Students reported that they enjoyed the challenge of attempting to respond to the questions in real time and speak Irish during an entire voice response session.

While students were communicating with their teacher in an asynchronous way, they liked how the system provided them with more anonymity – "I knew only the teacher would hear me so I wasn't as nervous." They reported feeling freer to speak out than they would have felt in class and that their confidence in speaking Irish had grown. Teachers noted that the system gave them the opportunity to hear students who were usually shy or retiring in class. They could listen to each student, which in turn led to the requirement for more time to do so. They noted students' improvement in competence and confidence in speaking over the five week period. Students' suggestions for improvements to the system included developing the option to chat with someone – something we took on board when scoping the FÓN project.

3.2.2 Student Motivation and Enjoyment

Students reported that the use of mobile phones for learning and practising Irish was more fun and enjoyable than more traditional methods. References were made to mobile phones being 'new' and 'different' as well as how it was enjoyable to use 'new age technology' for Irish. 95% of the students reported that they had enjoyed using the mobile phones for Irish. Some of the comments from the 5% reporting that they had not enjoyed using then, included that they preferred using the book and that they already knew what was being learnt or practised through the content within the MALL system. Further comments referred to the functionality of the system, which will be covered in Section 3.2.5.

Students reported feeling that they had more ownership over this process of learning and that they could practise whenever they pleased, indicating their realisation of more autonomous learning. 93% of the students reported that they thought the forthcoming second year class should also have the opportunity to take part in a similar project - "I enjoyed it and it would be cool if they could do it too."

Teachers felt that students were more 'switched on' to Irish and attributed this to the mobile phone use. They noticed students' increase in motivation and reported that it was making Irish more modern for them. They noted how students' technological know-how, far exceeded their own, something the students relished and the forward-looking teachers embraced. Through observation in class and listening back to student recordings through the system, teachers noticed students' enhanced self-esteem and confidence in speaking and using their Irish.

3.2.3 Student Competence and Learning

67% of the students felt that they had improved their Irish competency as a result of using the MALL system. They referred to increases in their vocabulary, improved comprehension, more fluency in that they could respond more readily, improved pronunciation and feeling more adept at putting sentences together. They reported getting better grades in their class tests and feeling that their speaking skills had improved. Teachers referred to the positive shift from teacher-led learning to more student-led learning. Teachers observed that students were enabled to set their own pace through the differentiation in the question levels. Teachers also reported being able to move students on to the next level where they thought the students were ready, or mark certain questions for students to repeat, where they needed more practice. While the questions and levels within the voice-response system were designed by the three project teachers with their classes in mind, they reported that as they monitored students over time, they would like to have had the opportunity to add more questions and differentiate further for students.

3.2.4 Monitoring and Assessment

Teachers were very positive about the feature which enabled them to listen back to students' recorded responses. As referred to above, teachers felt they had more 'ear time' for every student and had the opportunity to hear the usually quiet or more withdrawn student. The teachers felt that the additional time required to listen back to each student, may not be too practical if all of their classes were participating in the project. The additional time requirement for teachers was also a finding in Hsu, Wang and Cormac's study (2008) where the teacher participating felt that it could be difficult to listen back to all students' recordings over time (in this instance, they were students' audioblogs). The teacher suggested that a class of ten would be more feasible than the class of 20 which participated in the study.

Teachers referred to the novel way of providing feedback on students' oral recordings, and how it was "more interesting and interactive than sitting with a red pen." They also referred to how it was possible to monitor students more generally, in their grammar and syntax, as well as their oral production.

Students' responses to open-ended questions indicated their own realisation that the MALL system provided them with opportunities to monitor themselves and self-assess. One student said "it improved my oral because I could hear what I said after I said it and I could hear my pronunciation." They made reference to the feature that allowed them to re-record their answers, noting that they could do this until they were happy with their answer – "you could speak any time and you could always revise your answer." Another student said "you could hear what you said and you could expand on what you had said before by pressing 2." Other references were made to their realisation about how much Irish they already knew when they could respond readily to questions.

3.2.5 Logistics and Practicalities

Features and shortcomings of the MALL system were noted for evaluation, so that any further project could benefit from what had been learnt.

The MALL system had the capacity to accept up to 30 incoming calls at any one time. As all three participating classes were second year classes, they were scheduled to have their Irish lessons at the same time. If all teachers decided to use the system at the same time, students often had difficulty gaining access to the system. Students did not experience any access problems when they used the system outside of Irish class times. Some students experienced dropped calls – where they would be cut off mid-session. This was also attributed to the 30-call maximum that the system could cater for.

Where all students were using their mobile phones in an Irish class, they needed to spread out as much as the classroom size would allow. They tended to use lowered voices, so they did not disturb their classmates too much and equally, so that they could not be heard too distinctly by their classmates. Students complained about the quality of the recorded questions in the system, and reported that they were sometimes difficult to hear, that the speaker spoke too fast or that the volume was too low. The questions had been recorded with native Irish speakers via the phone system so that although the quality of the language was not an issue, the quality of the sound recording was.

Teachers felt that they needed to be more prepared for each lesson – technically, and with a backup plan in case the technology failed (this was the case for the online text chat which was not successfully integrated due to the intermittent school internet connection).

No cost was incurred by the participating teachers or students, as the project partners took over the running costs to ensure equality of access among participants. This prevented any associated lack of use of certain features due to costs to participants, as seen in Section 2.2 within other studies. The set-up costs as a ratio to the running costs were 3:2. Set-up costs included software development, hardware purchase, teacher training, server costs and line costs. The running costs included the mobile phone line rental and call costs. A longer active project period would prove more cost effective, as the initial set-up costs would be absorbed across a longer term. Teachers and students also requested that any extension to the pilot project be run for a longer period, so that they could spread the use of the system throughout the full school year.

The vocabulary SMSs were sent each morning before school. One of the project teachers was required to log-in online to send the message out in bulk each time to students. For this reason, one recommendation included developing an option to set a schedule of SMS deliveries for students which would automatically send the SMSs at specified times and intervals. Again, this was a recommendation taken on board for the FÓN project. The SMS component was so successful during the MALL project that students requested that they be sent to their personal mobile phones during the summer holidays. The mobile phone screen size did not present any problem for students. SMS content was restricted to the standard length of 160 characters, which students were used to scrolling through within their personal use of SMS.

The implications of the findings from the MALL project and the FÓN project will be discussed in tandem in Section 3.5.

3.3 FÓN Project

The FÓN project built on the work and findings of the MALL project, involving more students, more schools and a cross border element. The software system itself was also developed to be more sophisticated, providing more options for teachers to customise the content their students would access and an additional online student interface.

Six schools, three in the Republic of Ireland and three in Northern Ireland, 16 teachers and a total of 393 students participated in the FÓN project. The original MALL school and teachers, now housed in their new school building, continued to work on the project into the FÓN phase. The project ran for a total of eight months from October 2008 to the end of May 2009. The active project periods were divided among the project cohort, into two terms of work, so that at any time only half of the cohort was using the phones. This reduced running costs, as 210 mobile phones were active per term, rather than 420 for the duration. It also allowed us to compare user and non-user groups, while maintaining equality of access for all. The active period per term was twelve weeks in duration, preceded by teacher training and familiarisation period. Mobile phones were again provided for all participants and costs paid by the partner organisations.

The content for the system was again designed by some of the teachers involved. This time, it encompassed content for all themes and topics within the Junior Certificate Irish syllabus, which also reflected the content of the Northern Ireland GCSE syllabus. A bank of voice-response questions was scripted, according to 10 themes and three levels of difficulty. A bank of SMS vocabulary items was also scripted, to complement the questions. To overcome the difficulties encountered during the MALL project with the quality, volume and speed of recorded questions, all questions for the FÓN voice response element were recorded in a recording studio. National celebrities volunteered their voices for a few hours each, so a variety of speakers, dialects, and Irish levels of ability were recorded. This added to the motivation for students as well as ensuring that they were exposed to a variety of speaker types through recorded questions, as they would be in real life encounters.

During MALL, students worked through the content and levels of voice response questions that their teachers designated for them. The same carried through this time for FÓN, with the extra facility for teaches to pick and choose questions from the bank of recorded questions, or to record their own questions, in order to make up a customised session for their students. An additional feature was the creation of a student practice area – where students could practise any theme/topic, which was tailored to their level of ability, as often as they wanted to. This practice area accessed the bank of pre-recorded voice response questions, as well as being topped up by any teacher-recorded questions that were saved to the system. Again, students used an individual student id and PIN to log into the system, so that their recorded responses were available for later access online by their own teacher only.

While the voice response system became more sophisticated and could be tailored by teachers, an additional aspect of a mobile phone's call functions was added - voice to voice chat. This worked in a similar way to the voice response system to begin with – students dialled the same number, logged into the system and were presented with a number of options. One of these was to chat with someone else. Naturally, this component required that another person was logged into the system at the same time and had opted for a voice to voice chat. Chat sessions were initiated by teachers, where they defined which classes would be able to access the chat session and what the students would be requested to chat about. Teachers could invite other classes to participate, and if they accepted, they were added to the classes which could access that particular chat session. Once paired up to chat, students were presented with the teacher-decided role play scenario, for example, student A is a shopkeeper and student B would like to purchase some fruit, or both students are sitting in a dentist's waiting room and they should chat about why they are visiting the dentist. Both parts of the conversation were recorded for later access by the students' own teachers. So while both teachers could hear both parts of the conversation, each teacher could only provide feedback on their own student's production.

The voice to voice chat facilitated inter-class, inter-school and intra-school chats among students. This feature really opened up opportunities for students to chat with other Irish speakers, who were not in their own class or school and who potentially spoke a different dialect of Irish. Suddenly Irish became a living language for students, where meaningful communication was a facet of what had been just a school subject to them – "it has helped me talk to people in Irish in a real life situation. It changes your views on speaking Irish, it gets better."

SMS was again integrated to deliver vocabulary words, phrases or quizzes to students. This time, teachers were presented with options for setting up and disseminating SMSs to students. They could set up a pool of words, which could be drawn from the system bank or entered themselves, and schedule messages to be delivered from the custom pool they created. They could also decide to send messages relating to scheduling, course reminders or words from the vocabulary pool straight away. Some difficulty was experienced with the delivery of SMSs to students, as the restrictions applied to phones by the network provider, were overly restrictive – and failed to permit the receipt of SMS. The process of trying to resolve this issue across the 210 project mobile phones lasted approximately three months. By the time phones were receiving messages as envisaged, teachers had already adopted their method of use of the mobile technologies, and they did not include the SMS component to same extent as it had been used through MALL.

The teacher online interface included new features that had not been in the MALL version. Teachers could now set up custom voice response sessions for their students, record their own questions online to add to their available pool of questions, set up schedule deliveries for SMS and set up voice chat sessions. They could also view, listen back to and provide feedback on students' voice response recordings and voice chat recordings. Additionally, teachers could mark any student answer or recording as a sample answer, which was then made available to other students in that class. FÓN also introduced a new student online interface, where they could access all of the recorded responses and chat sessions they completed in the context of the questions and role play scenarios they were in response to, compare their answers to teacher-marked sample answers, access teacher feedback associated with their recordings and view all SMSs which had been sent to them.

Teachers tended to integrate the technologies in different ways. Some teachers mostly used the voice response system in class after preparing students for a new topic. They then roved around the classroom, helping students with any difficulties or unknown vocabulary encountered. These words and phrases then fed into a round up session nearing the end of the class. Other teachers employed the voice response system largely as a homework activity. They followed this homework assignment, at times, by logging into their online teacher area the following day using their overhead projector and screen, and playing back samples of students' recordings (with their permission) to discuss the recordings as a group.

The voice to voice chat component was usually carried out as a homework activity, especially where the pairing allowed inter-school conversations. This facilitated the conversations taking place, where Irish class schedules did not match. Students were told to log in at 7pm in the evening and to chat with their partner around the scenario they were presented with.

3.4 Findings from the FÓN Project

Preliminary findings from the FÓN project reinforce and extend findings from the MALL project. A mixed methods approach was also used during this phase of work to gather qualitative and quantitative data about the project. Student and teacher pre- and post-project questionnaires were used, on site observation visits and informal interviews, semi-structured post-project interviews with teachers and students and teacher reflective diaries.

The response rate for the teacher pre-project questionnaires was 75% (12/16) and post-project questionnaires was 50% (8/16).[11] The response rate for the student pre-project questionnaires was 80% (293/368) and post-project questionnaires was 72% (265/368). As the second term of the project finished at a busy time of year, it is hoped that the outstanding questionnaires can be gathered after schools start back in September. As before, the data gathered from the numbers of teachers participating is not sufficient for sophisticated statistical analysis. However, data is reported here in percentages for ease of reading.

3.4.1 Student Oracy and Communication

The majority of students, 74%, reported that they felt that their Irish had improved as a result of participating in the project. Students referred to how they had more words to choose from, the increased ability to provide fuller answers to questions over time and a better grasp of understanding different Irish accents and dialects. This perception matched what teachers reported about student competence. Teachers noticed an increase in students' competence which they attributed to FÓN. They made reference to improved comprehension skills, including alternative accents and dialects, increased vocabulary, better fluency and more use of appropriate tenses and verbs. When listening back to students' responses from across the active term, teachers noticed a marked improvement in students' competency and fluency when comparing the initial weeks to the final weeks.

From the pre-project questionnaire to the post-project questionnaire, teachers' perceptions of students' spoken and listening abilities in Irish improved. Before the project, 27% of teachers reported that their students' speaking skills were *very good* and 36% *good.* After the project, 75% of teachers reported that their students' speaking skills were *very good* and 25% *good.* In listening, scores also increased from 9% *very good* and 18% *good* to 50% *very good* and 50% *good* respectively. Interestingly, teachers' perceptions of students' abilities in reading and writing also improved.

Students' own perceptions of their level of Irish across the four skills were also shown to improve between pre- and post- project questionnaires. Before the project, 8% and 44% of students considered their speaking skills to be *very good* and *good* respectively, with these percentages increasing to 22% and 47% for *very good* and *good.* Similarly, perceived listening skills showed a small improvement. As for teachers, reading and writing also showed improvements between the pre- and post- project questionnaires. Students reported that they felt more comfortable speaking Irish than they had previously, a finding which was echoed by their teachers. Teachers reported that students, across all levels or ability, exhibited improved confidence and competence in speaking Irish. 73% of the students reported that their competence in Irish had improved as a result of taking part in the project.

Students reported speaking more Irish than before, as they now had additional opportunities to speak Irish at home and with their friends. 78% of the students felt that they had spoken and used more Irish during the project than they had previously. All teachers (100%) agreed that students had used and spoken more Irish as a result of the project, than they had before.

As within MALL, teachers noted students' positive shift to more autonomous learning, where teachers felt they were now "facilitating students, rather than lecturing."

3.4.2 Student Motivation and Enjoyment

Students reported that learning and practising Irish with the assistance of the FÓN system made Irish more fun, interesting and enjoyable. Teachers noted an improved student attitude for learning Irish, which they attributed to their using the FÓN system. Teachers reported that students were very excited and positive about using the FÓN system and that there was an additional novelty factor for students as they were permitted to use their mobile phones, which must usually be kept out of sight in school, for learning Irish. One area of the FÓN system which caused some difficulty for students during the initial few days or week of its use, was learning how to navigate the instructions over their phone which were through Irish. The instructions provided students with the prompts for entering their user-numbers and PINS and directed students to press 1 for voice response or press 2 to chat with someone. Until students familiarised themselves with these instructions, they acted as somewhat of a barrier to their accessing and using the system.

Teachers also reported a marked improvement in students' enjoyment levels for Irish between the pre- and post-project questionnaires. Figures moved from 50% *really enjoy/enjoy learning Irish* to 100% *really enjoy/enjoy learning Irish.* A 100% of the teachers also reported that they thought their students had enjoyed learning Irish more as a result of using the FÓN system.

Within the pre- and post-project questionnaires, students were asked to indicate an Irish learning activity that they particularly enjoyed, and to provide a reason why they had enjoyed it. Through the pre-project questionnaires, students indicated a range of activities that included collaborative activities or active/ task-based and goal-orientated activities, as activities that they particularly enjoyed. The reasons cited for choosing these activities, related to working with others, and that they were fun and interesting activities. The post-project questionnaire saw the FÓN project as the most frequently mentioned activity (cited by 58% of students) that was enjoyed by students. When students were asked what they enjoyed about participating in the FÓN project, 25% mentioned that it was a "fun way of learning", 17% said that it was a new or different way to learn and 14% that they enjoyed talking to other people. A further 16% felt it had helped them to learn.

Teachers reported an element of competition among classes, where there was more than one class participating in a school, to see which class could transition through the topics fastest. One teacher in particular, when discussing student motivation, reported that "motivation went through the roof!" Teachers reported that 66% of students used the FÓN technologies more than requested, indicating that they enjoyed using them to some extent or felt that it was a useful activity.

3.4.3 Student Competency and Learning

As noted in Section 3.4.1, both teachers and students perceived improvements in students' abilities across the four skills – speaking, listening, reading and writing. Particular reference was made to students' improved ability to understand various accents and dialects of Irish. From the students' perspectives, they mentioned the accents/dialects as a challenge to begin with, but something that they could later understand without difficulty.

Teachers referred to their perception that students were mastering topics faster, through using the FÓN system, as they had the opportunity and impetus to practise and reinforce topics until they felt that they were fully competent in them. As within MALL, references were made to increased vocabulary, fluency, and better use of tenses and verbs. This was also voiced by students, who referred to how they were learning faster when using the FÓN technologies than without them - "you get it into your head quicker and learn quicker."

Students reported that using the FÓN system was a better way to learn Irish than using books and notes all of the time and that they only needed to remember their mobile phones while participating in the FÓN project, rather than needing to carry books home to complete homework activities. 73% of the students felt that their Irish had improved more than usual as a result of their participation. An advantage of the FÓN system cited by both teachers and students was the increased ability to speak Irish outside the classroom and for parents to get a sense of the learning process. Log-in records from the FÓN system showed students logging in as late as 10pm and 11pm at night to complete voice response sessions and as early as 6.43am. Positive parental feedback about the project was also evident during parent-teacher meetings and school open days.

3.4.4 Monitoring and Assessment

Students made references to their ability to assess their own level of competence or to exact some measure of quality about their performance. From the students' perspectives, they liked being able to listen back to themselves and correct their answer until they were happy with it - "I can hear my mistake and try to do better." Students also reported comparing their level of ability to students they had voice to voice chat sessions with, and how sometimes, they helped the other student along. Where teachers played back student recordings in class for the class to critically analyse, students made reference to how they could gauge their own level of ability against that of their classmates'. They also used the voice response sessions to test their knowledge of the content which had been learned in class.

References were made by students to how the FÓN system helped them to realise that they never fully knew a topic, even if they had learnt everything that had been covered in class. The variety in the FÓN system content meant that students could be presented with content that was not always phrased in the same way as their teacher phrased it, or with additional vocabulary to what they might have learnt in class. Teachers referred to this feature positively, in that students' realised that language was not as finite as what was learned in class and that there was always more to learn.

The sample answer feature offered teachers an opportunity to show students what a good answer would sound like and for students to compare their answers to marked sample answers within their online area. Teachers did not tend to use this facility to its full extent, perhaps as there were already many system features to familiarise themselves with. As a result, few sample answers were available for students to access.

Teachers reported that the system was suitable for all levels of learner, so they could monitor every learner, regardless of what level of competence they had. As discussed above, teachers referred to their ability to hear quieter students who would not usually be vocal in class. They mentioned how it was possible to pick up on recurring mistakes among students' answers and address them later in class. They were positive about the facility that enabled them to set up an oral exam session, which all students could take at one time and that they could return to when time allowed, to mark and provide feedback on. Until they participated in the FÓN project, all student oral assessment required scheduling class time or other available school-hour slots, to examine each student individually. Teachers felt the FÓN system offered a great way to monitor whether students were completing oral homework assignments, a feature which is usually difficult to gauge as there would not necessarily be any tangible output. All teachers reported positively on the FÓN system's potential to provide a medium for students to practise for an oral examination.

Some teachers felt a little overwhelmed by the volume of student recordings and felt they needed to provide feedback on each one. Having learnt from MALL that this could be a potential challenge for teachers, it was suggested during teacher training that teachers might provide feedback once per week or for particular sessions, rather than for every student recording. This would reduce student expectation for receiving feedback on every session they completed and ease the time burden on teachers to work through the many recordings. Teachers recognised that students were learning, even if they weren't providing feedback on each and every student recording.

3.4.5 Logistics / Practicalities

When the content for the FÓN system was being assembled, and volunteer celebrities sought to record the voice response questions, it proved difficult to acquire speakers with Ulster Irish, which is spoken in Northern Ireland. As a result, teachers in Northern Ireland felt they could not make as much use of the bank of questions available within the FÓN system, as the teachers in the Republic of Ireland could. The participating students in Northern Ireland had been learning Irish for a period of approximately three years, while those in the Republic of Ireland had been learning Irish for ten/eleven years, so teachers did not wish to overwhelm their students with alternative accents and dialects. This also implied that there was more content available for the Republic of Ireland schools, as students could progress through the levels of content. The students and teachers in Northern Ireland were more restricted to the beginner and introductory levels of content. This imbalance could be addressed to some extent, as teachers could record their own questions and add them to the content available to their classes. However, this also implied additional work for the teachers in Northern Ireland. Any potential extension to the pilot project would address this imbalance and ensure that all teachers and classes are more equally catered for.

Within MALL, the concurrent call capacity of the system was 30. For FÓN, this figure was increased to 100. As a result, there were no issues reported of dropped calls or difficulties in connecting to the system as a result of the FÓN system capacity. One local issue presented a connection problem – one school had three classes participating. The local network mast had a maximum call capacity of 50, which meant that when phones were being used collectively within the school (where Irish classes were scheduled at the same time), students sometimes experienced difficulty in connecting via the network.

Students had indicated that the quality of the sound recordings they heard through the voice response system in MALL tended to be a little low. Having recorded all content for the FÓN system in a recording studio, we were able to ensure the sound quality was at its optimum. Aside from comments referring to speaker accents and dialects, we received no comments relating to the sound quality of recordings.

Through each term of use within FÓN, 210 active mobile phones were in circulation among the teachers and students. An additional 20 phones were held by the project team for training purposes and to replace any breakages. The sheer scale of managing 210 mobile phones, allocating them to individual students and labelling them accordingly, as well as monitoring any reported problems, presented a large administrative load. This was especially evident during the changeover period from term one to term two, where mobile phones and chargers were collected, checked in, checked that they were functioning and reassigned to and labelled for a new student. The impact of supplying students with mobile phones is further discussed in the next section.

Where the FÓN system was used within class, students again needed to spread out to ensure that they could hear the prompts on the system and weren't disturbed by their peers. Interestingly, some teachers noted that students tended to be highly focussed when engaging with the FÓN system in class. They did not pay any heed to their peers if they tried to interrupt, until they had completed their own sessions first. Students also tended to pay little attention to what their peers were saying/recording around them.

Costs associated with the FÓN project were higher than the MALL project as the project ran for a longer period and the FÓN system development was a lot more sophisticated that it had been for MALL. The set-up costs as a proportion to the running costs were 5:2. Set-up costs included software development, purchase of hardware, server and call lines, teacher training and content development and recording. Running costs included mobile phone line rental and call charges. In this instance, teachers had a longer active time on the project of twelve weeks (rather than the five weeks within MALL). Set-up costs were more substantial than for MALL as the software developed was more sophisticated and a large amount of system content was developed and recorded. Teachers and students commented that a longer period would facilitate them in working with all components of the system and giving more time to the technologies overall. From a costs perspective, the split terms allowed the running costs to ultimately be halved, as well as allowing for a more scientific examination of the findings where user and non-user groups were present. Further investigation is required to weigh up the potential of a longer running term for all students within any potential expansion to the project.

Where SMS was integrated by teachers (mostly in term 2 following the network restriction issue), students reported that SMSs were arriving too early for them before they had woken up or while they were eating breakfast. This would confirm Kennedy and Levy's (2008) finding that students did not welcome the arrival of SMS too early in the morning. They reported that the optimum delivery time was between 10am and 1pm.

3.5 MALL and FÓN: Solutions, Recommendations and Future Research Directions

The findings of the MALL project fed into the scoping process for the FÓN project, as all findings across both projects will feed into any future work with schools.

While it was important to remove any cost from students, it is also important to reduce project set-up and running costs to make project scalability and feasibility more realistic. One potential avenue which could be explored to reduce costs involves students' use of their own mobile phones or landline from home. In order to address any issues of equality of access, a small pool of mobile phones could be assigned to each school to cover any student who does not own their own mobile phone. These actions would also alleviate the administrative burden of each student being assigned a mobile phone. Students would dial up to the system via a free phone number, so call costs are displaced to the project organisations. The majority of running costs of the MALL and FÓN projects was the monthly phone and line rental fee from the network. The call charges themselves were relatively small in comparison.

A further avenue which could be explored would be a partnership to include one of the mobile phone network providers, where they would absorb some/all of the phone rental and call charges. A longer-term project with equally as many or fewer students participating, could also remedy the requests from participants for a longer project term, without costs over and above those amounts already invested, being incurred. In the case of both projects, the start-up costs far outweighed the running costs. Hence, a longer running project would naturally prove more economical, as the start-up costs would be absorbed over a longer project running period.

The foci of use in these projects were on the call and SMS functions of the mobile phone. Using the natural communicative abilities of a phone and applying these to language learning to create real, meaningful and authentic contexts which are vital for effective language learning (Kiernan and Aizawa, 2008) resulted in increased student motivation, enjoyment and learning. Indeed, Willis (2007, no pagination) reports on research that suggests that "superior learning takes place when classroom experiences are enjoyable and relevant to students' lives, interests and experiences" (online). Uses of mobile phones which can be better served via other modes, for example internet browsing, were avoided.

In terms of developing students' oral abilities and providing opportunities for teachers to observe these, mobile phones and the FÓN system offer prospects for both to be achieved. These have previously proved difficult to achieve within the confines of the school timetable, where the target language is not the medium of instruction in the school.

From the partner organisations' perspectives, the projects were largely successful in achieving the aims they set out to achieve. Students spoke more Irish, beyond the school walls, they communicated with their teacher asynchronously and other students synchronously, they exhibited increased motivation for Irish through using mobile phones and the enjoyment that brought and they perceived and showed improvements in competency across the four language skills. From the perspective of assessment – students were facilitated in self-assessment, and teachers in formative and summative assessment of students.

4. CONCLUSION

"... a really great challenge to my stone age teaching methodologies – a renaissance"

(Teacher participating in the FÓN project)

Students can become disengaged from the learning process, as happens in the case of Irish, as they progress through school. Where students enjoy a learning activity or perceive it as worthwhile with attainable results, they are more likely to become or stay engaged. Couple these types of activities with some ICT which generally does hold students' attention for long periods of time – the mobile phone – and suddenly the learning potential is maximised.

Mobile phones have been integrated into language teaching and learning environments for a variety of purposes; vocabulary SMS, web-browsing to access language content and emails that contain language quizzes, to name but a few. The integration of mobile phones has shown to have benefits and challenges in all of the scenarios in which it has been integrated for language learning. Where the course of use is well planned to suit a mobile phone, the addition of the learning tool has been more positive than negative.

The MALL and FÓN projects have contributed to the growing literature about the integration of mobile phones for learning, practising and assessing languages, the benefits and challenges these technologies present and the practicalities of the presence of mobile phones in schools. It seems there can be a useful place for mobile phones in schools, which help to apply student motivation to the task at hand and certainly result in a learning *renaissance* for all involved.

5. REFERENCES

Brown, E. (Ed.). (2001). *Mobile learning explorations at the Stanford learning lab. Speaking of Computers, 55*. Stanford, CA: Board of Trustees of the Leland Stanford Junior University.

ComReg. (2008). Irish Communications Market: Quarterly Key Data Report. Document number 08/101. Commission for Communications Regulation. Retrieved June 10, 2009, from http://www.comreg.ie/_fileupload/publications/ComReg08101.pdf

Department of Education and Science. (2007). *Irish in the Primary School. Inspectorate Evaluation Studies*. Dublin: DES.

Dias, J. (2002). Cell phones in the classroom: Boon or bane? *C@lling Japan 10(2),* 16-22. The Newsletter of the JALT-CALL Special interest Group.

Dias, J. (2002a). Cell phones in the classroom: Boon or bane? [Part 2] *C@lling Japan 10(3),* 8-13. The Newsletter of the JALT-CALL Special interest Group.

Giollagáin, Ó. C., Mac Donnacha, S., Ní Chualáin, F., Ní Shéaghdha, A., & O'Brien, M. (2007). *Staidéar Cuimsitheach Teangeolaíoch ar Úsáid na Gaeilge sa Ghaeltacht: Príomhthátal agus moltaí.* Dublin: The Government Stationary Office.

Godwin, J. (2005). Messaging, gaming, peer-to-peer sharing: Language learning strategies and tools for the millennial generation. *Language Learning & Technology*, *9*(1), 17–22.

Government of Ireland. (1999). *Primary School Curriculum*. Dublin: The Government Stationary Office.

Harris, J. (2006). *Irish in Primary Schools: Long-Term National Trends in Achievement*. Dublin: The Government Stationary Office.

Hsu, H., Wang, S., & Cormac, L. (2008). Using audioblogs to assist English-language learning: An investigation into student perception. *Computer Assisted Language Learning*, *21*(2), 181–198. doi:10.1080/09588220801943775

Joint Council for Qualifications. (2008). *National Results: GCSE*. Retrieved June 10, 2009. from *Joint Council for Qualifications* website, http://www.jcq.org.uk/national_results/

Kennedy, C., & Levy, M. (2008). L'italiano al telefonino: Using SMS to support beginners' language learning. *ReCALL*, *20*(3), 315–330. doi:10.1017/S0958344008000530

Keogh, K., Koller, T., Ward, M., Úí Dhonnchadha, E., & van Genabith, J. (2004). *CL for CALL in the Primary School* eLearning for Computational Linguistics and Computational Linguistics for eLearning. International Workshop in Association with COLING 2004, Geneva, Switzerland.

Kiernan, P. J., & Aizawa, K. (2004). Cell phones in task-based learning. Are cell phones useful language learning tools? *ReCALL, 16*(1), 71–84. doi:10.1017/S0958344004000618

Kukulska-Hulme, A., & Shield, L. (2008). An overview of mobile assisted language learning: From content delivery to supported collaboration and interaction. *ReCALL, 20*(3), 271–289. doi:10.1017/S0958344008000335

Lee, T. (2006). Pleasurable Cities: A discussion paper. UK: FutureLab. Retrieved June 1, 2009, from http://futurelab.org.uk/resources/documents/project_reports/Pleasurable_Cities_discussion_paper.pdf

Levy, M., & Kennedy, C. (2005). Learning Italian via mobile SMS. In Kukulska-Hulme, A., & Traxler, J. (Eds.), *Mobile Learning: A handbook for educators and trainers* (pp. 76–83). London: Routledge.

Lu, M. (2008). Effectiveness of vocabulary learning via mobile phone. *Journal of Computer Assisted Learning, 24*(6), 515–525. doi:10.1111/j.1365-2729.2008.00289.x

Mac Gréil, M., & Rhatigan, F. (2009). *The Irish language and the Irish people: Report on the attitudes towards, competence in and use of the Irish language in the Republic of Ireland in 2007-'08.* Ireland: National University of Ireland, Maynooth.

Nah, K. C., White, P., & Sussex, R. (2008). The potential of using a mobile phone to access the Internet for learning EFL listening skills within a Korean context. *ReCALL, 20*(3), 331–347. doi:10.1017/S0958344008000633

National Council for Curriculum and Assessment. (2008). *Primary Curriculum Review, Phase Two: Final report with recommendations.* Research Report No. 7. Dublin: Author.

Pęcherzewska, A., & Knot, S. (2007). Review of existing EU projects dedicated to dyslexia, gaming in education and m-learning. WR08 Report to CallDysc project. June 2007. Retrieved June3, 2009 from http://www.calldysc.info/attachments/114_WR08_existingprojectreview.doc/

Prensky, M. (2004). *What can you learn from a cell phone? – Almost anything!* Retrieved June 3, 2009 from http://www.marcprensky.com/writing/

Shield, L., & Kukulska-Hulme, A. (2008). Editorial. *ReCALL, 20*(3), 249–252. doi:10.1017/S095834400800013X

Shudong, W., & Higgins, M. (2006). Limitations of Mobile Phone Learning. *The JALTCALL journal 2*(1), 3-14.

Smyth, E., Dunne, A., Darmody, M., & McGoy, S. (2007). *Gearing Up for the Exam? The Experiences of Junior Certificate Students*. ESRI: NCCA and the Gender Equality Unit funded research report. Dublin: The Liffey Press in association with the Economic and Social Research Institute.

Smyth, E., Dunne, A., McGoy, S., & Darmody, M. (2006). *Pathways through the Junior Cycle: The Experiences of Second Year Students*. ESRI: NCCA and the Gender Equality Unit funded research report. Dublin: The Liffey Press in association with the Economic and Social Research Institute.

Stockwell, G. (2007). Vocabulary on the Move: Investigating an intelligent mobile phone-based vocabulary tutor. *Computer Assisted Language Learning, 20*(4), 365–383. doi:10.1080/09588220701745817

Stockwell, G. (2008). Investigating Learner Preparedness for and usage patterns of mobile learning. *ReCALL*, *20*(3), 253–270. doi:10.1017/S0958344008000232

Taylor, R. P., & Gitsaki, C. (2003). Teaching WELL in a computerless classroom. *Computer Assisted Language Learning*, *16*(4), 275–294. doi:10.1076/call.16.4.275.23412

Thornton, P., & Houser, C. (2002). M-learning in transit. In Lewis, P. (Ed.), *The changing face of CALL* (pp. 229–243). Lisse, The Netherlands: Swets and Zeitlinger.

Thornton, P., & Houser, C. (2003). Using mobile web and video phones in English language teaching: Projects with Japanese college students. In Morrison, B., Green, C., & Motteram, G. (Eds.), *Directions in CALL: Experience, experiments and evaluation* (pp. 207–224). Hong Kong: English Language Centre, Hong Kong Polytechnic University.

Thornton, P., & Houser, C. (2005). Using mobile phones in English education in Japan. *Journal of Computer Assisted Language Learning*, *21*, 217–228. doi:10.1111/j.1365-2729.2005.00129.x

Twarog, L., & Pereszlenyi-Pinter. (1988). Telephone-assisted language study and Ohio University: A report. *Modern Language Journal*, *72*, 426–434.

Willis, J. (2007). The Neuroscience of joyful education. *Educational Leadership 64.* Retrieved May 30, 2009 from http://www.ascd.org

Wishart, J., Ramsden, A., & McFarlane, A. (2007). PDAs and Handhelds: ICT at your side and not in your face. *Technology, Pedagogy and Education*, *16*(1), 95–110. doi:10.1080/14759390601168106

ENDNOTES

1 The National Centre for Technology in Education provides advice, support and information on the use of information and communications technology (ICT) in education. See http://www.ncte.ie.

2 FÓN is also the Irish word for 'telephone'.

3 Foras na Gaeilge is an organisation responsible for the promotion of the Irish language throughout the whole island of Ireland. See http://www.gaeilge.ie.

4 Monica Ward has reported on the use of an Intelligent Computer-Assisted Language Learning tool in primary schools in Ireland for teaching and learning Irish (see Keogh, Koller, Ward, Úí Dhonnchadha and van Genabith, 2004).

5 The three stages of the communicative approach are (1) pre-communicative phase, (2) communicative phase and (3) post-communicative phase. When taught in full, the communicative approach embeds enjoyable and task-based activities but also grammar, structures and accuracy in language learning.

6 Hsu, Wang and Cormac (2008) reported on students' use of mobile phones to record and upload audio files to an audioblog in response to comprehension materials posted by their teacher.

7 WAP stands for Wireless Application Protocol. This can involve accessing the web from a mobile phone or refer to the format of webpages which can be viewed on a mobile phone.

8 For example, Thornton and Houser (2005) reported that their students sent an average of 200 emails per week via their mobile phones.

9 A blended learning environment is a setting where mixed methods are used in teaching. In this instance, traditional teaching methods and those facilitated by the MALL and FÓN systems.

10 Learnosity Ltd., formerly Websonic Ltd., was the company contracted to develop the software solution for both the MALL and FÓN projects, according to the specification supplied. http://www.learnosity.com

11 13 individual teachers participated in the project. Some teachers participated in term one and term two of the project, so were counted twice for evaluation purposes.

Chapter 14
The Pedagogy of Mobility

Kathy M. Stewart
Macquarie University, Australia

John G. Hedberg
Macquarie University, Australia

ABSTRACT

Direct student experience of the real, live organism, object, place or environment is recognised by teachers and other educators as having powerful potential for high quality learning. Rendering this rich experience into explicit learning contexts for students remains an on-going pedagogical challenge. The use of mobile technologies as a way for students to capture their experience as it happens in the real world holds great promise as a vehicle to support authentic learning. This chapter presents a series of cases where different school groups were challenged with different mobile learning activities. The first two cases are derived from research studies in the local environment. In these two cases students were involved in inquiry-based learning where they used data collected in the field with handheld computer and digital cameras to assess the health of their local environment. The second two cases are derived from complementary geography tasks in the field. In one case students worked in pairs to navigate a given route, and in the fourth engaged in a debate about an issue pertaining to urban planning using text- and picture- messaging technologies. The results of the cases are analysed with a view to generating some general principles around the technologies, the activities and the tools that comprise the pedagogies embodied in these cases. It is hoped that they will inform more effective professional practice, specifically when teaching spatially dependent tasks and, more generally, in mobile collaborative learning environments.

DOI: 10.4018/978-1-61692-849-0.ch014

INTRODUCTION

"The fact is that to achieve high quality learning inside the classroom you have to take students outside the classroom."

(Principal, Environmental Education Centre)

The pedagogy of mobility embraces the idea of thoughtful teaching practice that engages students in meaningful learning in environments beyond the classroom. Experiences gained and captured in the field form part of a sequence of learning activities embedded in a unit of classwork. Effective use of mobile technologies enables a seamless connection between learning environments where both teacher and students can access a wide range of activities and resources as well as their own materials to support learning across time and place. This chapter seeks to explore the complex issue of effective pedagogy relevant to this 21st century learning situation by exploring the practice of teachers and researchers as they scaffold student learning using mobile technologies in local outdoor environments.

The broad aspirations for high quality learning and teaching held by teachers and other education professionals are based on the application of social-constructivist perspectives drawn from the work of researchers and theorists such as Piaget, von Glasersfeld and Vygotsky. This perspective holds that learners, individually and as part of a social group, actively construct their own understandings of the world as a result of their experiences through processes of construction and re-construction of mental frameworks. Clearly, in the context of this perspective, there would be no 'one size fits all' teaching approach. Rather, sets of principles or guidelines that describe high quality learning have been developed based on research of effective teaching and learning practice (e.g., Department for Education and Skills, DES, 2004; NSW Department of Education and Training, NSWDET, 2003a).

Common elements of 21st century teaching and learning involve a learner-centred approach (DES, 2004; National Research Council, NRC, 1999) that is based on deep understanding of key concepts (NSWDET, 2003a; NRC, 1999) that are linked to the learner's real world contexts with access to real sets of data and information (DES, 2004, NSWDET, 2003a; NRC, 1999). Rather than a fixed body of information, knowledge encountered by students is seen as problematic and may be viewed from a range of perspectives. Students are encouraged to actively engage with their learning, being aware of teacher expectations and any assessment criteria that may be applied to learning activities. The role of the teacher becomes less of an authoritative source of information and disciplinarian to more of a facilitator of learning – a coach or a guide for students. Students clarify and build their understandings by engaging in sustained conversations around the ideas and concepts that they encounter in their learning activities (Laurillard, 2007; NSWDET, 2003a).

This chapter considers the nature and role of fieldwork within this context and discusses research of teaching practice or pedagogy associated with fieldwork with a view to locating practice within a 21st century learning context. Four case studies are used to demonstrate the use of mobile technologies to support student learning and the pedagogical principles associated with these cases are discussed.

Fieldwork and Mobility

Rickinson and his colleagues (2004) in their comprehensive review of research relevant to outdoor learning concluded that "well planned, competently delivered and effectively followed-up fieldwork can result in learning that supplements and enhances students' experiences in the classroom" (Rickinson et al., 2004, p. 52). The review acknowledged the increasing number and range of research studies conducted over the past two decades including accounts of work of and for practitioners specialising in outdoor education. However the review identified areas for further research especially in the areas of increasing understanding of the learning processes involved in fieldwork and the need for grounding of teaching practice in learning and education theory. This chapter aims to contribute to a more explicit stance on both the theoretical and practical elements of pedagogy in fieldwork. In much the same process as that demonstrated by Brody (2005) in the evolution of his theory of learning in nature, this chapter utilises Glaser and Strauss' grounded theory methodology (1967) to generate theory from practice while also drawing from established education theory.

Authentic Pedagogy

Hein (1998) considers that any educational stance (such as that of a fieldwork provider e.g. an environmental education centre, a museum or a botanic garden) requires a theory of knowledge, a theory of learning and a theory of teaching. A theory of knowledge considers epistemological questions about the nature of knowledge and from a social constructivist perspective, knowledge is constructed through *interaction* between the learner and their physical and social environment (e.g. Driver, Asoko, Leach, Mortimer & Scott, 1994; Rennie, Feher, Dierking & Falk, 2003;). Definitions of learning arising from this perspective can be seen in the work of Brody (2005) who states that "meaningful learning in nature is a result of direct experience(s) over time in which personal and social knowledge and value systems are created through complex cognitive and affective processes" (p. 610).

Over the past decade considerable research has endeavoured to illuminate characteristics of pedagogy that contribute to effective learning for all students. This work can be seen as the basis for a theory of teaching. The 'authentic pedagogy' movement (Newmann, 2000; Newmann and Associates, 1996) articulated "three broad criteria for authentic intellectual work:

1. Construction of knowledge: using or manipulating knowledge as in analysis, interpretation, synthesis and evaluation, rather than only reproducing knowledge in previously stated forms.
2. Disciplined inquiry: gaining in-depth understanding of limited topics, rather than superficial acquaintance with many; and using elaborated forms of communication to learn and to express one's conclusions.
3. Value beyond school: the production of discourse, products, and performances that have personal, aesthetic, or social significance beyond demonstration of success to a teacher." (Newmann 2000, p. 2)

This research has been adapted to an Australian context and articulated for schools in New South Wales as the Quality Teaching model (NSWDET 2003a; 2003b). This model "identifies generic qualities of pedagogy that have been successfully applied in a range of school contexts and are shown to lead to improved student learning." (NSWDET, 2003a, p. 5). Ballantyne and Packer (2009) utilised a similar model to interrogate teaching and learning in environmental education centres across Queensland and have proposed a pedagogical approach specifically designed for teaching in natural environments. They define experience-based learning as comprising five elements: learning by doing; being in the environment; real life learning; sensory engagement and local context. This pedagogical approach captures the active, hands-on aspect of fieldwork where students are encouraged to use their senses to interact with their environment. This conception of experience-based learning also incorporates student involvement in authentic tasks and problem solving activities based around real issues in their environment.

A further aspect of learning in the field identified by researchers (e.g. Baumgartner & Zabin, 2008; Brody, 2005; Dunphy & Spellman, 2009; Kent et al., 1997) is the social aspect of the activities. Dunphy and Spellman (2009) in their study of undergraduate geography students found that although students' learning styles appear not to affect their fieldwork experience, the social and inspirational aspects of fieldwork are highly valued. "While these aspects may seem superficial and academically unimportant, they clearly influence students and motivate them to learn, and certainly direct them to consider fieldwork extremely valuable" (p. 27).

Mobile Learning

In their review of fieldwork in geography teaching in higher education Kent, Gilbertson, and Hunt (1997) drew attention to the changing nature of field studies – especially with the introduction of emerging technologies at that time.

Old observation, surveying and sketching of landforms has been supplanted by hypothesis generation and testing, involving measurements of phenomena such as streamwater quality using a range of instrumentation... the use of portable computers record and provide instant analysis of project data while still in the field. (Kent et al., 1997, p. 314)

The use of 'portable computers' has expanded with increasingly smaller and powerful devices being used by educators to support fieldwork activities. This emerging phenomenon has resulted in a growing body of literature associated with description and analysis of practice (e.g. Naismith, Lonsdale, Vavoula, & Sharples, 2004; Sharples, 2007). Mobile learning implies the concept of the mobility of the learner supported by handheld devices that provide access to the affordances of the classroom including internet access, data logging facilities and seamless transfer of information across learning environments. Reflections by educators and researchers working in this area emphasise the "broader concept of learning taking place in the mobile age, rather than the use of the narrower term 'mobile learning'" (Taylor, 2007, p. 26).

Mobile technologies can support student-centred and inquiry-based pedagogies (e.g. Anastopoulou, Sharples, Wright, Martin, Ainsworth, Benford, Crook, Greenhalgh, & O'Malley, 2008; Stewart, 2007; Tan, Chen, Looi, Zhang, Seow, & Chan, 2007; Thompson & Stewart, 2007) and also both personalized and collaborative learning (e.g. Chen, Kao, & Sheu, 2003; Swan, van 't Hooft, Kratcoski, & Unger, 2005). Clearly the way in which mobile technologies are used has to be appropriate to the pedagogical approach and to the learning activities themselves. "Technology in itself won't make the difference; it's what students do with it that does" (Swan et al., 2005, p. 110). The following discussion describes the use of mobile technologies to support a range of fieldwork activities in science, environmental education and geography in primary and secondary schools in Australia and Singapore.

Pedagogy of Fieldwork

The pedagogy of fieldwork can be viewed from three related aspects: the activities designed for the event, the artefacts produced as a result of the event and the assessment strategies for the event. This discussion focuses on the teaching and learning associated with fieldwork event itself rather than detail its links to the wider school context. This wider context necessitates further considerations of issues such as the appropriate placement of fieldwork within a learning sequence; the administration and management of fieldwork; assessment of any risk associated with the activity.

Activities

Study of research in fieldwork reveals scant detail of the activities designed for learning in the field. Broad categories of activities are identified and these include: field identification of plants and animals; observation using the senses; development of fieldwork skills such as sketching, data collection; development of team work skills; problem solving activities (e.g. Baumgartner & Zabin, 2008; Chew, 2008; Cook, 2008; Halocha, 2005; Thompson & Stewart, 2007).

The activities for a real-world based learning event can be conceptualised by considering the ways in which the particular group of students can achieve a set of learning outcomes and experiences envisaged for the fieldwork. Articulating the purpose for the fieldwork serves as a lens to focus the teaching and learning activities students will encounter and experience during the event. Clear learning outcomes for fieldwork also inform the nature of assessment that can be used to determine if students have achieved these outcomes. The case studies reported in this chapter demonstrate the wide variety of reasons teachers hold for fieldwork. With each different pedagogical requirement teachers can select those technologies and the affordances they provide to support and enable particular learning activities (Bower 2008).

Pedagogical decisions involved in designing learning activities for fieldwork often include:

- positioning fieldwork within the teaching sequence of the topic being studied,
- the content and structure of the day's activities,
- selection of appropriate technologies to support student activities,
- organisation of student groups and student roles within the group.

The cases reviewed in this chapter explore the pedagogical decisions made in these four areas. Cases involve the use of a range of mobile devices, including hand held computers and mobile phones. It appears that each device has its own set of pedagogical issues that can be identified within the context of its selection for use in the field.

Artefacts

Baumgartner and Zabin (2008) stress the importance of producing "significant and useful artifacts" (p108) as part of the processes involving fieldwork activities. To record the processes of fieldwork experience, modern technology provides many options including high quality digital photographs, graphs, tables and charts.

Traditionally a teacher-designed paper worksheet completed by students during fieldwork served as the record of the event. The nature and effectiveness of this record can be greatly enhanced with the use of mobile technologies (e.g. Stewart, Thompson, Hedberg & Wong, 2009). Students are able to access permanent high quality representations of their fieldwork experience captured as they are engaged in the moment. These data and image artefacts can then be used as part of larger, richer tasks associated with the topic itself. Technologies and their associated pedagogical scaffolds can be employed to ensure effective completion of the fieldwork tasks and activities. The scaffold may take the form of a teacher or student created artefact saved on a hand-held computer or a web-based activity sequence accessed in the field. Pedagogical decisions about the nature of the artefacts of fieldwork include:

- determining the nature and extent of *scaffolding* required for students to achieve the learning outcomes of the fieldwork events in the context of the wider outcomes of the unit,
- determining the *nature of the product/record* of fieldwork, and
- selection of *appropriate technologies to underpin the production of the field re-*

cord and those that document the learning trajectory.

There is always tension between the need to provide all step-by-step instructions and support, and the desire to allow a completely student-designed solution. The designer's concern is to ensure that the learning activity is not wasted time or overly prescriptive circumscribing alternate processes or sequences. The choice of the wrong artefact might lead in turn to the loss of data or the loss of the place at which it was collected thereby limiting its use in some tasks.

Assessment

There is a close relationship between learning outcomes, the artefact and the final assessment. The desire is to ensure that assessment task is not trivialised to be a completed worksheet, but at the same time to capture process and product, strategy and solution. The choice of appropriate activity, artefacts and technology will resolve many of the assessment challenges, as the cases which follow illustrate. Several principles can help to guide the choices to be made by the learning designer. The cases illustrate these as:

- technology stamping of place and time.
- using tools that capture the message or piece of data every time they are collected reducing the risk of loss of evidence
- assessment strategies to capture evidence in visual, textual, numerical and temporal modes,
- assessment processes that demonstrate the sequences of solution and the choice of elements in the final solution.

Other assessment choices that are illustrated in these cases indicate that the written field report might be varied at times to emphasise the planned solution, the collected evidence, and the analysis of that evidence into a final product.

Thus the following four cases describe a variety of fieldwork events each employing different mobile devices. Each case describes the context of the project: the learning task, students involved, and operating principles for the project. In particular they explore the pedagogical practice that enabled the achievement of the activity goals through a discussion of the learning activities for the event, the artefacts produced as a component of the learning activity and assessment strategies that indicated the achievement of the learning outcomes.

Case 1: Using Tablet PCs to Support Primary Students Involved in Fieldwork at their Local Creek

A class of 30 upper primary students and their teacher from a suburban primary school located in the south-western part of Sydney, Australia, formed part of a school-university project to explore the potential of mobile technology to enhance learning. The affordances of mobile technology were incorporated into a school-based unit of work on water focussing on increasing students' knowledge of and familiarity with their school's use of water and its position as part of the local water catchment area.

A main operating principle of this project was for a core team consisting of two classroom teachers and a university lecturer in science education to devise an inquiry-based learning activity incorporating fieldwork and the use of mobile technology. Specialist assistance with ICT was provided by a university-based support centre and the project liaised with the school-based ICT teacher. Information about the local environment was sought from the local council and from a local community-based amateur naturalist group. Hand held tablet personal computers (tablet PCs) provided mobile access to generic word processing software and device-specific drawing software. Digital cameras with macro facility were used to provide landscape and close-up images of environmental features.

As Figure 1 demonstrates, numerous fieldwork events were planned to be conducted throughout the unit. The purpose of field activity included:

- to locate and record waste water drainage points on the school grounds
- to use local government maps of storm water drainage to track the movement of rain water from the school, through suburban streets to a creek just over one kilometre from the school.
- to conduct a study of plants and animals at the creek assisted by members of the local community

Activities: What do the Students do?

Learning outcomes associated with the fieldwork activities clustered around increasing student knowledge of and experience in their local area and developing skills in fieldwork techniques and in using mobile technologies. A further set of skills related to the social aspect of the experience and included working as part of a team and effective interaction with knowledgeable adults. These outcomes can be identified specifically as follows:

- increasing student knowledge of waste water management in their school and in the local area, and specific local plants and animals
- increasing student fieldwork skills in: making accurate observations; recording observations; and organising field data
- increasing student skill in using the technologies; transferring data; and selection and use of mobile electronic devices
- developing student skills in collaborative teams and interacting with adults to obtain information

Figure 1. Unit of work

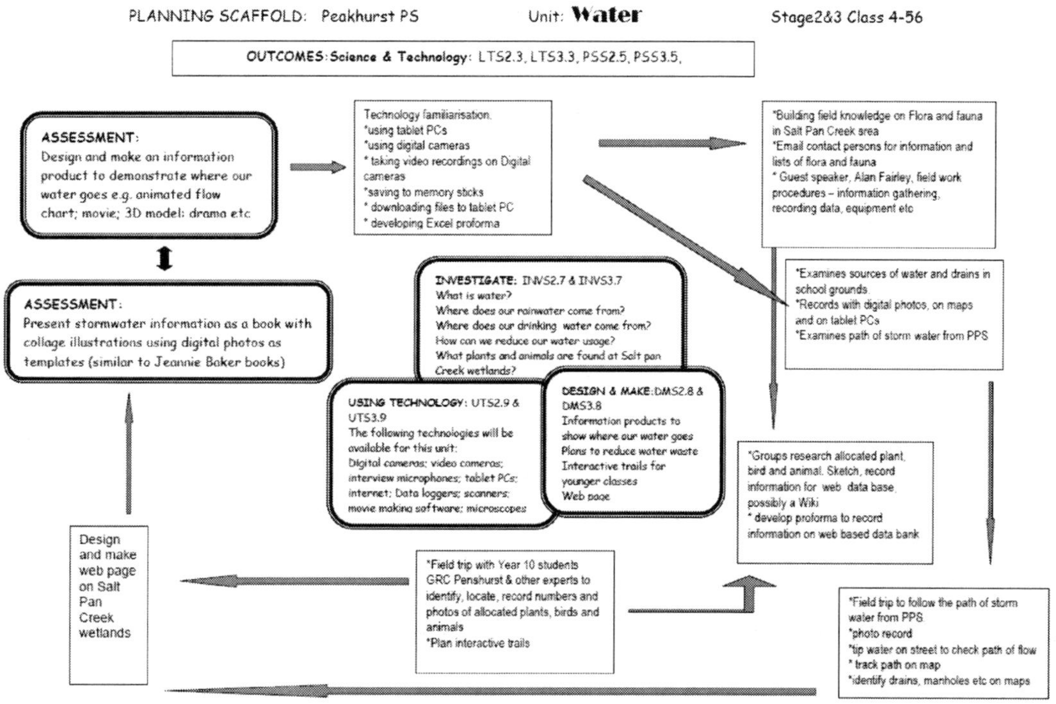

- increasing student awareness of environmental issues: need to minimise street litter to avoid its movement into the local waterways; diversity of local plants and animals; urban infrastructure and the maintenance role of local councils; and the role of volunteers in the local community

Articulation of this quite substantial set of outcomes reflects the rich learning potential of fieldwork, providing a broad palette of opportunities for the classroom teacher to develop more explicitly as part of the broader teaching and learning sequence of the unit of classwork. In this particular case successive field trips provided a number of opportunities for achievement of these outcomes.

Students working in groups of four using a tablet PC and digital camera per group were involved in the following activities as part of their fieldwork: using a map to locate drains and underground pipes; photographing physical aspects of the urban landscape; locating and identifying specific plants and birds; photographing plants and animals; making an audio recording of information from adults, interpretive signs; or bird calls; keeping records of their observations; and transferring digital information between devices.

Artefacts: What do Students Produce by the End of the Activity?

Students involved in this project created their own digital scaffolds to record their field observations. These scaffolds included the use of tables, hand written descriptions incorporating the use of both generic and device-specific software. For example one group linked each storm water drain from the school to the creek to their photographs and an assessment of the area. Students captured a large quantity of high quality digital photographs including images taken through a binocular microscope. These digital records created in the field were transferred to the school's computer system and used to create student presentations about their assessment of the health of the local catchment.

Assessment/Evaluation of these Products

The permanent nature of the products of the fieldwork enabled the project team to view students' work. A number of aspects of students' work were outstanding. Firstly the digital photographs produced by students were, without exception, of a very high quality. These student-produced resources were then used to support a wide variety of learning activities ranging from creating digital artworks to illustration of information reports about local plants and animals and fieldwork reports assessing the quality of the local environment. A second area where students excelled was in the nature of the scaffold they created to capture and record their fieldwork. The best examples of this student-directed task incorporated the use of tables and photographs as well as written information such as the date, members of the group, fieldwork procedures and observations.

This case study demonstrates that, as mobile learners, students are more than capable of incorporating the use of powerful computing capabilities into their own investigations of their local environment. In this pedagogical context the teacher does not have to be a content or ICT specialist but Baumgartner and Zabin (2008) comment that for some teachers the role of co-learner may be 'counterintuitive' to more traditional teaching approaches. The role of the teacher becomes that of a facilitator of student learning, managing student access to scientific expertise and training in the use of devices and project software. Clearly the educational 'richness' of learning situations that place students in some control of their own learning, that locate learning in environments beyond the classroom can be more accessible with the increasing availability of portable computers.

Case 2: Using Handheld Devices and Web 2.0 Technology to Support Mixed Groups of High School and Primary School Students Monitor their Local Environment

Ninety students from middle year classes across a comprehensive high school and its local primary schools were combined in mixed groups to work collaboratively to assess the health of their local environment. The rationale for the wider project (of which fieldwork formed a component) was to create links between schools to focus on the design and implementation of a multi-disciplinary, inquiry-based unit of work focusing on sustainability. The schools involved in the project are located in Sydney's northern beach-side suburbs with two classes of 60 students from the final year in the primary school combining in smaller groups with a class of 30 first year high school student mentors.

The project was supported by a joint school-university funded ICT centre which provided specialist ICT expertise and the mobile devices. Closely linked to the project schools was the local environmental education centre which provided teaching expertise and access to information about local ecosystems. These ICT specialists and environmental educators worked with student groups in the field to use project technologies to capture and record information about the biodiversity of the local area.

The devices used in the project included hand held computers with data logging capabilities as well as word-processing and drawing software compatible with more generic software used in the school environment. Digital cameras were used in their capacity to produce quality landscape and macro images of environmental features. Web 2.0 technology (a wiki) was used as a tool to support a range of collaborative activities including support of a discussion forum between teachers, between teachers and students and between the students themselves. The wiki served as a site for each group of to students to record, store and access their fieldwork data from a number of locations including their school, home and ICT centre.

Activities: What do the Students do?

The operating principle of this project involved a core group of classroom teachers designing, conducting and evaluating the multi-discipliary unit of work. The expertise of the ICT and environmental educator support teams was called upon in the planning stages and during the teaching sequence of the unit of classwork. The fieldwork activities were located in the 'skill building' phase of the inquiry-based unit. In this early stage of the teaching sequence students were engaged in building their understandings and experiences in a range of areas including: understanding of local environmental and sustainability content, expertise in use of the project technologies, teamwork and collaboration, mentoring, thinking and organisational skills. These understandings and skills could be assessed by the project before students moved on to demonstrating their understandings by creating a group presentation around their interpretation of human impact on the local environment.

Artefacts: What do Students Produce by the End of the Activity?

Before the commencement of the unit, project teachers attended the ICT centre for professional development in the use of the project devices and the project wiki. After the commencement of the unit, staff from the ICT centre visited the high school to train the secondary students in the use of the handheld computer and its accompanying data loggers as well as the digital camera. These students then trained the primary members of their team in the use of this technology during the two fieldwork events associated with the project.

The first fieldwork event was held at the ICT centre. Mixed groups of students from the high school and two primary schools worked together with ICT centre, environmental educators and their teachers to practice fieldwork techniques by conducting a group of environmental monitoring activities at the site. The second fieldwork event was located in the local environment chosen by the project schools. The same groups of students, teachers and specialists conducted a set of similar activities at this remote site.

Data collected in the field was transferred from the mobile devices into a project wiki on both fieldwork days. The days were busy with a range of outcomes to be met. Typically students were involved in a set of monitoring activities each one resulting in a digital product. These products were then transferred from the device onto the host computer system and then onto a wiki. Figure 2 is an example of a resulting group wiki entry which is a permanent, high quality record of one of the day's activities.

Assessment/Evaluation of these Products

Wiki records of students fieldwork served a range of assessment purposes. A check list of skills associated with the fieldwork was compiled and included items such as being able to record a graph of temperature using the datalogger; produce a digital image of a plant using the macro facility of the device etc. This list enabled both students

Figure 2. Wiki record produced by a student group

Plant Transect

Time :

Date : 30th March 2009.

Location : On the banks of Macquarie University Lake.

Student Researchers:[Student names]

(a) Insert your Nova Drawing in the cell below | **(b) Insert your Photo of the Transect in the cell below**

(c) Additional photos in this column | **Notes about your additional photos**

(i) Pittosporum.

(ii) Geebung Tree under a blue gum tree

(iii) Wattle

and teachers to monitor both the development of skills and the output of the group. The final wiki site of the project serves as a permanent record of student monitoring of environmental features at both the university site and the local environment of the school.

Case 3: Using Mobile Phones to Support Fieldwork in Secondary Geography

This case explores how mobile phones can be used as tools for collaborative learning around a geographical task. The learning activities were designed to provide insight into how adolescents perceive their local environment and in particular, select evidence to support land use planning decisions.

One hundred grade nine students were taken in small groups to a variety of different and unfamiliar neighbourhoods and asked to complete an orienteering activity, through using only the exchange of text- and picture- messages (Lim, Hedberg & Chatterjea, 2004). The task was to determine how well they could follow verbal instructions about direction and how they could communicate their insights about the real world context. Each group of two or three students were given broad directions about key elements on their journey. They set off to follow the directions and to share them with a following group who were only equipped with a mobile phone.

Activities: What do the Students do?

The learning outcomes were to develop collaborative and language skills as students explored their local area. In particular, how they constructed their language and communication options when using mobile phones. It was to explore how well the leading group could identify geographical landmarks that served to efficiently identify the students' paths and how effectively could the students share text and image by messaging to ensure that both leading and following team made it through to the end of the task.

The task was an orienteering task that comprised the following steps:

- Both pairs of students were brought to a neighbourhood that was unfamiliar to them; Team A proceeded to navigate to the first checkpoint, using conventional means (following direction clues on a worksheet). Along the way, the team members were to keep a pictorial record, using a camera phone, of any significant landmarks or exemplars of land-use;
- Once sufficient time had elapsed, Team B proceeded to navigate to the first checkpoint. The only means Team B had at its disposal was to engage in an exchange of text- and picture-messaging with Team A about the correct route to follow. The exchange of messages was saved for subsequent examination;
- The steps were repeated for additional checkpoints – routes were typically between 1.6 km to 2.0 km.

Artefacts: What do Students Produce by the End of the Activity?

In the context of learning, the students are sending messages and images which are captured and in effect create a sequence of records of the decisions made and evidence collected. The pieces of evidence show time, sequence and position, such that the choices and omissions can be used to form a coherent story of the group's pathway.

Assessment/Evaluation of these Products

The approach to analysis of Sinclair and Coulthard (1975) was employed as the basis of analysing the discourse modes in text-messaging. In analysing classroom discourse, they devised a hierarchical structure comprising acts (elicitations, directives and informatives), which in turn structure moves (framing, focusing, opening, answering and follow-up), which in turn realise exchanges. Examples of the latter include 'teacher inform', 'teacher direct', 'teacher elicit', 'student elicit', 'student inform', 'check' and 're-initiation'.

The analysis of discourse provided the evidence for how the mobile phone was used as a tool for collaborative learning around the geographical task.

Figure 3 presents a typical transcript of the dialogue between the lead and following teams in the orienteering task with the statements coded for the analysis.

For the orienteering task, twenty-six of the twenty-eight groups of students were successful in completing it within the allocated time. That is to say, twenty-six of the twenty-eight 'leading' pairs were able to successfully guide their associated 'following' pairs along the prescribed route, using only a combination of text- and picture-messaging.

One result, in particular, from the orienteering task was that there was a significant gender difference with respect to the propensity to use

Figure 3. Sample discourse between pairs (Lim et al, 2004)

Time	**Leading team** Lavina & Ziwei (0010)	**Following team** Michelle & Poh Sian (0013)
1519	*Instruction* Then reach the wats0ns..Turn left then walk straight where u wil pas by c0ld st0rage..U wil c 'better h0mes 4 us'	
1521		*Acknowledgment* K. *Question seeking to verify attainment* Holand rd shpin centre huh
1522		*Procedural question* After tat hw
1523	*Exploratory question* Where r u nw? *Question seeking to verify attainment* U c the better home for us already? *Instruction* That wil b ur next destinati0n after the dental surgery..*Check 4 understanding* U gt it?	
1524	Where r u nw? U c the better home for us already? That wil b ur next destinati0n after the dental surgery..U gt it?	
1526		*Reporting attainment* Beter hm 4 us found le . *Procedural qn* Then
1527		*Qn verifying attainment* Upgradin huh
1529		*Urgency statement* Faster
1529	*Instruction* After that, l0ok 4 senior citizens c0rner. It's under an hdb bl0ck..	
1530		*Qn seeking elab* Block wat
1531		*Acknowledgment* K .*Procedural qn* Then
1532	*Admission of fallibility* I never n0tice leh..*Re-instruction* Bt after that better h0me u need to turn right..*Directed tip* It's near i mean the distance	
1532		*Reporting attainment*  (i)
1534		*Procedural question* Then nex
1536	*Question seeking to verify attainment, using the learner's own terms* What is the mms u send? *Re-instruction* U reach the c0rner then..Is u walk al0ng the shelter then walk up the stairs then u wil c the fitness c0rner, *Directed tip* u have to l0ok out 4 the signb0ard leg lift..	
1540	*Check 4 understanding* Gt it? I mean the leg lift..*Instruction* After reach that, l0ok out for blk 10 n 11..Between that 2 bl0cks..G0t semicircle shapes, l0t of them..*Request for report on attainment* Reply tel us where u r..	
1543		*Reporting attainment* 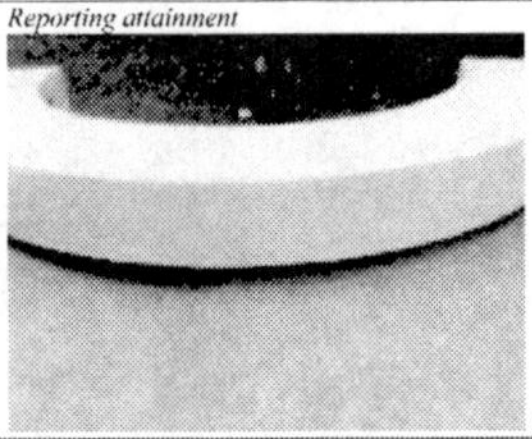 (ii)
1547	*Instruction* Nw walk straight then turn right..D0nt cros rd..Bt when u are near the traffic light u wil c a cl0ck in the air, *Directed tip* meaning u have to l0ok up to sp0t it	

pictures in communication as opposed to text. A comparison of the twenty-one pairs of female students and the twenty-two pairs of male students revealed that among the former, they sent, on average, one picture-message for every 5.0 text-messages; the corresponding figure for the male students was one for every 4.0 text-messages. Among the 'following' pairs, those who were successful sent on average one text-message every 3.66 minutes, compared to only one every 4.42 minutes for those who could not complete the route in time. Thus confirmation and constant communication resulted in more efficient and successful completion of the task. In effect, the importance of mobility was important to collecting evidence but both teams had to work actively at the communication when the physical connection was not present.

One pair thought that they might circumvent the need for detailed instructions and key landmarks, they drew a map and sent a picture of it to the following pair. However, given the small size of the screens on the phones and the inability to zoom in, reading the map was useless and resulted in several minutes being lost as the team waited to send the image in a message.

There was a significant correlation between the time taken to complete the orienteering task, and the use by leading-pairs of discourse modes including 'affirmations', 'statements of assurance' and 'edifications'. More of such statements needed to be sent by leading pairs as following pairs grew more anxious with the passage of time.

Similarly, there was a significant correlation between the time taken to complete the orienteering task, and the use by following-pairs of either 'utterances of desperation' (which did not add to the communication) and 'searches for information' (where the resulting communication resulted in meaningful interchanges and kept the pair on task). As time passed, the following pairs grew increasingly concerned about their chances of success. Those who persevered used questions seeking elaboration, exploratory and procedural questions, and statements or questions seeking to either verify or report attainment.

The use of the mobile phones indeed created a greater evidential basis for the learning task, and the process of capturing all the messages enabled not only the students to review their processes but also the nature of their evidence. In this instance the basic structure was scaffolded through the provision of general instructions so that while the task was still a challenge for the following pairs the evidence and the thinking about what would work best could be focused upon and not other skills such as map reading or cryptic clues as to the best choice of route.

Case 4: Real-Time Debates Using Spatial Data

The learning task in this case required students to adopt, collect evidence, and defend non-congruent social perspectives about land use in the same neighbourhood. Teams of students explored a bounded area, looking for sources of evidence that they could use to support non-congruent points of view. They recorded these pieces of evidence pictorially, again using mobile phones, and exchanged these pictures and texts in real time while still in the field, but physically separated from each other. By analysing the pictorial exchange, the researchers were able to identify which particular aspects of their local environments the students perceived to be relevant to the given geographical themes.

Johnson and Johnson (1979) have suggested that *Structured Academic Controversies* can be employed to stimulate intellectual conflict by designing a highly structured learning task where one student's ideas, information, conclusions, theories, and opinions are incompatible with those of another. The different perspectives can then be used as the basis of a dialogue through which the positions are potentially reconciled. In fact they suggest the proponents seek to reach a consensual agreement by engaging in Aristotelian 'deliberate discourse' (Johnson, Johnson and Smith, 1997).

Garrison and Anderson (2003) suggest that learning facilitated online through network technologies can result in a 'community of inquiry' and the need to resolve multiple and differing perspectives can encourage higher-order learning.

Activities: What do the Students do?

The task in this case required students to adopt and defend non-congruent social perspectives about the same neighbourhood. The model followed the design below:

Team A	Exploring an area, gathering evidence to support a given point-of-view	Engaging in a Structured Academic Controversy regarding the optimal land-use of the given area
Team B	Exploring the same area, gathering evidence to support another point-of-view	

Teams of students explored a bounded area, looking for pieces of evidence that they could use to support non-congruent points-of-view. They recorded these pieces of evidence pictorially, using mobile phones, and exchanged these pictures in real time while still in the field, but physically separated from each other. Students used the evidence to explore given geographical issues within the bounded area. As the students sought evidence for their task and their potential solution, they started piecing together quite complex text and images to support it.

Artefacts: What do Students Produce by the End of the Activity?

In a similar way the students can collect evidence and submit it in real time. In this case each team had supporters back in the classroom, who collated the evidence as it was collected. By comparing one group's evidence against another it was possible to send out for additional evidence if there were omissions while the team was out in the field.

By analyzing the pictorial and text exchange, it is possible to identify which particular aspects of the local environments the students perceived to be relevant to the given geographic challenge of land use and planning. Current mobile phones also attach the geo-spatial position at which the image was taken automatically enabling a reference link between the evidence and its positional source. Analyses of the results enabled the extent of the assumed contribution of multimedia-messaging to students' powers of observation, cognitive mapping abilities and appreciation of multiple points-of-view to be determined.

Messages were collected at all stages of the learning task, the exploration for evidence, the integration of evidence and the resolution of position phases. The unit of transcript analysis is blurred when applied to the analysis of text-messages as they are limited to 160 alphanumeric characters and most were even shorter. Some example messages are shown in Figure 4.

Assessment/Evaluation of These Products

An analysis of the messages sent during the controversy tasks showed that of a total of 806 messages, 483 (60%) were sent during what Garrison, Anderson and Archer (2000) term the exploration phase. Fahy's (2002) study was also similar with a figure of 62%. For the integration phase, 241 (30%) messages were sent, and for the resolution phase, 82 (10%) messages were sent. It would appear therefore that the application of the controversy in the context of a field-based mobile learning environment does result in a worthwhile outcome for the students, as they collect real-world evidence and seek to support their position. The relatively high number of messages sent during the integration phase (compared with the previous studies) can be accounted for by the structure imposed by the pedagogical task, and is particularly encouraging, given that Garrison et al, (2000) also highlight that integration is "more challenging than

Figure 4. Examples of the types of text and images used to support arguments

Group A	Group B	Group C
 (i)	There is no railway nearby so it's very inconvenient. The streets are very dirty and dusty because of the heavy traffic. Roads very congested. Apartment construction takes very long. Supported by the photos we sent just now. 3 yrs.	a lot of clinics provided for people to use... We can find many food stalls therefore people can find a place to eat very easily. Here we also dry and wet markets.
There are lots of wet markets, bus stops, banks recreation centre, provisions shop, coffee shop, mini mart, a community centre, playground, upgrading on apartments.	 (ii)	Senior citizen corner provide a place for elders to gather. Fitness corner and swimming pool allow residents to exercise to stay healthy.. Community club allows different races to build relationship.

exploration" for learners, and as a consequence, "students will be more comfortable remaining in a continuous exploration mode" (p. 89).

Garrison et al, (2000) explain that cognitive presence is 'operationalized' by what they term the 'practical inquiry model'. In practical inquiry, there are four phases: namely a triggering event, in which teaching presence is employed to explicitly introduce an issue or problem, in terms of learning tasks, to the learners; exploration, in which the learners are required to understand the nature of the problem, and then move through the social exploration of ideas to a fuller exploration and subsequent selection of relevant information; integration, in which learners begin to assess the applicability of ideas in terms of how well they describe the issue; and finally resolution, in which learners work towards building a consensus. It should be noted that these four stages of the practical inquiry model correspond very well to the respective stages in a Structured Academic Controversy.

DISCUSSION

After consideration of these four cases, a number of issues and commonalities become apparent. There is a wide range of purposes for fieldwork and a corresponding wide range of pedagogies implemented with technologies employed to achieve these outcomes. The cases in this chapter reveal purposes ranging from collection and recording of environmental data, effective communication of geographical position, through to collection of data to support a point of view.

Purpose/Pedagogy Framework

Purpose and pedagogy may be aligned by consideration of the framework provided by Jergen Habermas' theory of 'knowledge-constitutive interests' (Habermas 1972). He describes three areas of cognitive interest: technical, practical and emancipatory. Each of these areas represents three corresponding ways of knowing: the empirical-analytical view, the historical-hermeneutic view and the critical-reflective view. Grundy (1987) applies this approach to curriculum design and in turn this can be used to frame and align the purpose and pedagogy of mobility.

1. Technical Interest

Real-world activities designed with a technical interest have purposes related to students locating and recording specific content and information from a site. Activities that include collecting and recording data, monitoring environmental features and finding out information are characteristic of technical knowing. The pedagogy of technically based fieldwork is often asymmetrical where there are experts who know site specific information and novices who seek out this specific information. The products of the learning activity include original evidence in the form of information reports, databases or field reports which can be assessed for their accuracy in reproducing and interpreting the information from the site.

The technology selected to support this type of activity would be expected to provide a permanent, digital record of data that can be seamlessly transferred from the field to the classroom and to other environments where it may be required. The digital records can then be transduced into other forms of representation to assist in the interpretation of the phenomena under study. In this chapter mobile phones, hand held computers in tandem with digital cameras were utilised to fulfil these requirements. Specific monitoring activities can be loaded onto the device or accessed via the internet. Students may also access remote experts or fellow students via the Internet. Web 2.0 technologies can be utilised to store and build a collaborative record and interpretation of the evidence.

2. Practical Interest

Learning activities with a practical interest have purposes related to students negotiating and communicating their understanding of their own or another's point of view. The activities operating with a practical interest might include seeking a different cultural perspective about an issue or a location from a member of that cultural group, or communicating a personal or group perspective about a site. The pedagogy of this interest involves creating opportunities for discussion, negotiation and formulation of consensus. The products of practically oriented activities would include the script of the shared communication, a representation of a particular point of view.

Use of technologies like the mobile phone has been incorporated into activities that have a practical interest. The messaging capability of the phone facilitates capture of student to student communication strategies and their modes of representation.

3. Emancipatory Interest

Learning activities with an emancipatory interest seek to change the behaviour or attitudes of participants as a result of the experience. This purpose in effect was achieved by challenging the students to 'see' what the land uses were currently and how they might be improved to facilitate a policy or human rights initiative. Changing students' behaviours may take extended involvement with a longer activity which might require action related to the emerging evidence and point of view of the students. This might result in designing, creating and caring for a garden or involvement in caring for a particular societal group such as the elderly. The pedagogy of emancipation is symmetrical with the teacher and students operating as co-learners engaged in an enactive learning context. The product of this interest area is evidence of change in students' behaviour and attitudes.

The technologies to support such interest, store, communicate and provide the temporal record of decisions made and evidence reviewed. These cases provide simple and more complex ways of assessing these outcomes for the students. They also illustrate the students' engagement with challenges that have value and meaning in a real way and are not simply an academic exercise abstracted to complete a syllabus topic.

Further research of the application of the purpose /pedagogy framework generated by this study will determine its potential as a predictive tool for guiding the planning and evaluation of pedagogical approaches incorporating the use of fieldwork supported by mobile technologies as part of high quality learning.

CONCLUSION

"From both the teachers and students involved this was an exciting project, mainly because it enabled the students to have many meaningful experiences using different forms of technology. Also, they were using the technology to assist their learning in real life situations for a real purpose and not just for the sake of ticking the technology box." (Classroom primary teacher, 2006)

This chapter has explored both theoretical and practical considerations of the use of mobile technologies to support student learning in outdoor environments. The social constructivist perspectives embedded in theoretical expositions of characteristics of high quality learning in the 21st century require a more student directed pedagogy using inquiry and problem-based approaches located within student experience and in the students' physical environment. Further research in the area will illuminate the application and relevance of the frameworks proposed here to guide pedagogy that can scaffold learning for our mobile teachers and students.

ACKNOWLEDGMENT

Director, Macquarie ICT Innovations Centre, Debbie Evans; school principals Chris Geelan, Peter Cusworth; curriculum consultant Chris Painter; classroom teachers Jenny Hoskin, Lizzie Smith; Macquarie ICT Innovations Centre team: Andreas Utomo Kuswara, John Willis, Jenny Madson, Karen Blackwell, Pam Kelly; environmental educators Steve Papp, Cate Frederickson, Sharon Kinisson; Coco Lab team: Peter Reimann, Kate Thompson, Wai Yat Yi.

REFERENCES

Anastopoulou, S., Sharples, M., Wright, M., Martin, H., Ainsworth, S., & Benford, S. (2008). Learning 21st century science in context with mobile technologies. In Traxler, J., Riordan, B., & Dennett, C. (Eds.), *The bridge from text to context. Proceedings of mLearn 2008, Wolverhampton, UK* (pp. 12–21). Wolverhampton, UK: University of Wolverhampton.

Anderson, T., Rourke, L., Garrison, D. R., & Archer, W. (2001). Assessing teacher presence in a computer conferencing context. *Journal of Asynchronous Learning Networks*, *5*(2), 1–17.

Ballantyne, R., & Packer, J. (2009). Introducing a fifth pedagogy: Experience-based strategies for facilitating learning in natural environments. *Environmental Education Research*, *15*(2), 243–262. doi:10.1080/13504620802711282

Baumgartner, E., & Zabin, C. J. (2008). A case study or project-based instruction in the ninth grade: a semester-long study of intertidal biodiversity. *Environmental Education Research*, *14*(2), 97–114. doi:10.1080/13504620801951640

Bower, M. (2008). Affordance analysis – matching learning tasks with learning technologies. *Educational Media International*, *45*(1), 3–15. doi:10.1080/09523980701847115

Chen, Y. S., Kao, T. C., & Sheu, J. P. (2003). A mobile learning system for scaffolding bird watching learning. *Journal of Computer Assisted Learning*, *19*, 347–359. doi:10.1046/j.0266-4909.2003.00036.x

Chew, E. (2008). Views, values and perceptions in geographical fieldwork in Singapore schools. *International Research in Geographical and Environmental Education*, *17*(4), 307–329. doi:10.1080/10382040802401565

Cook, V. (2008). The field as a 'pedagogical resource'? A critical analysis of students' affective engagement with the field environment. *Environmental Education Research, 14*(5), 507–517. doi:10.1080/13504620802204395

Department for Education and Skills (DES). (2004). *Pedagogy and practice. Creating effective learners: Using ICT to enhance learning*. London: DES.

Dewey, J. (1916). *Democracy and education*. New York: Free Press.

Driver, R., Asoko, H., Leach, J., Mortimer, E., & Scott, P. (1994). Constructing scientific knowledge in the classroom. *Educational Researcher, 23*(7), 5–12.

Dunphy, A., & Spellman, G. (2009). Geography fieldwork, fieldwork value and learning styles. *International Research in Geographical and Environmental Education, 18*(1), 19–28. doi:10.1080/10382040802591522

Faculty of Education & Social Work, University of Sydney. (2009). *Supporting mobile learning from the ground up*. University of Sydney. Retrieved on July 21, 2009 from http://coco.edfac.usyd.edu.au/mobilelearning/peakhurstPS.php

Fahy, P. J. (2002). *Assessing critical thinking processes in a computer conference*. Retrieved November 23, 2009, from http://cde.athabascau.ca/softeval/reports/mag4.pdf

Garrison, D. R., & Anderson, T. (2003). [*st century: A framework for research and practice*. London: Routledge Falmer.]. *E-learning*, 21.

Garrison, D. R., Anderson, T., & Archer, W. (2000). Critical inquiry in a text-based environment: Computer conferencing in higher education. *The Internet and Higher Education, 2*(2/3), 87–105.

Garrison, D. R., Anderson, T., & Archer, W. (2001). Critical thinking, cognitive presence and computer conferencing in distance education. *American Journal of Distance Education, 15*(1), 7–23. doi:10.1080/08923640109527071

Gaughran, B. (1996). *The graphics code, visualization and CAD*. Flexible automation and intelligence manufacturing conference. Atlanta, GA.

Glaser, B. G., & Strauss, A. L. (1967). *The discovery of grounded theory*. Chicago: Aldine.

Grundy, S. (1987). *Curriculum: Product or praxis?* Lewes, East Sussex: The Falmer Press.

Habermas, J. (1972). *Knowledge and human interests* (2nd ed.). London: Heinemann.

Halocha, J. (2005). Developing a research tool to enable children to voice their experiences and learning through fieldwork. *International Research in Geographical and Environmental Education, 14*(4), 348–355. doi:10.1080/10382040508668370

Hein, G. (1998). *Learning in museums*. London: Routledge.

Johnson, D. W., & Johnson, R. T. (1979). Conflict in the classroom: Controversy and learning. *Review of Educational Research, 49*, 51–61.

Johnson, D. W., Johnson, R. T., & Smith, K. A. (1997). *Academic controversy: Enriching college instruction through intellectual conflict*. Washington, DC: George Washington University.

Kent, M., Gilbertson, D., & Hunt, C. O. (1997). Fieldwork in geography teaching: A critical review of the literature and approaches. *Journal of Geography in Higher Education, 21*(3), 313–332. doi:10.1080/03098269708725439

Laurillard, D. (2007). Pedagogical forms for mobile learning: Framing research questions. In Pachler, N. (Ed.), *Mobile learning: Towards a research agenda. Occasional papers in work-based learning 1* (pp. 15–24). London: WLE Centre for Excellence.

Lim, K. Y. G., Hedberg, J. G., & Chatterjea, K. (2004). Adolescent usage of multimedia messaging in the negotiation, construction, and sharing of meaning about local environments. In L. Cantoni, & C. McLoughlin (Eds.), *Proceedings of Ed-Media 2004 World Conference on Educational Multimedia, Hypermedia and Telecommunications* (pp. 1179-1186). Lugano, Switzerland: Association for the Advancement of Computing in Education.

Lipman, M. (1991). *Thinking in education*. Cambridge, UK: Cambridge University Press.

Naismith, L., Lonsdale, P., Vavoula, G., & Sharples, M. (2004). *Literature review in mobile technologies and learning*. Bristol: NESTA Futurelab. Retrieved November 22, 2007 from http://www.futurelab.org.uk/download/pdfs/research/lit_reviews/futurelab_review_11.pdf

National Research Council. (1999). *How people learn: Brain, mind, experience and school*. Washington, DC: National Academy Press.

Newmann, F. (1996). *Authentic achievement: Restructuring schools for intellectual quality*. San Francisco: Jossey-Bass.

Newmann, F. (2000). Authentic intellectual work: What and why? University of Minnesota. Retrieved November 21, 2009, from http://education.umn.edu/CAREI/Reports/Rpractice/Fall2000/newmann.html

NSW Department of Education and Training (NSWDET). (2003a). *Quality teaching in NSW public schools: Discussion paper*. Sydney: NSWDET.

NSWDET. (2003b). *Quality teaching in NSW public schools: An annotated bibliography*. Sydney: NSWDET.

Rennie, L. J., Feher, E., Dierking, L. D., & Falk, J. H. (2003). Toward an agenda for advancing research on science learning in out-of-school settings. *Journal of Research in Science Teaching, 40*(2), 112–120. doi:10.1002/tea.10067

Rickinson, M., Dillon, J., Teamey, K., Morris, M., Choi, M. Y., Sanders, D., & Benefield, P. (2004). *A review of research on outdoor learning*. London: National Foundation for Educational Research and King's College London.

Sharples, M. (Ed.). (2007). *Big issues in mobile learning*. Nottingham: Learning Sciences Research Institute.

Sinclair, J. McH., & Coulthard, R. M. (1975). *Towards an analysis of discourse: The English used by teachers and pupils*. Oxford, UK: Oxford University Press.

Stewart, K. (2007). Mobile learning: Designing the learning context. In A. Norman, & J. Pearce (Eds.), *Making the connections*. Proceedings of the 6th Annual International Conference on Mobile Learning, Melbourne (pp. 253–258). Parkville, Vic: University of Melbourne. Available online at http://www.mlearn2007.org/files/mLearn_2007_Conference_Proceedings.pdf

Stewart, K., Thompson, K., Hedberg, J., & Wong, W. (2009). Using technology to support quality learning for school activities involving field studies. In Caballe, S., Xhafa, F., Daradoumis, T., & Juan, A. A. (Eds.), *Architectures for distributed and complex m-learning systems: Applying intelligent technologies*. Hershey, PA: IGI Global.

Swan, K., van 't Hooft, M., Kratcoski, A., & Unger, D. (2005). Uses and effects of mobile computing devices in K-8 classrooms. *Journal of Research on Technology in Education, 38*(1), 99–113.

Tan, N., Chen, W., Looi, C.-K., Zhang, B., Seow, P., & Chan, A. (2007). Handheld computers as cognitive tools: An environmental learning project in Singapore. In T. Hirashima, U. Hoppe & S. Shwu-Ching Young (Eds.), *Supporting learning flow through integrative technologies* (162, pp. 377–384). Amsterdam: IOS Press.

Taylor, J. (2007). Evaluating mobile learning. In Sharples, M. (Ed.), *Big issues in mobile learning* (pp. 26–28). Nottingham: Learning Sciences Research Institute.

Thompson, K., & Stewart, K. (2007). *The mobile jigsaw: A collaborative strategy for mlearning about the environment.* Proceedings of the 6th Annual International Conference on Mobile Learning, Melbourne (pp. 248–252). Parkville, Vic: University of Melbourne. Available online at http://www.mlearn2007.org/files/mLearn_2007_Conference_Proceedings.pdf

Chapter 15
Formal and Informal Use of Handhelds by Australian and British Students: A Comparative Case Study

Wan Ng
University of New South Wales, Australia

Stamatina Anastopoulou
University of Nottingham, UK

ABSTRACT

In this chapter, we discuss the capacity of mobile technology in facilitating young people's ubiquitous learning and socializing, both formally and informally. We report on a study of how young adolescents (12-13 year olds) in Australia and UK use handheld computers in formal (school) and informal (home) situations and their perceptions of the usefulness of these devices for their learning. The data show that entertainment activities with the handhelds dominate both school and home for both groups of students and that there is little continuity between the activities carried out in the school and activities in the home. We argue that schools have a responsibility to bridge the home-school learning and to support these students to become self-directed learners for lifelong learning.

INTRODUCTION

The mobile device that is the focus of this chapter is the handheld computer. The chapter presents a comparative study of how Year 7 (12 year old) students in Australia and UK with personal ownership of handheld computers used them formally in the classrooms and informally at home. According to the manufacturers' labels and specifications of the devices used, the Australian students referred to the handheld computers as pocket PCs (Personal Computer) or PDAs (Personal Digital Assistance) while their UK counterparts called them EDAs (Educational Digital Assistance). In this chapter, we will refer to them as handheld computers or just handhelds.

Handheld computers are ubiquitous tools that have the potential to contribute to an individual's seamless learning and social lives. They are not as powerful as laptops or tablet PCs but their

DOI: 10.4018/978-1-61692-849-0.ch015

small size means that they can be easily carried around in the pocket, school- or hand-bags. In educational institutions, wireless technology enables the handhelds to access information on the Internet anytime for students' research purposes. Reference and other learning materials such as simulations, lecture notes, e-books, encyclopedia and worksheets could be installed in the students' handhelds or they could be transferred via Bluetooth from the teacher's computer to the students' handhelds. At the school level, handhelds alleviate the necessity to book computer laboratories or queue up to get access to technology resources such as the Internet and other learning materials.

Schools with handheld programs where students purchase and own their own handhelds will be aiming at the integration of these devices into the students' education and other aspects of their daily lives. For example, they are able to replace their notebooks and diaries by using the calendar and Note functions on their handhelds to manage learning and social schedules as well as their learning tasks (Sharples, 2003). Nicholas and Ng (2009, p.2174) have suggested that 'having access to a handheld all the time is like having the pens, text books, camera, calculator, voice recorder, clock and Internet access all in one'. They have identified a list of uses that personalised handhelds could offer that provides 'immediacy' to learning (Ogata and Yano, 2004) and learning in context. These uses include: (i) accessing the Internet via wireless technology to research for information, access emails and social networks (ii) organising their daily activities and time management using the calendar, the clock and alarm (iii) record keeping and writing, for example, by taking notes, audio recording and drawing figures using the Note function; writing short paragraphs or keeping reflective logs using Word Mobile where students either transcribe (the writing recognition software will change the writing to text) or tap on the keyboard with the stylus (iv) communicating and collaborating with peers via Bluetooth in class for the exchange of files or via the Internet to access emails and other social networking websites such as Facebook, Flickr and blogs (v) learning from software packages e.g. foreign languages, dictionary and encyclopedia, Periodic Tables and other subject-related software (vi) collecting data with handheld sensors and probewares or taking pictures and video recording (v) displaying, calculating and analyzing data using Excel Mobile and the calculator (vi) accessing educational resources and software such as e-documents, simulations and multimedia materials, movies with Media Player, interactive worksheets exercises e.g. those created with Hot Potatoes software (vii) creating PowerPoint presentations using Pocket Slides and (viii) self-education and entertaining with installed or downloaded games, music, podcasts, lectures and interviews.

The integral access to a wide ranging set of applications that support ubiquitous learning enables the handheld to become a natural and immersed part of the student's learning to learn seamlessly where the boundaries of time and place are blurred. In this regard, learning with handhelds is blended learning that takes place both formally in classrooms and informally outside of school and on the move, for example, in buses or trains.

The aim of this chapter is to present an investigative study on the attitudes of Year 7 students in Australia and the UK toward personal ownership of handhelds, how they made use of their handhelds in formal (school) and informal (home) situations and their perceptions of the usefulness of these devices for their learning. The broad research question is: How do Year 7 Australian and UK students with ownership of handheld computers make use of them in schools and at home? Underlying this question are the sub-questions of:

1. What are the similarities and differences in trends between Australian and UK students' usage of their handhelds at home and in school?

2. What are their attitudes toward the use of these devices and their perceptions of their usefulness in helping them learn?

Informal and Self-Directed Learning

Inculcating the desire to keep learning is the key attitude to be developed in people (Dewey, 1938). A large percentage of people continue to learn beyond their formal school years and gain knowledge through self-directed learning informally, for example, at home or at work (Drucker, 2004; Sargant, 1991). In this regard, work, leisure, learning and collaboration within communities are intertwined activities situated within contexts (Lave & Wenger, 1991) for lifelong learners. Schools and tertiary institutions should be preparing students to become more critical thinkers by providing opportunities to synthesise what they learn in many settings throughout their private and work life outside formal education (Fisher & Scharff, 1998). Being able to self-direct their own learning is one of the major steps in achieving this goal.

A definition of self-directed learning is the development of autonomous learners who are able to take control of and responsibility for their own learning, including seeking help when difficulties are encountered (Garrison, 1992; Tough, 1967). Smith (1996) suggested two characteristics of self-directed learning as (i) a continuous activity where the individual has 'authentic control' and decides all aspects of the learning, for example, planning, doing and evaluating the learning and (ii) the ability to access and select from a range of appropriate resources. Hence in educating students to be self-directed learners, and to cater for differences in background and expertise, there needs to be flexibility in both the diversity of resources offered and freedom for the learners to select learning materials (Sinitsa, 2000). Other skills that need to be promoted for self-directed learning include both critical thinking skills (seeking, selecting and utilizing appropriate resources from the Web) and social and co-operative skills to assist with knowledge construction individually or collaboratively to solve problems or complete set tasks. In this regard, technology and, in particular, handhelds play a vital role since they offer greater freedom to learn ubiquitously both formally and informally.

Waycott, Jones, & Scanlon (2005) promoted handheld computers as lifelong learning tools. These tools allow the learner to be mobile, interacting with a variety of contexts so that learning can take place both formally and informally. They offer the potential for ubiquitous learning in ways that are independent and self-paced. Learning becomes personalised so that it fits an individual's learning path and their social networks. A personal learning environment using technology, such as a PDA as a ubiquitous learning device, recognizes that learning is continuous and not provided by a single learning provider (Attwell, 2007; Madia, 2007). In order for school students to make use of handhelds to self-direct learning ubiquitously, schools should teach them how to maximize the utilization potential of these devices and the resources available to these devices for learning.

In this chapter, the kinds of opportunities that school and home environments offer for learning with handhelds for the Year 7 students participating in the study are explored. By indentifying the similarities and differences between home and school use of their personal handhelds, we might begin to understand the mobile nature of adolescents' learning with the use of handhelds and how the learning in formal settings and skills associated with this learning are translated to learning in informal settings and vice versa. We will discuss what makes handhelds become useful learning tools and provide some recommendations on how they could be used in the classrooms to promote self-directed learning that bridges home-school learning.

The Study

The study compares students' use of handhelds across two countries to provide a more expanded view of what adolescents do with them in schools and at home. In the literature, there are numerous reports of schools adopting the use of handheld computers in the classrooms. Many of these reports are associated with funded projects where the funds are able to assist, technically and pedagogically, with the introduction of the devices into classrooms. Examples of these projects include the US Palm Education Pioneers Program (Vahey & Crawford, 2002), UK's Becta's ICT research on pocket PCs in schools (Hartnell-Young, 2008; Perry, 2003; McFarlane, Triggs & Yee, 2008; 2005Twining, Evans and others, 2005!), Europe's MobiLearn project (Brugnoli, Morabito, Bo & Murelli, 2007) and MyArtSpace (Vavoula, Sharples, Rudman, Meek & Lonsdale, 2009). Under normal circumstances, where there are no external funds, schools adopting the use of handhelds are required to plan carefully their strategy of uptake to ensure successful implementation. The adoptions of handhelds in the schools participating in the study described in this chapter are not part of externally-funded projects and hence issues they face would have to be solved within the school community and with existing available resources. These issues include infrastructure and financial considerations, teacher professional training and identifying effective pedagogies related to handheld usage in the classrooms. These issues will inevitably impact on how students learn using their personal handhelds for learning.

Participants and Background Information

As shown in Table 1, 58 Australian and 59 English Year 7 (12-13 year-old) students participated in the study. The Australian students were spread across three classes of Year 7 and the English students were spread across five. Both the Australian and English students were from government schools. The socio-economic location of the schools however differed, with the Australian school located in a middle-class socio-economic area while the English school was located in a lower working class area. Before discussing the details of the study, we will describe how the devices became available to the students and the role that the teachers took in becoming familiar with the device and its uses.

Table 1. Demographics of the two schools

	Australian students	English students
Number of year 7 participants	58	59
Type of school	Government	Government
Location of school	Eastern metropolitan region of Victoria	East Midlands, inner city school
Socio-economic status	Middle class	Lower working class

The Australian school started a pocket PC program in 2008 where parents were invited to enrol their Year 7 students into the program. Students electing to be in the program were required to purchase their own Ipaq HP pocket PC. In this regard, they have ownership of the device and could use it ubiquitously (anytime and anywhere). The Australian teachers were inducted into the program by being provided with the device and two in-school professional learning conferences. At these conferences, a group of teachers who undertook initial exploration and usage of the handhelds in their classes shared their experiences and showed how the devices could be used in the classrooms. Novice teachers were given the opportunity to practise using the handhelds and work with more advanced teachers in a collegial manner.

The English school started the EDA program in 2007 where parents were invited to enrol their Year 7 children into the program. Students electing to be in the program had to contribute a small amount to purchase their own Fujitsu Siemens

EDA, with a built-in camera. The English teachers were not inducted into the program and they were not given devices to familiarize with or explore usages. There was no in-school training for the teachers to promote the use of the handhelds in the classroom or for homework. Furthermore, according to informal accounts from the teachers, students were not allowed or provided with the technical support from the school to connect their devices to the Internet or to the school's intranet.

Method

The research instrument used in this study is a paper-based questionnaire (see Appendix) containing a set of 30 Likert-scale statements to elicit (i) computer ownership and technical skills and (ii) attitudes toward the use of handhelds, and perceptions of their usefulness for learning, doing homework and personal use. The scale for the Likert-scale statements was: strongly agree = 5; agree = 4; not sure = 3, disagree = 2 and strongly disagree = 1. Students were also asked to indicate the types of handheld usage at home and at school and the frequencies associated with these usages. The open question asked for general comments and recommendations on how to improve the handheld programs at their schools.

For each of the 30 questionnaire statements, the Likert-scale responses were analysed for means and percentage responses in the categories of strongly agree/agree (SA/A), not sure (NS) and disagree/strongly disagree (D/SD). The open responses were analysed for themes that emerged from the responses.

A further set of statements was used to elicit the types and frequency of usage in school and at home. The frequency of responses reflects the number of students ticking these activities on the questionnaire according to how frequently they undertook the activities. The frequencies were coded: R=rarely (< once a week); S=sometimes (1-3 times a week); F=frequently (4-6 times a week); V=Very frequently (>7 times a week). As the sample size of N=58 was too small to spread the analysis over the four categories, the responses were combined into two categories of R/S (i.e. infrequent use) and F/V (more frequent use).

Results and Discussions

The 30 Likert-scale statements are grouped into 5 categories and analysis of means and frequencies will be undertaken under these categories. Since the data indicated a relatively high proportion of 'not sure' responses for many of the statements which will inevitably affect the mean score, a focus on the distribution of frequencies for most of the statements will be undertaken. Similarities and differences between the Australian and UK students' responses will be discussed.

Computer Literacy of the Year 7 Students

Over 90% of the Australian and English students have computers at home with 86% Australian and 72% of English students being able to access the Internet easily. Both groups of students appear to be computer literate and technically proficient, with:

- 83% of Australian and 72% of English students indicating that they did not need training
- 85% of Australian and 69% of English students indicating that handhelds were easy to use and
- 95% of Australian and 78% of English students indicating that they knew how to use them easily

There are however, other technical (physical) obstacles unrelated to skills that both groups have identified. For example, 43% and 68% of Australian and English students respectively indicated that it was not easy for them to access the internet with the handhelds (item 2). More than a third of both groups indicated that it was not easy to send messages (see Table 5, item 10) to their peers via Bluetooth or via the Internet. In their responses to the open questions in the questionnaire, the British students mentioned that they did not find it easy to access the Internet mainly because they did not receive any technical support from the school. Students and staff could access the Internet only if they could sort out network problems themselves. While their Australian counterparts had technical support provided to them, the support was insufficient or not readily available when it was needed. Furthermore, the infrastructure was inadequate, for example, there were insufficient access points in certain areas of the school. As a result the Australian students indicated that accessing the school's Internet was frequently met with technical problems. The difficulties encountered by both groups of students in accessing online resources presented an obstacle in their development as self-directed learners.

Attitudes toward Learning with Handhelds

As shown in Table 3, the majority of the students (more than 60%) from both groups liked learning with technology (item 23). In addition, 52% of Australian students and 42% of English students indicated that handhelds made learning fun (item 20). Nearly 60% of both the English and Australian students agreed with the statement that with a handheld it was easy to change what they had written (item 24) even though 64% of the Australian students disagreed or strongly disagreed with the statement that writing with a handheld is easier than writing by hand (item 21). The views of their English counterparts about this statement differed substantially with only 34% disagreeing or strongly disagreeing. A possible reason for this discrepancy could be found in the responses to item 18 which indicates that Australian Year 7 students have used handhelds more (62% Australian versus 41% English students). The more usage by the Australian students meant that they could have written more using the handhelds and had experienced frustrations with tapping on the small keyboard or slowness in transcribing hand-writing into text. The increase in usage by Australian students have, however, led to 54% of them saying that they were more excited about using handhelds (item 7) compared with only 25% of their English counterparts in this regard. The

Table 2. Computer ownership and literacy

Statement	Aust mean N=58	UK mean N-59	Aust % SD/D	UK % SD/D	Aust % NS	UK % NS	Aust % A/SA	UK % A/SA
1. I need special training to use a handheld	1.90	1.93	83	72	9	16	8	14
2. It is easy to access the Internet with handhelds	2.81	2.27	43	68	14	12	43	20
14. handhelds do the same things as a desktop computer	2.74	2.84	43	41	28	24	29	34
26. handhelds are easy to use	4.04	3.88	5	10	10	20	85	69
27. I can easily use the Internet at home	4.50	3.91	9	22	5	5	86	72
28. I know easily how to use a handheld	4.28	4.09	5	3	0	17	95	78
30. I have a computer (desktop or laptop) at home	4.88	4.59	0	5	1	3	99	92

more frequent use by the Australian students meant that they were provided with insights into what the handheld was capable of doing, hence generating excitement. There is also more Australian students (55% compared with 33% of English students) who indicated that handhelds made learning more interesting (see item 15, Table 3).

Table 3 also shows that a higher percentage of Australian students indicated that they enjoyed using handhelds for schoolwork (45% versus 31% for item 11). The lack of enjoyment by the English students could be due to students using the devices for routine type of work such as calculating with the calculator and organising time with the calendar rather than being instructed to use a variety of features to solve problems or complete set tasks. The English teachers did not embrace the use of handhelds in their classroom since they were not trained to use them nor did they have access to a pool of 'good practice' examples that could facilitate the handhelds' inclusion in the class activities. In most cases they were not given the opportunity to borrow a device for exploratory use.

There was general agreement with the statement that it would not be worth using a handheld unless they could take it home (item 22). Over 50% of the students from both countries agreed or strongly agreed with this statement. It could be suggested from this response that students valued the mobile technology because they owned it which is consistent with findings of other studies where students owning the technology felt empowered to find resources relevant to their schoolwork (e.g. Vahey & Crawford, 2002), and when they did not own the devices, they were reluctant to devote time and effort in personalising and extending them (Corlett, Sharples, Chan & Bull, 2005)

Table 3. Attitudes toward learning with handhelds

Statement	Aust Mean N=58	UK mean N=59	Aust % SD/D	UK % SD/D	Aust % NS	UK % NS	Aust % A/SA	UK % A/SA
7. I am excited about using a handheld	3.57	2.63	19	46	28	30	54	25
11. I enjoy using the handheld for schoolwork	3.26	2.71	22	46	33	24	45	31
12. There are no disadvantages to using handhelds in the classroom	2.72	2.81	48	36	21	39	31	25
15. Handhelds make learning more interesting	3.45	2.90	26	40	19	28	55	33
17. handhelds at school are a waste of time	2.75	2.97	47	36	26	31	26	33
18. I have used a handheld lots already	3.66	3.03	17	41	21	17	62	41
20. Handhelds make learning fun.	3.13	3.10	27	36	31	22	52	42
21. Writing with a handheld is easier than writing by hand	2.26	3.14	64	34	17	24	19	41
22. If I couldn't take the EDA home, it wouldn't be worth using it in school	3.59	3.53	19	19	29	29	52	53
23. I like using technology for learning a lot	3.86	3.54	9	22	21	17	70	61
24. With a handheld it is very easy to change what I have written	3.54	3.64	16	10	26	33	59	57
25. I learn better with technology	3.50	3.78	16	12	27	24	57	64

The responses to item 17 in Table 3 shows that nearly 50% of Australian students and 36% of English students did not think that the use of handhelds in schools was a waste of time. Despite this, a similar number of students did not think that there were no disadvantages in using handhelds in the classroom (item 12). These students recognized the limitations of handhelds but were still optimistic about their use in the classrooms. A larger percentage of English students were unsure of these two statements than their Australian counterparts, suggesting again that the less frequent use of the devices in the classrooms created uncertainty in these students. This reason could be similar for the 26% of Australian students who indicated uncertainty for the statement that 'handhelds at schools are a waste of time'. In the open responses to the questionnaire, one of the three classes in the handheld program at the school indicated strongly the lack of usage in their classes.

Benefits of Learning with Handhelds

The students from both countries did not regard handhelds as beneficial for their learning. The evidence supporting this claim is based on their perceptions, indicated in Table 4, of the statements 'my marks have been better with handhelds', 'handheld has helped me learn all subjects a lot better' and 'handheld has helped me learn better'. There is a fairly consistent pattern in the responses with over 50% of English students and about 40% of Australian students responding negatively to all the three statements. The percentages of 'not sure' responses for these three statements are fairly high for both groups of students, ranging from 27% to 43%. While only 16% and 12% of Australian and English students agreed/strongly agreed that their marks have been better with handhelds (item 1), these numbers are doubled when it came to beliefs that the handhelds have helped them learn better and easier (see responses in the last two columns of Table 4 for items 4, 5 and 6). An explanation for the responses described above could be that for these students, it was difficult to measure the impact of the handhelds on their own achievements. It is difficult to know if better marks are due to using the handhelds for learning or other factors, for example, teacher instructions, social discourses with peers or the nature and level of difficulty of the topic under study.

Despite the above responses, and as discussed in the previous section, 47% of the Australian students and 36% of UK students did not think that handhelds were a waste of time. About 30% of the students from both the countries were unsure of this. Apart from the lack of use of these devices (discussed in the previous section), this result could also be due to the lack of specific enough objectives for handheld use. For example, the most prominent school objective is often about getting better marks. Even though learners did

Table 4. Handhelds and learning

Statement	Aust Mean N=58	UK Mean N=59	Aust % SD/D	UK % SD/D	Aust % NS	UK % NS	Aust % A/SA	UK % A/SA
3. My marks have been better with handheld	2.60	2.27	41	54	43	34	16	12
4. Handheld has helped me learn all subjects a lot better	2.85	2.46	41	53	24	24	34	24
5. Handheld has helped me learn better	2.90	2.46	36	51	31	27	33	22
9. Handheld has made learning a lot easier.	3.05	2.69	28	41	36	34	36	25

Table 5. Other personal uses of handhelds

Statement	Aust Mean N=58	UK Mean N-59	AU % SD/D	UK % SD/D	A % NS	UK % NS	A % A/SA	UK % A/SA
10. With a handheld, it is easy to send messages to my friends	2.86	2.71	35	44	30	24	35	32
29. I enjoy using the handhelds for games	4.33	4.10	3	2	9	17	88	81
19. I have lots of music on my handheld	3.53	3.10	22	41	16	10	62	49
13. handhelds help me organise my time better	3.28	2.83	26	39	23	34	51	27

not see their marks having improved, they may have realised that they have acquired new skills from using and exploring the handhelds which were not captured by assessment. Therefore the use of handhelds was not considered as a waste of time.

The results in Table 4 suggest that with new ways of learning we need new ways of assessment. Since handhelds can allow seamless transition between school and the home (Waycott at el., 2005), they are used to help learners extend educational practices beyond the school hours and open up further opportunities to try different activities that were not plausible before. Such innovations would need to be assessed differently on the basis of the long-term benefits and life-long learning skills that learners acquire rather than short term, content specific assignments. For example, if students use handhelds to research on the Internet about healthy diet, they may have a broader understanding of the term, such that the word 'diet' is not only used for weight-loss or weight gain situations. Even though such activity is plausible from a desktop computer, the probability for the desktop to be used for extending school activities is low (Kerawalla & Crook, 2005). Another example could relate to the activities that new learning tools allow for them to be carried around. Students may, for example, learn about alternative ways of getting the required nutrients (e.g. pulses and meat are good sources of iron and protein, milk and yogurt are good sources of calcium). If assessment only focused on the need to and reason why people should receive protein and minerals among other nutrients (which is what the curriculum requires), it will not capture their expanded understanding of where the different sources are to obtain these nutrients. Research using handhelds as students meet with different sources of food would be more motivating and likely to succeed with learning. Therefore, new learning tools could push for new educational activities that could influence a change on what it is assessed.

Homework with Handhelds

The students from the two countries differed considerably in their thinking about the ease of transferability of schoolwork to home with the handhelds (item 8) and being able to do their homework anytime and anywhere (item 6). As shown in Figure 1, the mean scores of these items are 3.47 (item 6) and 3.43 (item 8) for Australian students and 2.51 (item 6) and 2.64 (item 8) for their English counterparts. In frequency terms, this translated to about 60% of Australian students agreeing or strongly agreeing with both these statements while only about 25% of the English students thinking similarly. It is likely that the English students had little opportunities to explore these experiences due to a lack of use of the handhelds in their classrooms and the dif-

Figure 1. Mean scores for items 6 and 8 for the two groups

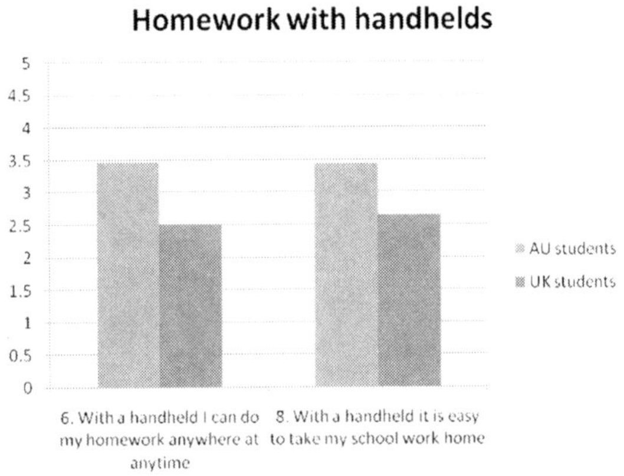

ferent degree of the teachers' involvement. Here it needs to be noted that the English students were not requested by their teachers to do homework with their handhelds. The students who did use it did so out of their own personal initiative.

Cooper, Robinson and Patall (2006) indicated that homework activities have multiple instructional purposes such as a) to practice or review material that has already been presented in class, b) to introduce material to help students obtain the maximum benefit when the new material is covered in class (preparation assignments), c) to involve the transfer of previously learned skills to new situations (extension homework) and d) to integrate separately learned skills and concepts by undertaking book reports, science projects, or creative writing. As a result, the provision of technology to take home work could be seen as an opportunity to improve on not only academic achievement but also the degree of engagement with school-related tasks. Homework as schoolwork that is done at home could be supported by making learning materials available in both contexts, e.g. by using the same software as used in class so that students are familiarised with the activities (Kerawalla, O'Connor, Underwood, duBoulay, Holmberg, Luckin, Smith and Tunley (2007). This seems to be the case for the Australian students but not for the English ones.

Homework however is conceptualised differently from other learning activities taking place at home because it is initiated, set by the teacher and is usually not optional (Kerawalla & Crook, 2005) and not assessed. Technology and the activities that it affords could play a role in blurring the boundaries between home learning and school so that it is a continuous learning experience. Learning materials therefore could allow for space for student-initiated activities that are fun and engaging.

Other Personal uses of Handhelds

The data in Table 5 shows that more than 80% of the students from the two countries enjoy using the handheld for playing games. About 50% of the UK students and 60% of the Australian students agreed or strongly agreed that they have lots of music in their handhelds too. This data concords with the data in Table 6 where students from both countries indicated that playing games and listening to music are the top five uses of the handhelds both at school and at home. In the home environment, these two activities remain as the top two uses of the handhelds by students from both countries, supporting the assertion by Kerawalla and Crook (2005) that when devices are used at home (which is primarily a place for relaxation and leisure), students spent more time playing entertainment games than being involved in educational activities.

A notable difference between the Australian and English students in terms of personal usage of the handhelds is the use of these devices to help them organise their time better. From the results in Table 5, it can be seen that over 50% of the Australian students indicated that the handhelds helped them organised their time better compared to only 27% of English students who indicated this. As shown in Table 6, using the calendar and diary both at school and home appear to be activities that helped with the students' time organisation. The higher positive response by the

Table 6. Handheld use by year 7 Australian and UK students at home and school: types and frequencies.

Home					School			
Australian students		English students			Australian students		English students	
1	Playing Games	1	Playing Games		1	As a calculator	1	Playing Games
2	Listening to music	2	Listening to music		2	Playing Games	2	As a calculator
3	As a calculator	3	As a calculator		3	Accessing the Internet	3	Taking down notes
4	Accessing the Internet	4	Accessing the Internet		4	Sharing Games	4	Listening to music
5	Downloading Music	5	Downloading Music		5	Listening to music	5	Sharing Games
6	Using the Calendar	6	Taking down notes		6	Using the Calendar	6	Excel spreadsheet
7	Downloading Games	7	Using the Calendar		7	As a Diary	7	Accessing the Internet
8	Sharing Games	8	Sharing Games		8	Completing worksheets	8	Downloading Games
9	As a Diary	9	Downloading videos, e.g. YouTube		9	Taking down notes	9	As a Diary
10	Completing worksheets	10	Facebook, Bebo, MySpace, etc		10	Excel spreadsheet	10	PowerPoint Presentation
	Facebook, Bebo, MySpace, etc		Email			Email		Using Maps
	Taking down notes		Downloading Games			Listening to Podcasts		Downloading Music
	Email		PowerPoint Presentation			Creating Podcasts		Email
	Downloading videos, e.g. YouTube		Excel spreadsheet			PowerPoint Presentation		Completing worksheets
	Excel spreadsheet		As a Diary			Downloading Games		Using the Calendar
	Listening to Podcasts		Using Maps			Downloading Music		Reading eBooks
	Reading eBooks		Creating Podcasts			Facebook, Bebo, MySpace, etc		Downloading videos, e.g. YouTube
	Creating Podcasts		Listening to Podcasts			Using Maps		Creating Podcasts
	PowerPoint Presentation		Completing worksheets			Reading eBooks		Listening to Podcasts
	Using Maps		Reading eBooks			Downloading videos, e.g. YouTube		Facebook, Bebo, MySpace, etc

Australian students would be due largely to their school implementing a policy where parents have to sign their children's diary. For these students with handhelds, they were required to use the devices as diaries. On a daily basis, their parents were required to 'sign' in the handhelds which were then checked by the classroom teachers. A substantial number of students (23% and 34% of Australian and English students respectively) from both countries were, however, unsure if the devices helped with organising their time better. It could be that these students have not explored the use of these features sufficiently or consistently to see an impact.

Item 10 in Table 5 referred to the ease of sending messages to friends with handhelds. The responses to this item are mixed with both groups of students. The responses of the Australian students were split a third each way for 'agree/strongly agree', 'not sure' and 'disagree/strongly disagree'. There were more English students who disagreed/strongly disagreed with the statement that 'with a handheld, it is easy to send messages to my friends'. Forty-four percent of English students responded in this category with about a third indicating that sending messages to friends was easy. The reasons for the mixed responses could be that students have not used Bluetooth or the Internet to send messages or that they have encountered technical difficulties, as discussed in section 'Computer literacy of the Year 7 students' above. The research is unable to distinguish between whether the technical difficulties occurred at home or in school.

TRENDS IN HANDHELD USAGE BY YEAR 7 AUSTRALIAN AND ENGLISH STUDENTS AT HOME AND SCHOOL: TYPES AND FREQUENCIES OF USAGE

Table 6 shows that handheld uses by Australian and English students at home and school show similar trends. Both groups of Year 7 students ranked entertainment and accessing the Internet and activities related to it, for example downloading music, games and videos as the most frequent activities that they engaged their handhelds with. These activities are ranked within the top 10 most frequently conducted activities with handhelds, both at home and in school. Using the handhelds as a calculator and for organisational purposes (calendar and as a diary) also ranked as frequent activities both at home and at school. Other similarities of uses by both groups of students at school are the frequent use of Excel spreadsheets, and Note and/or Word for taking down notes.

A difference between the two groups of students is in the use of the handhelds for completing worksheets. The Australian students appear to use their handhelds for completing worksheets quite frequently at school and at home (presumably as homework). This data concords with their perceptions of the usefulness of handhelds for homework purposes (see discussion on section 'Homework with handhelds' above), suggesting that the Australian students had more frequent usage in class and at home for academic work than their UK counterparts.

Table 6 also indicates a lack of frequent use of Web 2.0 technologies by the Year 7 students. Reasons for this could be due to the students' young ages where they may not need to share information over the Internet yet and concerns about risks of using Web 2.0 (Byron, 2007). Even though there are concerns about these risks, there are some striking benefits. In particular, when engaging in online collaboration and social networking, the development of skills required to prosper in the 21st century is supported, such as creativity, idea generation, presentation, leadership, team building, confidence, communication, innovation, initiative, critical awareness in information gathering, and ability to evaluate, question and prioritize information (Green and Hannon, 2007). To overcome the new digital divide between the Web confident and Web-restricted, combined efforts by policymakers, local authorities, teachers, parents and students are required where a necessary prerequisite is to balance the discussion of e-safety and child protection with that of Web entitlement and child development (Sharples, Graber, Harrison & Logan, 2009).

Students Open Comments and Recommendations

The open questions sought for comments and recommendations from the students on the use of handhelds in their schools. The responses to the open questions drew out two main themes from the students of both countries. These are:

a) There was insufficient use in the classrooms, hence a number of them viewed the handhelds as an expensive gaming consoles. Students indicated strong usage for entertainment – playing games, listening to music or downloading games and music into their handhelds. Indication of other usages include for research, organising (calendar) and the use of Excel, Word and Note functions. Examples of comments that the students wrote are shown below:

I use it to play games only and maybe twice a week for school work. A waste of money. A very expensive gaming console (Australian student)

We should use it in every class, because a lot of the time we don't get to use it (Australian student)

Use them in lessons because teachers won't let us use them (English student)

To use it more often (English student)

b) There were expressions of frustrations in terms of the technical aspects of the use of the handhelds, particularly in their inability to access the Internet properly.

In terms of recommendations, 30 Australian and 27 English students suggested recommendations. In accordance with the comments indicated above, their recommendations were strongest in 'more usage' and 'improving the technical aspects of the handhelds', notably better Internet access. As shown in Table 7, more than 50% of the students from both countries wanted more use of the handhelds and over 20% wanted better Internet access. Comments to support these recommendations include:

Table 7. Students' recommendations for the handheld program at their schools

Recommendations	Australian students %	English students %
More use of the handhelds	57	52
Better Internet access	20	22
Others e.g. improving battery life, have more software, have phone function	23	26

Teachers have more activities to do with them to make the subjects more fun (Australian student)

Try to improve the pda's internet things cos sometimes it lags (Australian student)

Teachers have software to use it (English student)

Teachers to be able to send homework to us (English student)

CONCLUSION AND IMPLICATIONS

This chapter compared students' use of handhelds across two countries to provide a view of the students' attitudes toward handheld usage and what these adolescents do with them in schools and at home. It explored the kinds of opportunities that school and home environments have to offer for students in relating out-of-school activities to experiences within the classroom. An aim of the research was to understand the mobile nature of adolescents' learning using handhelds. We recognise that how students learn with handhelds in school would depend on multiple factors like infrastructure and financial considerations, teacher professional training and identifying effective pedagogies related to handheld usage in the classrooms. The direct exploration of the full impact of these factors on learning was not the focus of this study.

Despite the schools' locations in socio-economic regions of different status, the research showed that students from both countries were very computer literate but they highlighted difficulties with connectivity to the Web. They held positive attitudes towards learning with technology and thought that they learn better with technology.

They also indicated that it was only worthwhile using handhelds at schools if they had ownership of the devices which they could take home. They did not, however, in this study consider it beneficial to learn with handhelds since they did not perceive better marks or better learning due to the technology. This finding raises implications related to the need to consider assessment when learning with such technologies and how best to capture improvements in learning outcomes and skills developed from the learning experiences with mobile devices.

Completing homework with handhelds was perceived more positively with the Australian students. This finding draws attention, particularly in the English situation, to the need for teachers to promote the use of the handhelds by allocating activities that take advantage of the device and which could bridge home-school learning. In this regard, the role of homework and other learning activities that occur at home but are not related to schoolwork also need to be considered.

The research found that the most frequent use of the handhelds by students from both countries was associated with games and music downloading and sharing. This finding concords with Kerawalla and Crook's (2005) assertion that since relaxation and fun activities are the main focus of young people, and for technology to be used, it would need to provide access to leisure activities.

A significant conclusion that emerged from the findings of this study is that the more frequent use of handhelds in their classes has resulted in the Australian students to adopt a more positive attitude toward the device, presumably through having gained a better understanding of the devices' potential for learning. The motivational aspect of mobile technologies on students' learning has been well documented (for example McFarlane, Triggs & Yee, 2008; Ng, 2008; Ng & Nicholas, 2009a; Perry, 2003; Twining et. al, 2005; Vahey & Crawford, 2002). Teachers should capitalise on this aspect of learning with mobile technologies to engage students to learn better. What was not captured by this study however, was the development of skills relevant to self-directed learning. An implication of this study for future research is the investigation of how students perceive 'authentic control' of the classroom activities and their own learning, and the development of critical thinking, co-operative and collaborative knowledge construction skills. These are issues that are of great importance and it shows that the school initiative to provide students with technology could go wrong if technology is not paired with meaningful activities that take advantage of the opportunities that the technology affords. Research should provide guidelines on what makes handhelds become useful learning tools so that schools can apply such findings in practice without the support from the academic community.

Responsibilities of Schools

This chapter has provided an overview of how young adolescents used new technologies in the home and in school. Mobile technology in particular can facilitate students' learning and socializing in both formal and informal settings and teachers should take advantage of the mobility of handhelds to provide students with opportunities to develop lifelong and independent learning skills. In introducing mobile technologies to students, schools have the responsibility to inculcate these skills in their students. These skills include critical thinking and research-based skills such as being able to search, locate, decide, assess (analyse and evaluate) and select for appropriate information, and design and synthesise new ideas or solutions to problems (McKenzie, 1998; Ng, 2008; Ng & Nicholas, 2009b). The teacher plays a central role in providing scaffolding activities to assist students with developing these skills. Setting tasks such as WebQuests is one way in which a teacher could scaffold learning and overcome the difficulties faced with the open and unstructured information seeking nature of an open research project (Dodge, 2007; March, 2003; Ng, 2006).

Tasks that are relevant and more personalized are a means of fostering self-directed learning skills. For example, the activity outlined in Appendix II where students have to investigate their own pulse rates during rest and after exercises, and to compare them with those of their peers (and other animals), is promoting research and analytical skills that are both experimental (see instructions 7-10 in Appendix II) and information seeking (see 13 and 14 in Appendix II). Learning in this context is authentic and meaningful as students are able to relate the learning to themselves, their peers and the surroundings that they are familiar with, for example, identifying the amount of blood pumped per heart pulse with the aid of a coca-cola bottle. The activity is also meaningful in that it traverses the learning disciplines of Mathematics, Science, Literacy and Information Technology, enabling students to apply skills from the different discipline areas to solve the problems required by the activity. In collecting the pulse rates of their peers, the handhelds enable students to Bluetooth their own pulse rates to their peers. Alternatively, the mobility of the handhelds allows the students to move around physically to obtain the data face-to-face and enter them directly into the handhelds so that they can generate the graphs on Excel later. The ubiquitous learning opportunities that handhelds offer mean that the students could take the data home and continue the investigation with family members.

In concluding, mobile technology is a tool that is able to bridge students' learning between formal and informal situations. More research needs to be conducted into (i) how young adolescents use and learn with these technologies in these situations and (ii) the types of learning activities that are pedagogically sound to bridge meaningful home-school learning and to develop lifelong learning skills.

REFERENCES

Attwell, G. (2007). Personal Learning Environments - the future of eLearning? *eLearning Papers, 2*(1). Retrieved July 10, 2009. from http://www.elearningeuropa.info/out/?doc_id=9758&rsr_id=11561

Brugnoli, M. C., Morabito, F., Bo, G., & Murelli, E. (2007). Augmented itineraries: Mobile services differentiating what museum has to offer. *PsychNology Journal, 4*(3), 311–335.

Byron, T. (2007) *Safer Children in aDigitalWorld: The Report of the Byron Review.* Department for Children, Schools and Department for Culture, Media and Sport, London. Retrieved 12 July, 2009 from http://www.dfes.gov.uk/byronreview/.

Cooper, H., Robinson, J. C., & Patall, E.A. (2006). Does Homework Improve Academic Achievement? A

Corlett, D., Sharples, M., Chan, T., & Bull, S. (2005). A Mobile Learning Organiser for University Students. *Journal of Computer Assisted Learning, 21*(3), 162–170. doi:10.1111/j.1365-2729.2005.00124.x

Dewey, J. (1938). *École et société*. Paris: PUF.

Dodge, B. (2007). *The Webquest page*. Retrieved July 7, 2009. from http://webquest.org/index.php

Drucker, P. F. (1994). The Age of Social Transformation. *Atlantic Monthly*, (November): 53–80.

Fischer, G., & Scharff, E. (1998). Learning Technologies in Support of Self-Directed Learning. *Journal of Interactive Media in Education, 98*(4). Retrieved July 10, 2009, from http://www jime.open.ac.uk/98/4.

Garrison, D. R. (1992). Critical Thinking and Self-Directed Learning in Adult Education: An Analysis of Responsibility and Control Issues. *Adult Education Quarterly, 42*(3), 136–148.

Green, H., & Hannon, C. (2007) *TheirSpace: Education for a Digital Generation.* Demos, London. Retrieved July 12, 2009 from http://www.demos.co.uk/publications/theirspace.

Hartnell-Young, E. (2008). Mobile phones for learning in mainstream schooling: resistance and change.

In, J. Traxler, B.Riordan, C. Dennett (eds) *Proceedings of the mLearn 2008 Conference: The bridge from text to context*, 7-10 October 2008, Ironbridge Gorge, Shropeshire, UK,pp. 160-167

Kerawalla, L., & Crook, C. (2005). From Promises to Practices: the fate of educational software in the home. *Technology, Pedagogy and Education, 14*(1), 107–126. doi:10.1080/14759390500200195

Kerawalla, L., O'Connor, J., Underwood, J., duBoulay, B., Holmberg, J., & Luckin, R. (2007). Exploring the Potential of the Homework System and Tablet PCs to Support Continuity of Numeracy Practices between Home and Primary School. *Educational Media International, 44*(4), 289–303. doi:10.1080/09523980701680904

Lave, J., & Wenger, E. (1991). *Situated learning: Legitimate peripheral participation*. Cambridge, UK: Cambridge University Press.

Madia, M. S. (2007). Selection interviews using mobile technology. In Norman, A. & Pearce, J. (Eds) *Proceedings of the 6th International Conference on Mobile Learning* (pp140-150), Melbourne, Australia: Melbourne University Press.

March, T. (2003). *Webquests for learning.* Retrieved July 10, 2009, from http://bestwebquests.com/

McFarlane, A., Triggs, P., & Yee, W. (2008). *Mobile learning: Research findings.* Coventry, UK: Becta. Retrieved May 2, 2009 from http://partners.becta.org.uk/uploaddir/downloads/page_documents/research/mobile_learning.pdf.

McKenzie, J. (1998). Grazing the Net. [from http://www.fno.org/text/grazing.html.]. *Phi Delta Kappan, 79*(September), 26–31. Retrieved July 9, 2009.

Ng, W. (2006). Web-based Technologies, Technology Literacy and Learning in Tan, L.W.H. and Subramaniam, R. (Eds). *Handbook of Research on Literacy in Technology at the K-12 Level* (pp. 94-117) Hershey, PA: Idea Group Publishing.

Ng, W. (2008). Self-directed learning with web-based sites: How well do students' perceptions and thinking match with their teachers? *Teaching Science, 54*(2), 26–30.

Ng, W., & Nicholas, H. (2009a). Introduction of pocket PC in schools: attitudes and beliefs in the first year. *Computers & Education, 52*, 470–480. doi:10.1016/j.compedu.2008.10.001

Ng, W., & Nicholas, H. (2009b). Self-directed learning with web-based technologies. In Rogers, P., Berg, G., Boettcher, J., Howard, C., Justice, L., & Schenk, K. (Eds.), *Encyclopedia for online and distance Learning* (pp. 1847–1852). Hershey, PA: IGI Global Publishers.

Nicholas, H., & Ng, W. (2009). Ubiquitous learning and handhelds: An overview of theory and pedagogy. In Rogers, P., Berg, G., Boettcher, J., Howard, C., Justice, L., & Schenk, K. (Eds.), *Encyclopedia for online and distance learning* (pp. 2171–2176). Hershey, PA: IGI Global Publishers.

Ogata, H., & Yano, Y. (2004). Context-Aware Support for Computer Supported Ubiquitous Learning, *Proceedings of the 2nd IEEE International Workshop on Wireless and Mobile Technologies in Education*, (pp.27-34).

Perry, D. (2003). *Handheld computers (PDAs) in schools*. Coventry, UK: Becta.

Sharples, M. (2003). Disruptive Devices: Mobile Technology for Conversational Learning. *International Journal of Continuing Engineering Education and Lifelong Learning, 12*(5/6), 504–520. doi:10.1504/IJCEELL.2002.002148

Sharples, M., Graber, R., Harrison, C., & Logan, K. (2009). E-safety and Web 2.0 for children aged 11–16. *Journal of Computer Assisted Learning, 25*, 70–84. doi:10.1111/j.1365-2729.2008.00304.x

Sharples, M., Taylor, J., & Vavoula, G. (2007). A theory of learning for the mobile age. In Andrews, R., & Haythornthwaite, C. (Eds.), *The SAGE Handbook of e-learning research* (pp. 221–224). London: Sage.

Sinitsa, K. (2000). Learning individually: a lifelong perspective. *Journal of Educational Technology & Society, 3*(1), 17–23.

Smith, M. (1996). *Self-direction. The encyclopaedia of informal education.* Retrieved July 9, 2009 from http://www.infed.org/biblio/b-selfdr.htm

Synthesis of Research. 1987–2003. *Review of Educational Research, 76*, 1–62.

Tough, A. M. (1967). *Learning Without a Teacher. A study of tasks and assistance during adult self teaching projects*. Toronto: Ontario Institute for Studies in Education.

Twining, P., & Evans, D. with Cook, D., Ralston, J., Selwood, I., Jones, A., et al. (2005). Should there be a future for tablet PCs in schools? *Journal of Interactive Media in Education, 20*, 1–13. Retrieved July 7, 2009, from http://jime.open.ac.uk/2005/20/twining-2005-20-paper.html.

Vahey, P., & Crawford, V. (2002). *Palm Education Pioneers Program Final Evaluation Report*. Retrieved July 7, 2009, from http://www.palmgrants.sri.com/PEP_Final_Report.pdf

Vavoula, G., Sharples, M., Rudman, P., Meek, J., & Lonsdale, P. (2009). Myartspace: Design and evaluation of support for learning with multimedia phones between classrooms and museums. *Computers & Education, 53*(2), 286–299. doi:10.1016/j.compedu.2009.02.007

Waycott, J., Jones, A., & Scanlon, E. (2005). PDAs as lifelong learning tools: an activity theory based analysis. *Learning, Media and Technology, 30*(2), 107–130. doi:10.1080/17439880500093513

APPENDIX A: THE RESEARCH INSTRUMENT

Please place a tick in the box that best matches how you feel for each statement.

	Strongly Agree	Agree	Not Sure	Disagree	Strongly Disagree
I know easily how to use a EDA					
EDAs help me learn all subjects a lot better					
I like using technology for learning a lot					
EDAs are easy to use					
EDAs make learning a lot easier.					
I am excited about using a EDA					
I have lots of music on my EDA					
I need special training to use a EDA					
EDAs make learning fun.					
I learn better with technology					
With a EDA I can do my homework anywhere at anytime					
There are no disadvantages to using EDAs in the classroom					
I have used a EDA lots already					
EDAs make learning more interesting					
I enjoy using the EDAs for games					
I have a computer (desktop or laptop) at home					
I can easily use the Internet at home					
EDAs help me organise my time better					
EDAs do the same things as a desktop computer					
Writing with a EDA is easier than writing by hand					
If I couldn't take the EDA home, it wouldn't be worth using it in school.					
The screen on the EDA makes it difficult to do my homework					
It is easy to access the Internet with EDAs					
With a EDA it is very easy to change what I have written					
With a EDA, it is easy to send messages to my friends					
With a EDA it is easy to take my school work home					
EDAs at school are a waste of time					
I enjoy using the EDA for schoolwork					
EDA has helped me learn better					
My marks have been better learning with EDA					

So far, what have you used the pocket PC for? Please tick next to the letter on 'usage'.

Use	Tick here		Approximately how many times a week do you do it?	
Sending work to teacher				
Taking work home				
Use	**Tick here**		**Approximately how many times a week do you use it?**	
	home	school	home	school
Accessing the Internet				
Reading eBooks				
Creating Podcasts				
Listening to Podcasts				
Playing Games				
Sharing Games				
Taking down notes				
Completing worksheets				
Downloading Games				
Downloading Music				
Downloading videos, e.g. YouTube				
Listening to music				
Facebook, Bebo, MySpace, etc				
PowerPoint Presentation				
Using Maps				
As a calculator				
Excel spreadsheet				
Using the Calendar				
Email				
As a Diary				
Others Please specify:				

1. What recommendations, if any, would you make for the pocket PC/EDA program at your school?
2. Any other comments?

APPENDIX B: INVESTIGATING THE HEART AND PULSE RATES

Conduct this activity using your pocket PC only. Use resources such as stop clock in the pocket PC to assist you with completing this activity. Present data collected and answer all questions clearly on the pocket PC. You will work in pairs for task 1, 4 and 13.

1. Determine your resting pulse rate.
   Each time the heart beats, it sets up a wave of pressure which travels along the main arteries. This is called a pulse wave and if you put your finger on your skin just above the artery in your wrist or neck, you can feel it as a slight throb. Sitting down, determine your pulse rate by counting the number of times your pulse beats in one minute. Record your result. Do this twice more and find your average 'resting' pulse rate in 'beats per minute'.
2. How much work does the heart do?
   Each time the heart beats it pumps about 80 ml of blood into the arteries. Calculate how much blood your heart pumps in 1, 5 and 10 minutes.
3. How many 2 litre coca cola bottles could your heart fill in 5 minutes?
4. Determine your pulse rate after one minute of exercises.
   Stand up and do some vigorous exercises for one minute e.g. bending exercises or go for a run. Determine your 'after exercising' pulse rate. Record it as 'beats per minute'.
5. How does the 'after exercising' pulse rate compare with the 'resting' pulse rate.
6. Use your pocket PC to obtain all the 'resting' and 'after exercising' pulse rates from the rest of the class.
7. Record all the data collected and find the average class pulse rate for 'resting' and 'after exercising'.
8. What is the range of the 'resting' and 'after exercising' pulse rates of the class?
9. How does your 'resting' and 'after exercising' pulse rates compare with the average of the class?
10. Construct a graph or histogram for the number of students with pulse rates in each of these categories:
    71-80; 81-90; 91-100; 101-110; 111-120; 121-130
11. Where does all the blood that the heart pumps go to?
12. Show the connections between these 8 keywords: heart, blood vessels, veins, arteries, blush, heart attack, cholesterol, food. Choose whatever way you like, for example, you could draw a flowchart or write a short paragraph or voice record a description of how the 8 keywords are connected etc.
13. Internet search. Look up the resting heart rates of 2-3 small animals (e.g. rat) and 2-3 large animals (e.g elephant). Show the heart rate next to an image of the animal. What do you notice about heart rate and size of an animal?
14. Which animal's heart rate is closest to yours?

Compilation of References

Abas, Z. W., Chng, L. P., & Mansor, N. (2009). A study on learner readiness for mobile learning at Open University Malaysia. In *Proceedings of IADIS International Conference Mobile Learning,* (pp. 151-157).

Abas, Z. W., Lim, T., & Woo, T. K. (2009). Mobile learning initiative through SMS: A formative evaluation. *ASEAN Journal of Open Distance Learning, 1*(1).

Abfalter, D., Mirski, P. J., & Hitz, M. (2004). Mobile learning - Knowledge enhancement and vocational training on the move. In *Proceedings of the 5th European Conference on Organizational Knowledge, Learning and Capabilities*, Innsbruck, Austria. Retrieved June 25, 2009 from http://delphi.jura.uni-sb.de/texts/mirski2.pdf

Adam, D. M., & Hamm, M. E. (1990). *Cooperative learning critical thinking and collaboration across the curriculum.* Charles C. Thomas.

AIMIUA. (2009). *Australian Mobile Phone Lifestyle Index.* 5th Edition – Special topic: Mobile commerce, September 2009. Retrieved October 21, 2009, from http://www.aimia.com.au/i-cms?page=1.32.1491.2841

Ainsworth, S. (1999). The functions of multiple representations. *Computers & Education, 33*(2/3), 131–152. doi:10.1016/S0360-1315(99)00029-9

Ainsworth, S., & Iacovides, I. (2005, August). *Learning by constructing self-explanation diagrams.* Paper presented at the 11th Biennial Conference of European Association for Research on Learning and Instruction, Nicosia, Cyprus.

Albanese, M. A., & Mitchell, S. (1993). Problem-based learning: A review of literature on its outcomes and implementation issues. *Academic Medicine, 68*, 52–81. doi:10.1097/00001888-199301000-00012

Anastopoulou, S., Sharples, M., Wright, M., Martin, H., Ainsworth, S., & Benford, S. (2008). Learning 21st century science in context with mobile technologies. In Traxler, J., Riordan, B., & Dennett, C. (Eds.), *The bridge from text to context. Proceedings of mLearn 2008, Wolverhampton, UK* (pp. 12–21). Wolverhampton, UK: University of Wolverhampton.

Anderson, J. R. (1983). *The architecture of cognition.* Cambridge, MA: Harvard University Press.

Anderson, T., Rourke, L., Garrison, D. R., & Archer, W. (2001). Assessing teacher presence in a computer conferencing context. *Journal of Asynchronous Learning Networks, 5*(2), 1–17.

Andronache, A., Brust, M. R., & Rothkugel, S. (2007). Hycast- podcast discovery in mobile networks. In *Proceedings of the 3rd ACM Workshop on Wireless Multimedia Networking and Performance Modeling,* Crete Island, Greece, WMuNeP '07, (pp. 27-34).

Archibald, J. (2007). *Cellphones as an Educational Tool.* Retrieved July 9, 2009 from http://www.tectonic.co.za/?p=1396

Ardito, C., Costabile, M. F., & Lanzilotti, R. (2009). Enhancing user experience while gaming in archaeological parks with cellular phones. In *Proceedings of the 8th International Conference on Interaction Design and Children*, Como, Italy, pp. 270-271.

Artemeva, N., Logie, S., & St-Martin, J. (1999). From page to stage: How theories of genre and situated learning help introduce engineering students to discipline-specific communication. *Technical Communication Quarterly*, *8*(3), 301–316.

Attwell, G. (2007). Personal Learning Environments - the future of eLearning? *eLearning Papers, 2*(1). Retrieved July 10, 2009. from http://www.elearningeuropa.info/out/?doc_id=9758&rsr_id=11561

Australian Government. (2009). *State & Territory Privacy Laws*. Office of the Privacy Commissioner. Retrieved August 19, 2009, from http://203.210.106.3/privacy_rights/laws/

Bachfischer, A., Lawrence, E., & Litchfield, A. Dyson, L. E., & Rabran, R. (2008). Student perspective about using mobile devices in their studies. *2008 IADIS International Conference on Mobile Communication Proceedings* (pp. 287-295), Algarve, Portugal: ACM.

Ballantyne, R., & Packer, J. (2009). Introducing a fifth pedagogy: Experience-based strategies for facilitating learning in natural environments. *Environmental Education Research*, *15*(2), 243–262. doi:10.1080/13504620802711282

Bandura, A. (1989). Human agency in social cognitive theory. *The American Psychologist*, *44*(9), 1175–1184. doi:10.1037/0003-066X.44.9.1175

Banks, M. (2001). *Visual methods in social research*. London: Sage.

Bardini, C., Pierce, R., & Stacey, K. (2004). Teaching linear functions in context with graphics calculators: students' responses and the impact of the approach on their use of algebraic symbols. *International Journal of Science and Mathematics Education*, *2*, 353–376. doi:10.1007/s10763-004-8075-3

Batane, T. (2006). When computers are Scarce: a case of Botswana Schools. In *Proceedings of the 4th IEEE International Workshop on Technology for Education in Developing Countries TEDC '06* (pp. 93-98).

Bateson, G., & Mead, M. (1947). *Balinese character: A Photographic Analysis*. New York: New York Academy of Science.

Baumgartner, E., & Zabin, C. J. (2008). A case study or project-based instruction in the ninth grade: a semester-long study of intertidal biodiversity. *Environmental Education Research*, *14*(2), 97–114. doi:10.1080/13504620801951640

Becta, (Britian Educational Communications and Technology Agency) (2003). *Becta ICT Research Online*. Retrieved July 30, 2009, from http://www.becta.org.uk

BECTA. (2006). *Safeguarding children in a digital world: Developing a strategic approach to e-safety*. Retrieved January 5, 2009, from http://publications.becta.org.uk/display.cfm?resID=25933.

Benfield, J. A., & Szlemko, W.J. (2006). Internet-Based Data Collection: Promises and Realities. *Journal of Research Practice*., 2 (2), Article D1.

Benteley, C. (2007). The OLPC Laptop: Educational Revolution or Devolution? *Proceedings of World Conference on ELearning in Corporate, Government, Healthcare and Higher Education* (pp. 647-652). Quebec, Canada

Bermejo, S. (2005). Cooperative electronic learning in virtual laboratories through forums. *IEEE Transactions on Education*, *48*(1), 140–149. doi:10.1109/TE.2004.837045

Berners-Lee, T., & Fischetti, M. (1999). *Weaving the Web: The Past, Present and Future of the World Wide Web by its Inventor*. Britain: Orion Business.

Berson, I. R., & Berson, M. J. (2003). Digital literacy for effective citizenship. *Social Education*, *67*(3), 164–167.

Beyer, H., & Holtzblatt, K. (1998). *Contextual Design, Defining Customer-centred systems*. San Francisco: Morgan Kaufmann Publishers, Inc.

Bower, M. (2008). Affordance analysis – matching learning tasks with learning technologies. *Educational Media International*, *45*(1), 3–15. doi:10.1080/09523980701847115

Bradley, C., Haynes, R., & Boyle, T. (2005). Adult multimedia learning with PDAs - The user experience. In *Proceedings of mLearn 2005, 4th World conference on m-Learning*, Cape Town, South Africa.

Bravo, C., van Joolingen, W. R., & de Jong, T. (2006). Modeling and Simulation in Inquiry Learning: Checking Solutions and Giving Intelligent Advice. *Simulation*, *82*(11), 769–784. doi:10.1177/0037549706074190

Brown, J. S., Collins, A., & Duguid, P. (1989). Situated cognition and the cultural of learning. *Educational Researcher*, *18*(1), 32–42.

Brown, E. (Ed.). (2001). *Mobile learning explorations at the Stanford learning lab. Speaking of Computers, 55*. Stanford, CA: Board of Trustees of the Leland Stanford Junior University.

Brown, R., Ryu, H., & Parsons, D. (2006). Mobile helper for university students: a design for a mobile learning environment. In Kjeldskov, J. & Paay, J. (Eds.), *Proceedings of the 18th Australia Conference on Computer-Human interaction: Design: Activities, Artifacts and Environments*, Sydney, Australia.

Brugnoli, M. C., Morabito, F., Bo, G., & Murelli, E. (2007). Augmented itineraries: Mobile services differentiating what museum has to offer. *PsychNology Journal*, *4*(3), 311–335.

Bruner, J. S. (1960). *The process of education*. Boston: Harvard University Press.

Bruner, J. (1986). *Actual Minds, Possible Worlds*. Cambridge, MA: Harvard University Press.

Bruner, J. (1990). *Acts of Meaning*. Cambridge, MA: Harvard University Press.

Bruner, J. (1996). *The Culture of Education*. Cambridge, MA: Harvard University Press.

Burgard, W., Cremers, A. B., Fox, D., Hähnel, D., Lakemeyer, G., & Schulz, D. (1999). Experiences with an interactive museum tour-guide robot. *Artificial Intelligence*, *114*, 3–55. doi:10.1016/S0004-3702(99)00070-3

Businessweek (2000). *The highs and lows of Napster*. Retrieved April 21, 2009 from: http://www.businessweek.com/2000/00_33/b3694003.htm

Byron, T. (2007) *Safer Children in aDigitalWorld: The Report of the Byron Review*. Department for Children, Schools and Department for Culture, Media and Sport, London. Retrieved 12 July, 2009 from http://www.dfes.gov.uk/byronreview/.

Cake, M. (2006). Deep dissection: motivation students beyond rote learning in veterinary anatomy. *Journal of Veterinary Medical Education*, *33*(2), 266–271. doi:10.3138/jvme.33.2.266

Canalys.com. (2008). *Global smart phone shipment rise 28% - Nokia retains lead, but Apple moves into number two position*. Retrieved December 14, 2008 from http://www.canalys.com/pr/2008/r2008112.htm

Caron, A. H., & Caronia, L. (2007). *Moving cultures. Mobile communication in everyday life*. Montreal: McGill-Queens University Press.

Caron, A. H., & Caronia, L. (2008). The cultural dimensions of "IPODS" in higher education *IADIS International Journal on WWW/Internet*, *6*(2), 73-89.

Caronia, L. (2002). *La socializzazione ai media. Contesti, interazioni e pratiche educative*. Milano, Italy: Guerini.

Caronia, L., & Caron, A. H. (2004). Constructing a specific culture: Young people's use of the mobile phone as a social performance. *Convergence (London)*, *10*(2), 28–61. doi:10.1177/135485650401000204

Caronia, L., & Caron, A. H. (2009). Mobile learning in the digital age: a clash of cultures. In Kleinman, S. (Ed.), *The culture of efficiency: technology in everyday life* (pp. 191–211). New York: Peter Lang.

Carroll, J., Neale, D., Isenhour, P., Rosson, M., & McCrickard, D. (2003). Notification and awareness: synchronizing task-oriented collaborative activity. *Journal of Human-Computer Studies*, *58*(5), 605–632. doi:10.1016/S1071-5819(03)00024-7

Carvin, A. (2007). The iPod of the Beholder: Can MP3 Players Enhance Learning? *PBS Teachers*. Retrieved July 20, 2009 from http://www.pbs.org/teachers/learning.now/2007/05/the_ipod_of_the_beholder_can_m.html

Chan, A., & Lee, M. J. W. (2005). *An MP3 a day keeps the worries away*. Student Experience Conference 2005. Charles Stuart University. Retrieved June 20, 2009 from: http://www.csu.edu.au/division/studserv/sec/papers/chan.pdf

Chen, Y. S., Kao, T. C., & Sheu, J. P. (2003). A mobile learning system for scaffolding bird watching learning. *Journal of Computer Assisted Learning*, *19*(3), 347–359. doi:10.1046/j.0266-4909.2003.00036.x

Chen, Y. S., Kao, T. C., & Yu, G. J. (2004). A mobile butterfly-watching learning system for supporting independent learning. In *Proceedings of the 2nd International Workshop on Wireless and Mobile Technologies in Education*, pp. 11-18.

Chen, Y. S., Kao, T. C., Sheu, J. P., & Chiang, C. Y. (2002). A Mobile Scaffolding-Aid-Based Bird Watching Learning System. In *Proceeding of International Workshop on Wireless and Mobile Technologies in Education*, IEEE Computer Society, pp.15-22.

Chew, E. (2008). Views, values and perceptions in geographical fieldwork in Singapore schools. *International Research in Geographical and Environmental Education*, *17*(4), 307–329. doi:10.1080/10382040802401565

Christiansen, N., & Maglaughlin, K. (2003). Crossing from physical workplace to virtual workplace: be AWARE! In *Proceedings of HCI International Conference on Human Computer Interaction*. Lawrence Erlbaum Associates, NJ, pp.1128 – 1132.

Chu, H.-C., Hwang, G.-J., Huang, S.-X., & Wu, T.-T. (2008). A knowledge engineering approach to developing e-libraries for mobile learning. *The Electronic Library*, *26*(3), 303–317. doi:10.1108/02640470810879464

Cicourel, A. (1982). Interviews, surveys, and the problem of ecological validity. *The American Sociologist*, *17*, 11–20.

CISCO. (2008). *Multimodal Learning Through Media: What the Research Says*. Retrieved October 31, 2009, from http://www.cisco.com/web/strategy/docs/education/Multimodal-Learning-Through-Media.pdf

CNET News. (2005). *Apple WWDC Keynote Steve Jobs*. Retrieved April 2009 from: http://news.cnet.com/2100-1047_3-5748045.html

Colburn, A. (2000). An Inquiry Primer. *Science Scope*, *23*(6), 42–44.

Cole, H., & Stanton, D. (2003). Designing mobile technologies to support co-present collaboration. *Personal and Ubiquitous Computing*, *7*(6), 365–371. doi:10.1007/s00779-003-0249-4

Colella, V. (2000). Participatory simulations: Building collaborative understanding through immersive dynamic modeling. *Journal of the Learning Sciences*, *4*(9), 471–500. doi:10.1207/S15327809JLS0904_4

Collier, J., & Collier, M. (1986). *Visual Anthropology: Photography as a Research Method*. Albuquerque: University of New Mexico Press.

Collingwood, R. G. (1994). In Jan van Der Dussen (Ed.), The idea of history [1946] revised edition with lectures 1926–1928. Oxford: Clarendon Press.

Colombo, F., & Scifo, B. (2005). Social shaping of the new mobile devices. Representations and uses among Italian Youth. In Haddon, L., Mante, E., Sapio, B., Kommonen, K.-H., Fortunati, L., & Kant, A. (Eds.), *Everyday innovators. Researching the role of users in shaping ICTs* (pp. 86–103). London: Springer.

ComReg. (2008). Irish Communications Market: Quarterly Key Data Report. Document number 08/101. Commission for Communications Regulation. Retrieved June 10, 2009, from http://www.comreg.ie/_fileupload/publications/ComReg08101.pdf

Condie, R., Munro, B., et al. (2007). The Impact of ICT in Schools – a landscape review, *BECTA ICT Research*. Retrieved July 20, 2009 from http://partners.becta.org.uk/page_documents/research/impact_ict_schools.pdf

Conroy, S. (2007). Labor's plan for cyber-safety (fact sheet). Retrieved January 5, 2009 from http://www.alp.org.au/download/now/labors_plan_for_cyber_safety.pdf.

Constructionist learning theory. (n.d.). Retrieved June 18, 2009, from Wikipedia: http://en.wikipedia.org/wiki/Constructionist_learning

Cook, T. D., & Campbell, D. T. (1979). *Quasi-experimental design and analysis issues for field settings*. Boston: Houghton Mifflin Co.

Cook, V. (2008). The field as a 'pedagogical resource'? A critical analysis of students' affective engagement with the field environment. *Environmental Education Research*, *14*(5), 507–517. doi:10.1080/13504620802204395

Cooper, H., Robinson, J. C., & Patall, E.A. (2006). Does Homework Improve Academic Achievement? A

Corlett, D., Sharples, M., Chan, T., & Bull, S. (2005). A Mobile Learning Organiser for University Students. *Journal of Computer Assisted Learning*, *21*(3), 162–170. doi:10.1111/j.1365-2729.2005.00124.x

Cortez, C., Nussbaum, M., Santelices, R. A., Rodríguez, P., Zurita, G., Correa, M., & Cautivo, R. (2004). *Teaching Science with Mobile Computer Supported Collaborative Learning (MCSCL)*. Proceedings of IEEE International Workshop on Wireless and Mobile Technologies in Education (WMTE 2004), (pp. 67-74).

Cox, B., Macharia, R., Short, N., & Whittlestone, K. (2008) in Salmon, G., Ediringsha, P.,(Eds), Podcasting for Learning in Universities. Glasgow: McGraw Hill.

Craik, F. I. M., & Lockhart, R. S. (1972). Levels of processing: A framework for memory research. *Journal of Verbal Learning and Verbal Behavior*, *11*, 671–684. doi:10.1016/S0022-5371(72)80001-X

Cross, J. (2006). *Informal Learning: Rediscovering the Natural Pathways That Inspire Innovation and Performance* (Essential Knowledge Resource), Jossey Bass, San Fancisco: Pfeiffer.

Cui, Y., & Bull, S. (2005). Context and learner modeling for the mobile foreign language learner. *System*, *33*(2), 353–367. doi:10.1016/j.system.2004.12.008

Curtis, M., Luchini, K., Bobrowsky, W., Quintana, C., & Soloway, E. (2002). Handheld Use in K-12: A Descriptive Account. In U. Hoppe, M. Milrad, & Kinshuk (Eds.), *Proceedings of IEEE International Workshop on Wireless and Mobile Technologies in Education* (pp. 23-30). New York: IEEE Computer Society Press.

Davis, F. D. (1986). A technology acceptance model for empirically testing new end-user information systems: Theory and results. Cambridge, M.A: MIT Sloan School of Management.

Deal, A. (2007). A teaching with technology white paper, *Educause Quaterly*. Retrieved September 6, 2007, from: http://connect.educause.edu/files/CMU_Podcasting_Jun07.pdf

Delphian eLearning. (2007). iPodagogy: Using iPods and Podcasting for Learning, *Department of Education and Training*, Victoria. Retrieved July 20 2009 from http://www.delphian.com.au/images/documents/ipodagogy%20research%20report_final1.pdf

Delphian eLearning. (2008). iPod Touch Research Report, *Department of Education and Early Childhood Development*, Victoria. Retrieved July 20 2009 from http://www.eduweb.vic.gov.au/edulibrary/public/ict/ipodtouchresearch2009.doc

Delphian eLearning. (2009). Global Mobile Learning Research Report 2009, *The Etech Group*, Australia. Retrieved July 20 2009 from http://www.delphian.com.au/images/documents/mobile%20learning%20research%20report_etech2009_final.pdf

Department for Education and Skills (DES). (2004). *Pedagogy and practice. Creating effective learners: Using ICT to enhance learning*. London: DES.

Department of Education and Science. (2007). *Irish in the Primary School. Inspectorate Evaluation Studies*. Dublin: DES.

Derbentseva, N., Safayeni, F., & Canas, A. J. (2007). Concept maps: experiments on dynamic thinking. *Journal of Research in Science Teaching*, *44*(3), 448–465. doi:10.1002/tea.20153

Dewey, J. (1916). *Democracy and education*. New York: Free Press.

Dewey, J. (1938). *École et société*. Paris: PUF.

Dey, A. K., & Abowd, G. D. (1999). *Towards a better understanding of context and context-awareness*. GVU Technical Report GIT-GVU-99-22, College of Computing, Georgia Institute of Technology, Atlanta, Georgia.

Dias, B., & Brewer, Eric. (2009). How Computer Science Serves the Developing World. *Communications of the ACM*, *52*(8), 74–80. doi:10.1145/1516046.1516064

Dias, J. (2002). Cell phones in the classroom: Boon or bane? *C@lling Japan 10(2),* 16-22. The Newsletter of the JALT-CALL Special interest Group.

Dias, J. (2002a). Cell phones in the classroom: Boon or bane? [Part 2] *C@lling Japan 10(3),* 8-13. The Newsletter of the JALT-CALL Special interest Group.

Dicks, B., Mason, B., Coffey, A., & Atkinson, P. (2005). *Hypermedia Ethnography*. London: Sage.

Dicks, B., Soyinka, B., & Coffey, A. (2006). Multimodal ethnography. *Qualitative Research Journal*, *6*(1), 77–96. doi:10.1177/1468794106058876

Dillenbourg, P., Baker, M., Blaye, A., & O'Malley, C. (1996). The evolution of research on collaborative learning. In Spada, E., & Reiman, P. (Eds.), *Learning in Humans and Machine: Towards an interdisciplinary learning science* (pp. 189–211). Oxford, UK: Elsevier.

Dodge, B. (2007). *The Webquest page*. Retrieved July 7, 2009. from http://webquest.org/index.php

Dogbey, J. (2007). Using iPods for Instruction, *The Principals Partnership*. Retrieved July 20, 2009 from http://www.principalspartnership.com/iPods.pdf

Dourish, P., & Bellotti, V. (1992). *Awareness and coordination in shared workspaces. Proceedings of Computer Supported Cooperative Work '92* (pp. 107–114). ACM Press.

Dourish, P., & Bly, S. (1992). Portholes: Supporting Awareness in a Distributed Work Group. Proceedings of the *CHI '92 Conference on Human Factors in Computing Systems*, pp. 541-547.

Driver, R., Asoko, H., Leach, J., Mortimer, E., & Scott, P. (1994). Constructing scientific knowledge in the classroom. *Educational Researcher*, *23*(7), 5–12.

Drucker, P. F. (1994). The Age of Social Transformation. *Atlantic Monthly*, (November): 53–80.

Duffy, T. M., & Jonassen, D. H. (1992). Constructivism: New implications for instructional technology. In Duffy, T. M., & Jonassen, D. H. (Eds.), *Constructivism and the technology of instruction: a conversation* (pp. 1–16). Hillsdale, NJ: Lawrence Erlbaum Associates.

Dunphy, A., & Spellman, G. (2009). Geography fieldwork, fieldwork value and learning styles. *International Research in Geographical and Environmental Education*, *18*(1), 19–28. doi:10.1080/10382040802591522

Edelson, D. C., Gordin, D. N., & Pea, R. D. (1999). Addressing the Challenges of Inquiry-Based Learning Through Technology and Curriculum Design. *Journal of the Learning Sciences*, *8*(3-4), 391–450. doi:10.1207/s15327809jls0803&4_3

Edgar, D., & P. (2008).Television, Digital Media and Children's Learning, *Victorian Curriculum and Assessment Authority*. Retrieved October 30, 2009 from http://www.vcaa.vic.edu.au/earlyyears/Edgar_Paper_20081223.pdf

Edirisingha, P., Rizzi, C., Nie, M., & Rothwell, L. (2007). Podcasting to provide teaching and learning support for an undergraduate module on English language and communication. Turkish online journal of distance education. *8*(3). Article 6.

El-Bishouty, M. M., Ogata, H., Ayala, G., & Yano, Y. (2010). Context-aware support for self-directed ubiquitous-learning. *Int. J. Mobile Learning and Organisation*, *4*(3), 317–331. doi:10.1504/IJMLO.2010.033558

El-Bishouty, M. M., Ogata, H., & Yano, Y. (2007). PERKAM: Personalized Knowledge Awareness Map for Computer Supported Ubiquitous Learning. *Journal of Educational Technology & Society*, *10*(3), 122–134.

El-Bishouty, M. M., Ogata, H., & Yano, Y. (2008). A Model of Personalized Collaborative Computer Supported Ubiquitous Learning Environment. *Proceedings of the 8th International Conference on Advanced Learning Technologies (ICALT08)*, Spain, IEEE Computer Society, pp. 97-101.

Ellis, A. K. (2002). *Teaching and Learning Elementary Social Studies*. Boston: Allyn and Bacon, Inc.

Engestrom, Y. (1994). *Training for change: New approach to instruction and learning in working life*. Geneva: International Labour Office.

Engeström, Y. (1987). *Learning by expanding: An activity theoretical approach to developmental research*. Helsinki: Orienta-Konsultit.

Engeström, Y. (1999). Activity theory and individual and social transformation. In Engeström, Y., Miettinen, R., & Punamaki, R. (Eds.), *Perspectives on Activity Theory*. Cambridge, UK: Cambridge University Press.

Eshet-Alkalai, Y. (2004). Digital literacy: A conceptual framework for survival skills in the digital era. *Journal of Educational Multimedia and Hypermedia*, *13*(1), 93–106.

Facer, K., Joiner, R., Stanton, D., Reid, J., Hull, R., & Kirk, D. (2004). Savannah: Mobile gaming and learning? *Journal of Computer Assisted Learning*, *20*(6), 399–409. doi:10.1111/j.1365-2729.2004.00105.x

Faculty of Education & Social Work, University of Sydney. (2009). *Supporting mobile learning from the ground up*. University of Sydney. Retrieved on July 21, 2009 from http://coco.edfac.usyd.edu.au/mobilelearning/peakhurstPS.php

Fahmy, M. (2008). *If u cn rd this quickly, gd 4 u*. Retrieved October 31, 2009 from http://www.reuters.com/article/technologyNews/idUSTRE4B90YA20081210?rpc=64

Fahy, P. J. (2002). *Assessing critical thinking processes in a computer conference*. Retrieved November 23, 2009, from http://cde.athabascau.ca/softeval/reports/mag4.pdf

Finn, M., & Vandenham, N. (2004). The handheld classroom: Educational implications of mobile computing. *Australian Journal of Emerging Technologies and Society*, *2*(1), 1–15.

Fischer, G. (2001). User Modeling in Human-Computer Interaction. [UMUAI]. *Journal of User Modeling and User Adapted Interaction*, *11*(1/2), 65–86. doi:10.1023/A:1011145532042

Fischer, G., & Konomi, S. (2005). Innovative Media in Support of Distributed Intelligence and Lifelong Learning. In *Proceeding of the International Workshop on Wireless and Mobile Technologies in Education*, Japan, IEEE Computer Society, pp. 3-10.

Fischer, G., & Konomi, S. (2007). Innovative socio-technical environments in support of distributed intelligence and lifelong learning. *Journal of Computer Assisted Learning*, *23*, 338–350. doi:10.1111/j.1365-2729.2007.00238.x

Fischer, G., & Scharff, E. (1998). Learning Technologies in Support of Self-Directed Learning. *Journal of Interactive Media in Education*, *98*(4). Retrieved July 10, 2009, from http://www jime.open.ac.uk/98/4.

Ford, M., & Botha, A. (2007). MobilED-An Accessible Mobile learning Platform for Africa? *Proceedings of the IST-Africa Conference. Maputo Mozambique*. Retrieved December 1, 2009 from http://researchspace.csir.co.za/dspace/bitstream/10204/1777/1/Ford1_2007.pdf

Fortier, P. (2008). Improving student nurses clinical care experience through the use of a computerized mobile hand-held decision support system. In *Proceedings of IADIS International Conference Mobile Learning*, (pp. 58-72).

Fothergill, J. (2007, January). *Profcasts: supporting online learning of campus-based students*. Paper presented at the Learning Futures Conference, University of Leicester, UK.

France, D., & Ribchester, C. (2004). Producing websites for assessment: a case study from a Level 1 fieldwork module. *Journal of Geography in Higher Education, 28*, 49–62. doi:10.1080/0309826042000198639

Franklin, D. L. (1986). Mary Richmond and Jane Addams: From Moral Certainty to Rational Inquiry in Social Work Practice. *The Social Service Review, 60*(4), 504–525. doi:10.1086/644396

Garcia, A. C., Standlee, A. I., Bechkoff, J., & Cui, Y. (2009). Ethnographic approaches to the Internet and computer-mediated communication. *Journal of Contemporary Ethnography, 38*, 52–84. doi:10.1177/0891241607310839

Gardner, H. (2000). Can technology exploit our many ways of knowing? In Gordon, D. T. (Ed.), *The Digital Classroom: How Technology is Changing the Way We Teach and Learn* (pp. 32–45). Cambridge, MA: Harvard Education Letter.

Garland, V. (2006). Digital Literacy and the use of wireless portable computers, planners and cell phones for k-13 education. In Tan, L. W. H., & Subramaniam, R. (Eds.), *Handbook of Research on Literacy in Technology at the K-12 Level* (pp. 309–321). Hershey, USA: Idea Group Publishing.

Garrison, D. R., & Anderson, T. (2003). [*st century: A framework for research and practice*. London: Routledge Falmer.]. *E-learning*, 21.

Garrison, D. R., Anderson, T., & Archer, W. (2000). Critical inquiry in a text-based environment: Computer conferencing in higher education. *The Internet and Higher Education, 2*(2/3), 87–105.

Garrison, D. R., Anderson, T., & Archer, W. (2001). Critical thinking, cognitive presence and computer conferencing in distance education. *American Journal of Distance Education, 15*(1), 7–23. doi:10.1080/08923640109527071

Garrison, D. R. (1992). Critical Thinking and Self-Directed Learning in Adult Education: An Analysis of Responsibility and Control Issues. *Adult Education Quarterly, 42*(3), 136–148.

Gaughran, B. (1996). *The graphics code, visualization and CAD*. Flexible automation and intelligence manufacturing conference. Atlanta, GA.

Geddes, S. (2004). Mobile learning in the 21st century: Benefit for learners. *The Knowledge Tree: An e-Journal of Learning Innovation* 6. Retrieved June 25, 2009 from http://knowledgetree.flexiblelearning.net.au/edition06/download/geddes.pdf

Ghaith, G. M. (2002). The relationship between cooperative learning, perception of social support, and academic achievement. *System, 30*, 263–273. doi:10.1016/S0346-251X(02)00014-3

Ghaith, G. M., & Yaghi, H. (1998). Effect of cooperative learning on the acquisition of second language rules and mechanics. *System, 26*, 223–234. doi:10.1016/S0346-251X(98)00005-0

Gillani, B., & O'Guinn, C. (2004). *Cognitive Theories and the Design of E-Learning Environments*. IOS Press.

Gillet, D., Ngoc, A., Nguyen, V., & Rekik, Y. (2005). Collaborative web-based experimentation in flexible engineering education. *IEEE Transactions on Education, 48*(4), 696–704. doi:10.1109/TE.2005.852592

Gilster, P. (1997). *Digital Literacy*. New York: Wiley.

Giollagáin, Ó. C., Mac Donnacha, S., Ní Chualáin, F., Ní Shéaghdha, A., & O'Brien, M. (2007). *Staidéar Cuimsitheach Teangeolaíoch ar Úsáid na Gaeilge sa Ghaeltacht: Príomhthátal agus moltaí.* Dublin: The Government Stationary Office.

Gladstone State High School. (2009). *Recording Device Policy.* Retrieved August 19, 2009, from www.gladstonshs.eq.edu.au/.../School%20Information/Recording%20Device%20Policy.pdf

Glaser, B. G., & Strauss, A. L. (1967). *The discovery of grounded theory*. Chicago: Aldine.

Glenn, M. (2009). *What does it mean to educated in the 21st Century*. Open EdTech 2008. Retrieved December 10, 2009 from http://www.slideshare.net/timbuckteeth/what-does-it-mean-to-be-educated-in-the-21st-century

Gobbin, R. (1998). Adoption or rejection: information systems and their cultural fitness. In Hasan, H., Gould, E., & Hyland, P. (Eds.), *Information systems and activity theory: tools in context, University of Wollongong Press* (pp. 109–124). Wollongong: N.S.W.

Godwin, J. (2005). Messaging, gaming, peer-to-peer sharing: Language learning strategies and tools for the millennial generation. *Language Learning & Technology*, *9*(1), 17–22.

Goffman, E. (1981). *Forms of talk*. Philadelphia: University of Pennsylvania.

Goffman, E. (1967). On facework: an analysis of ritual elements in social interaction. In Jaworski, A., & Coupland, N. (Eds.), *The Discourse Reader* (pp. 306–321). London: Routledge.

Gokhale, A. A. (1995). Collaborative Learning Enhances Critical Thinking. *Journal of Technology Education*, *7*(1), 22–30.

Google Trends. (2009). *Podcast Google Trends*. Retrieved December 10, 2009, from http://www.google.com/trends?q=podcast&ctab=0&geo=all&date=2004

Goolamally, N. Abas, Z. W., Yusof, S., & Singh, H. K. D. (2009). Implementation of m-learning: Pedagogical Perspectives. In *Proceedings of IADIS International Conference Mobile Learning*, pp. 311-315.

Gould, S. J. (1987). *Time's arrow; time's cycle: Myth and metaphor in the discovery of geographical time*. Cambridge: Harvard University Press.

Government of Ireland. (1999). *Primary School Curriculum*. Dublin: The Government Stationary Office.

Green, H., & Hannon, C. (2007) *TheirSpace: Education for a Digital Generation.* Demos, London. Retrieved July 12, 2009 from http://www.demos.co.uk/publications/theirspace.

Grundy, S. (1987). *Curriculum: Product or praxis?* Lewes, East Sussex: The Falmer Press.

Gumpert, G., & Drucker, J. (2007). Mobile Communication in the twenty-first century or "Everybody, everywhere, at any time". In Kleinman, S. (Ed.), *Displacing place. Mobile communication in the twenty-first century* (pp. 7–37). New York: Peter Lang.

Gutwin, C., Greenberg, S., & Roseman, M. (1996). *Workspace awareness in real-time distributed groupware: Framework, widgets and evaluation. In proceedings of* (pp. 281–298). HCI on People and Computers XI.

Gwebu, K. L., & Wang, J. (2007). The role of organizational, environmental and human factors in e-learning diffusion. *International Journal of Web-Based Learning and Teaching Technologies*, *2*(2), 59–78.

Habermas, J. (1972). *Knowledge and human interests* (2nd ed.). London: Heinemann.

Hackling, M. W., & Garnett, Patrick J. (1995). The Development of Expertise in Science Investigation Skills. *Australian Science Teachers Journal*, *41*, 80–85.

Halocha, J. (2005). Developing a research tool to enable children to voice their experiences and learning through fieldwork. *International Research in Geographical and Environmental Education*, *14*(4), 348–355. doi:10.1080/10382040508668370

Hamelin, D. (2004, June), *Searching the Web to Develop Inquiry and Collaborative Skills*. Annual Joint Conference Integrating Technology into Computer Science Education. Working group reports from ITiCSE on Innovation and technology in computer science education, United Kingdom.

Hanewald, R. (2008). Confronting the Pedagogical Challenge of Cyber Safety. *Australian Journal of Teacher Education*, *33*(3), 1–16.

Harper, D. (2002). Talking about pictures: a case of photo elicitation. *Visual Studies Journal*, *17*(1), 13–26. doi:10.1080/14725860220137345

Harper, D. (1998). An Argument for Visual Sociology. In Prosser, J. (Ed.), *Image-based Research: A Sourcebook for Qualitative Researchers* (pp. 24–41). London: Falmer Press.

Harris, J. (2006). *Irish in Primary Schools: Long-Term National Trends in Achievement*. Dublin: The Government Stationary Office.

Hartnell-Young, E. (2008). Mobile phones for learning in mainstream schooling: resistance and change.

Hasan, H., Gould, E., & Hyland, P. (1998). *Information systems and activity theory: tools in context, University of Wollongong Press*. Wollongong: N.S.W.

Hatano, G., & Inagaki, K. (1973). *Intellectual curiosity*. Chuo Koronsya Press.

HEFCE. (2009). *Enhancing learning and teaching through the use of technology: A revised approach to HEFCE's strategy for e-learning*. Retrieved July 17, 2009 from http://www.hefce.ac.uk/pubs/hefce/2009/09_12/

Hein, G. (1998). *Learning in museums*. London: Routledge.

Hennessy, S. (1999). *The potential of portable technologies for supporting graphing investigations. British Journal of Educational Technology, 30*(1), 57-60. (hybrid article full version retrieved October 20, 2009, from www.education.leeds.ac.uk/research/mathseducation/gcalc_hennessy.pdf).

Heppell, S., Chapman, C., Millwood, R., Constable, M., & Furness, J. (2004) *Building learning futures*. Retrieved November 26, 2009, from http://rubble.heppell.net/cabe/final_report.pdf

Heron, J. (1993). *Co-operative inquiry and the primacy of the practical; Adapted from Chapter 2, Co operative Inquiry*. London: Sage. Retrieved November 25, 2009 from http://www.humaninquiry.com/primprac.htm

Higgins, L. (1994). Integrating background nursing experience and study at the postgraduate level: an application of problem-based learning. *Higher Education Research & Development, 13*, 23–33. doi:10.1080/0729436940130103

Hinchcliffe, D. (2006). *All We Got Was Web 1.0, when Tim Berners-Lee Actually Gave Us Web 2.0*. Retrieved October 21, 2009, from http://web2.socialcomputingmagazine.com/all_we_got_was_web_10_when_tim_bernerslee_actually_gave_us_w.htm

Hine, C. (2000). *Virtual Ethnography*. London: Sage.

Hofstadter, D. R. (1998). Popular Culture and the Threat to Rational Inquiry. *Science, 281*(3576), 512–513. doi:10.1126/science.281.5376.512

Holtzblatt, K., & Jones, S. (1993). Contextual inquiry: A participatory design technique for system design. In Schuler, D., & Namioka, A. (Eds.), *Participatory design: Principles and practice*. Englewood Cliffs, NJ: Prentice-Hall.

Hooper, S. (1992). Effects of peer interaction during computer-based mathematics instruction. *Journal of Educational Research and Development, 85*(3), 180–189.

Hooper, S. (2003). The effects of persistence and small group interaction during computer-based instruction. *Computers in Human Behavior, 19*, 211–220. doi:10.1016/S0747-5632(02)00033-X

Hsu, H., Wang, S., & Cormac, L. (2008). Using audioblogs to assist English-language learning: An investigation into student perception. *Computer Assisted Language Learning, 21*(2), 181–198. doi:10.1080/09588220801943775

Huber, G. L. (2003). Processes of decision-making in small learning groups. *Learning and Instruction, 13*, 255–269. doi:10.1016/S0959-4752(01)00039-1

Hull, R., & Reid, J. (2003). Designing engaging experiences with children and artists. In Blythe, M. A., Overbeeke, K., & Monk, A. F. (Eds.), *Funology: From usability to enjoyment* (pp. 108–116). Dordrecht, Netherlands: Kluwer.

Huntsberger, M., & Statvistky, A. (2007). The New "podagogy": Incorporating podcasting into journalism education. *Journalism & Mass Communication Educator, 64*(4), 397–410.

Hurworth, R. (2003). Photo-Interviewing for Research. *Social Research Update*, 40, Spring. University of Surrey. Retrieved October 12, 2009, from http://sru.soc.surrey.ac.uk/SRU40.html

Hwang, G. J., Tsai, C. C., & Yang, S. J. H. (2008). Criteria, strategies and research issues of context-aware ubiquitous learning. *Journal of Educational Technology & Society*, *11*(2), 81–91.

Inkpen, K. M., Ho-Ching, W., Kuederle, O., Scott, S. D., & Shoemaker, G. B. D. (1999). "This is fun! We're all best friends and we're all playing.": Supporting children's synchronous collaboration. In *Proceedings of Computer Supported Collaborative Learning (CSCL) '99*. Stanford, CA, (pp. 252-259).

Inkpen, K., & Booth, K. Gribble, S., & Klawe, M. (1995). Give and Take: Children collaborating on One Computer. *Proceedings of the International Conference on Computer Human Interaction CHI* Denver, CO. (pp. 258-259).

Inoue, H., Naito, E., & Koshizuka, M. (1997). Mediacy: What it is? Where to go? *The International Information & Library Review*, *29*(3/4), 403–413. doi:10.1006/iilr.1997.0059

ITU. (2009). *Measuring the Information Society. The ICT Development Index*. Geneve, Switzerland: International Telecommunication Union ITU.

ITWeb. (2008, April). *Mobile phones to teach Maths*. Retrieved July 13^{h}, 2009 from http://www.itweb.co.za/sections/telecoms/2008/0804081343.asp?A=CEL&S=Cellular&O=FPPN

Jaffer, S., Ng'ambi, D., & Czerniewicz, L. (2006). The Role of ICTs in Higher Education in South Africa: One Strategy for addressing teaching and learning challenges. *Proceedings of Emerge Online Conference: Learning Landscapes in Southern Africa*, 10-21 July. Retrieved from http://emerge2006.net

Janssen, J., Erkens, G., Kanselaar, G., & Jaspers, J. (2006). Visualization of participation: Does it contribute to successful computer-supported collaborative learning? *Computers & Education*, *49*(4), 1037–1065. doi:10.1016/j.compedu.2006.01.004

Jobbings, D. (2005). *Exploiting the educational potential of Podcasting.* Retrieved July 20, 2009 from http://www.recap.ltd.uk/articles/podguide.html

Johnson, D. W., & Johnson, R. T. (1991). *Learning Together and Alone: Cooperative, competitive, and individualistic learning* (3rd ed.). Englewood, NJ: Prentice Hall.

Johnson, D. W., Johnson, R. T., & Smith, K. A. (1998). Cooperative learning returns to college: What evidence is there that it works? *Change*, *30*(4), 26–35. doi:10.1080/00091389809602629

Johnson, D. W., Roger, T., & Smith, K. A. (1991). *Active Learning: Cooperation in the College Classroom*. Edina, MN: Interaction Book Company.

Johnson, D. W., & Johnson, R. T. (1979). Conflict in the classroom: Controversy and learning. *Review of Educational Research*, *49*, 51–61.

Johnson, D. W., Johnson, R. T., & Smith, K. A. (1997). *Academic controversy: Enriching college instruction through intellectual conflict*. Washington, DC: George Washington University.

Johnson, B. (2008). The rise and rise of the YouTube generation, and how adults can help. *Guardian* retrieved July 17, 2009 from http://www.guardian.co.uk/technology/2008/oct/06/youtube.youngpeople

Joint Council for Qualifications. (2008). *National Results: GCSE*. Retrieved June 10, 2009. from *Joint Council for Qualifications* website, http://www.jcq.org.uk/national_results/

Jonassen, D. H., & Rohrer-Murphy, L. (1999). Activity theory as a framework for designing constructivist learning environments. *Educational Technology Research and Development*, *47*(1), 62–79. doi:10.1007/BF02299477

Jones, M., & Marsden, G. (2006). *Mobile Interaction Design*. London: John Wiley and Sons, Ltd.

Kagan, S. (1985). Co-op Co-op: A flexible cooperative learning technique. In In Slavin, R. (Eds.), *Learning to cooperate, cooperating to learn* (pp. 437–462).

Kam, M., Mathur, A., Kumar, A., & Canny, J. (2009). *Designing Digital Games for Rural Children: A Study of Traditional Village Games in India.* Paper presented at Computer Human Interaction CHI. Boston, MA. Retrieved December 1, 2009 from http://www.cs.cmu.edu/~mattkam/publications/CHI2009.pdf

Kam, M., Ramachandran, D., Devanathan, V., Tewari, A., & Canny, J. (2007). *Localized Iterative Design for Language Learning in Underveloped Regions: The PACE Framework.* ACM Conference in Human Interaction: Designing for Specific Cultures, San Jose, CA. Retrieved December 1, 2009 from http://www.eecs.berkeley.edu/~divya/CHI2007_pace.pdf

Karimuddin, A. (2008). *Research and markets: Malaysia and Philippines mobile report.* Retrieved July 16, 2008 from http://3gweek.net/2008/07/16/research-and-markets-malaysia-and-philippines-mobile-market/

Kasesniemi, E.-L. (2003). *Mobile messages: Young people and a new communication culture*. Massachusetts: Tampere University Press.

Kasesniemi, E., & Rautiainen, P. (2002). Mobile culture of children and teenagers in Finland. In Katz, J. E., & Aakhus, M. A. (Eds.), *Perpetual contact. Mobile communication, private talk, public performance* (pp. 170–192). Cambridge, UK: Cambridge University Press. doi:10.1017/CBO9780511489471.014

Katz, J. E., Lever, K. M., & Chen, Y. (2008). Mobile music as environmental control and prosocial entertainment. In Katz, J. E. (Ed.), *Handbook of Mobile Communication Studies* (pp. 367–376). MIT Press.

Keamy, R., & Nicholas, H. (2007). Personalised learning: can governments guarantee diversity for individuals? *International Journal on Diversity in Organisations. Communities & Nations*, *7*, 137–146.

Kekwaletswe, R. M. (2007). *Knowledge transformation in a mobile learning environment: An inquiry of context and social presence awareness.* (Unpublished PhD thesis). University of Cape Town.

Kekwaletswe, R. M. (2009). Conceptualizing ubiquitous learning through context-aware wired and wireless web services. *In proceedings of the IADIS Mobile Learning Conference*. Barcelona, Spain. 25 – 28 February.

Kekwaletswe, R. M., & Ng'ambi, D. (2006). Ubiquitous Social Presence: Context-Awareness in a Mobile Learning Environment. In Proceedings of the *IEEE International Conference on Sensor Networks, Ubiquitous, and Trustworthy Computing.* Taichung, Taiwan. 5-7 June. pp. 90-95.

Kennedy, G., Krause, K.-L., Judd, T., Churward, A., & Gray, K. (2006). First year students' experience with technology: Are they really digital natives? *Australian Journal of Educational Technology*, *24*(1), 108–122.

Kennedy, C., & Levy, M. (2008). L'italiano al telefonino: Using SMS to support beginners' language learning. *ReCALL*, *20*(3), 315–330. doi:10.1017/S0958344008000530

Kent, M., Gilbertson, D., & Hunt, C. O. (1997). Fieldwork in geography teaching: A critical review of the literature and approaches. *Journal of Geography in Higher Education*, *21*(3), 313–332. doi:10.1080/03098269708725439

Keogh, K., Koller, T., Ward, M., Úí Dhonnchadha, E., & van Genabith, J. (2004). *CL for CALL in the Primary School* eLearning for Computational Linguistics and Computational Linguistics for eLearning. International Workshop in Association with COLING 2004, Geneva, Switzerland.

Kerawalla, L., & Crook, C. (2005). From Promises to Practices: the fate of educational software in the home. *Technology, Pedagogy and Education*, *14*(1), 107–126. doi:10.1080/14759390500200195

Kerawalla, L., O'Connor, J., Underwood, J., duBoulay, B., Holmberg, J., & Luckin, R. (2007). Exploring the Potential of the Homework System and Tablet PCs to Support Continuity of Numeracy Practices between Home and Primary School. *Educational Media International*, *44*(4), 289–303. doi:10.1080/09523980701680904

Kessy, D., Kaemba, M., & Gachoka, M. (2006). The Reasons for Under Use of ICT in Education: in the context of Kenya, Tanzania and Zambia. In *Proceedings of the 4th IEEE International Workshop on Technology for education in Developing Countries TEDC*.

Keyser, M. W. (2000). Active learning and cooperative learning: understanding the difference and using both styles effectively. *Research Strategies*, *17*, 35–44. doi:10.1016/S0734-3310(00)00022-7

Kiernan, P. J., & Aizawa, K. (2004). Cell phones in task-based learning. Are cell phones useful language learning tools? *ReCALL*, *16*(1), 71–84. doi:10.1017/S0958344004000618

Kimber, K., & Wyatt-Smith, C. (2006). Using and creating knowledge with new technologies: a case for students as designers. *Learning, Media and Technology*, *31*, 19–34. doi:10.1080/17439880500515440

Kimura, M. (2006). English Language Learning by Multimedia Mobile phones. In E. Pearson, E. & Bohman,P. (Eds.), In *Proceedings of World Conference on Educational Multimedia, Hypermedia and Telecommunications 2006*. Chesapeake, VA: AACE. (pp. 2575-2578).

Kirschner, P. A., Sweller, J., & Clark, R. E. (2006). Why Minimal Guidance During Instruction Does Not Work: An Analysis of the Failure of Constructivist, Discovery, Problem-Based, Experiential, and Inquiry Based Teaching. *Educational Psychologist*, *41*(2), 75–86. doi:10.1207/s15326985ep4102_1

Klaus, F., & Rachel, W. (2000). *RFID Handbook: Radio-Frequency Identification Fundamentals and Applications*. NY: John Wiley & Sons.

Klingner, J. K., & Vaughn, S. (2000). The helping behaviors of fifth graders while using collaborative strategic reading during ESL content classes. *TESOL Quarterly*, *34*, 69–98. doi:10.2307/3588097

Klopfer, E., Squire, K., & Jenkins, H. (2002). Environmental Detectives: PDAs as a window into a virtual simulated world. *Proceedings of IEEE International Workshop on Wireless and Mobile Technologies in Education* (pp95-98). Vaxjo, Sweden: IEEE Computer Society.

Kolb, D. A. (1984). *Experiential learning: Experience as the source of learning and development*. Englewood Cliffs, NJ: Prentice-Hall.

Kolb, D. A. (1985). *LSI Learning-Style Inventory*. McBer & Company, Training Resources Group, 116 Huntington Avenue, Boston, MA.

Korpela, M., Mursu, A., & Soriyan, H. (2002). Information systems development as an activity. *Computer Supported Cooperative Work*, *11*, 111–128. doi:10.1023/A:1015252806306

Korpela, M., Mursu, A., Soriyan, A., Eerola, A., Hikkinen, H., & Toivanen, M. (2004). IS research and development by activity analysis and development: Dead horse or the next wave? In Kaplan, B., Truex D.P. III, Wastell D, Wood-Harper A.T., and DeGross J.I (eds). *Information Systems Research: Relevant Theory and Informed Practice, IFIP TC8/WG8.2 Working Conference,* (pp. 453-470), Manchester, UK, July 15-17. Boston, MA: Kluwer Academic Publishers.

Kozinets, R. V. (2006), Netnography 2.0, in Russell W. Belk (Ed) *Handbook of Qualitative Research Methods in Marketing* (pp. 129-142), Cheltenham, UN and Northampton, MA: Edward Elgar Publishing.

Kraemer, K., Dedrick, J., & Sharma, P. (2009). One Laptop Per Child: Vision vs. Reality. *Communications of the ACM*, *52*(6), 66–74. doi:10.1145/1516046.1516063

Kukulska-Hulme, A., & Shield, L. (2008). An overview of mobile assisted language learning: From content delivery to supported collaboration and interaction. *ReCALL*, *20*(3), 271–289. doi:10.1017/S0958344008000335

Kuutti, K., & Molin-Juustilla, T. (1998). Information system support for 'loose' coordination in a network organisation: an activity theory perspective. In Hasan, H., Gould, E., & Hyland, P. (Eds.), *Information systems and activity theory: tools in context* (pp. 73–92). Wollongong, NSW: University of Wollongong Press.

Kuutti, K. (1996). Activity theory as a potential framework for human-computer interaction research. In Nardi, B. A. (Ed.), *Context and consciousness: Activity theory and human computer interaction* (pp. 17–44). Cambridge, MA: MIT Press.

Kuutti, K. (1991). Activity theory and its implications to information systems research and development. In Nissen, H. E., Klein, H. K., & Hirschheim, R. A. (Eds.), *Information systems research* (pp. 529–549). Amsterdam: Elsevier Science Publishers.

Lai, C.-H., Yang, J.-C., Chen, F.-C., Ho, C.-W., & Chan, T.-W. (2007). Affordances of mobiletechnologies for experiential learning: the interplay of technology and pedagogical practices. *Journal of Computer Assisted Learning*, *23*, 326–337. doi:10.1111/j.1365-2729.2007.00237.x

Lane, C. (2008). *UW Podcasting: Evaluation of year one*. Washington, DC: Catalyst.

Lang, B., Lu, Y., Zhang, Z., & Li, W. (2003). *A flexible access control mechanism supporting large scale distributed collaboration*. In The eighth international conference on computer supported cooperative work in design proceeding, 1, 500-504.

Laurillard, D. (2007). Pedagogical forms for mobile learning: Framing research questions. In Pachler, N. (Ed.), *Mobile learning: Towards a research agenda. Occasional papers in work-based learning 1* (pp. 15–24). London: WLE Centre for Excellence.

Laurilliard, D. (2002). *Rethinking University Teaching: A conversational Framework for the Effective Use of Learning Technologies* (2nd ed.). London: RoutledgeFalmer. doi:10.4324/9780203304846

Lave, J., & Wenger, E. (1991). *Situated Learning: Legitimate Peripheral Participation*. Cambridge, England: Cambridge University Press.

LeClaire, J. (2005). *Apple including podcast library in iTunes 4.9*. Retrieved December 10, 2009 from: http://www.technewsworld.com/story/EIGWrJ6Jx65C9M/Apple-Including-Podcast-Library-in-iTunes-49.xhtml

Lee, M. J. W., & Chan, A. (2007). Pervasive, lifestyle-integrated mobile learning for distance learners: An analysis and unexpected results from a podcasting study. *Open Learning*, *22*(3), 201–218.

Lee, T. (2006). Pleasurable Cities: A discussion paper. UK: FutureLab. Retrieved June 1, 2009, from http://futurelab.org.uk/resources/documents/project_reports/Pleasurable_Cities_discussion_paper.pdf

Lehner, F., & Nosekabel, H. (2002). The Role of Mobile Devices in E-learning – First Experience with a E-learning Environment. In Milrad, M., Hoppe, H.U. & Kinshuk (Eds.), *IEEE International Workshop on Wireless and Mobile Technologies in Education*, (pp. 103-106).

Leinonen, T., Sari, E., & Aucamp, F. (2006). *Audio Wiki for Mobile Communities: Information System for the Rest of Us*. Paper presented at the Workshop on Speech in Mobile and Pervasive Environments, during the ACM International Conference on Human Computer Interaction with Mobile Devices and Services MobileHCI. Espoo, Finland.

Lenham, R. (1995). Digital literacy. *Scientific American*, *273*(3), 253–255.

Leont'ev, A. (1974). The problem of activity in psychology. *Social Psychology*, *13*(2), 4–33.

Leont'ev, A. (1978). *Activity, Consciousness, and Personality*. Englewood Cliffs, NJ: Prentice-Hall.

Leung, C. H., & Chan, Y. Y. (2003). *Mobile learning: A new paradigm in electronic learning*. 3rd IEEE International Conference on Advanced Learning Technologies(ICALT'03), 76-80.

Levy, M., & Kennedy, C. (2005). Learning Italian via mobile SMS. In Kukulska-Hulme, A., & Traxler, J. (Eds.), *Mobile Learning: A handbook for educators and trainers* (pp. 76–83). London: Routledge.

Lim, B. R. (2004). Challenges and issues in designing inquiry on the Web. *British Journal of Educational Technology, 35*(5). doi:10.1111/j.0007-1013.2004.00419.x

Lim, K. Y. G., Hedberg, J. G., & Chatterjea, K. (2004). Adolescent usage of multimedia messaging in the negotiation, construction, and sharing of meaning about local environments. In L. Cantoni, & C. McLoughlin (Eds.), *Proceedings of Ed-Media 2004 World Conference on Educational Multimedia, Hypermedia and Telecommunications* (pp. 1179-1186). Lugano, Switzerland: Association for the Advancement of Computing in Education.

Lin, N., Kajita, S., & Mase, K. (2007). A multi-modal mobile device for learning Japanese kanji characters through mnemonic stories. In *Proceedings of the 9th International Conference on Multimodal interfaces*, pp. 335-338.

Lin, Y., Swan, K., & Kratcoski, A. (2005). Situated Learning in a Ubiquitous Computing Classroom, *Journal of the Research Center for Educational Technology,* Fall 2005, from http://www.rcetj.org/?type=art&id=3530&.

Ling, R. (1999). « *C'est bien d'être joignable!* »: L'usage du téléphone fixe et mobile chez les jeunes norvégiens. *Reseaux*, (92-93): 261–291.

Lipman, M. (1991). *Thinking in education*. Cambridge, UK: Cambridge University Press.

Lobet-Maris, C. (2003). Mobile phone tribes. Youth and social identity. In Fortunati, L., Katz, J., & Riccini, R. (Eds.), *Mediating the Human Body: Technology, Communication and Fashion* (pp. 87–92). Mahwah: Lawrence Erlbaum Associates.

Looi, C. K. (1998). Interactive learning environments for promoting inquiry learning. *Journal of Educational Technology Systems, 27*(1), 3–22. doi:10.2190/L4Q4-8QMM-QPJ3-B5LJ

Lu, M. (2008). Effectiveness of vocabulary learning via mobile phone. *Journal of Computer Assisted Learning, 24*(6), 515–525. doi:10.1111/j.1365-2729.2008.00289.x

Mac Gréil, M., & Rhatigan, F. (2009). *The Irish language and the Irish people: Report on the attitudes towards, competence in and use of the Irish language in the Republic of Ireland in 2007-'08.* Ireland: National University of Ireland, Maynooth.

Madia, M. S. (2007). Selection interviews using mobile technology. In Norman, A. & Pearce, J. (Eds) *Proceedings of the 6th International Conference on Mobile Learning* (pp140-150), Melbourne, Australia: Melbourne University Press.

Maker, C. J. (1993). Creativity, intelligence, and problem solving: A definition and design for cross-cultural research and measurement related to giftedness. *Gifted Education International, 9*(2), 68–77.

Maker, C. J., & Schiever, S. (1991). Enrichment and acceleration: An overview and new directions. In Davis, G., & Colangelo, N. (Eds.), *Handbook of Gifted Education* (pp. 99–110). Boston: Allyn & Bacon.

Malan, D. (2007). *Podcasting computer science E-1. 38th SIGCSE Technical Symphosium on Computer Science Education* (pp. 389–393). Cambridge, Massachusetts: Harvard University.

Malaysian Communications and Multimedia Commission. MCMC *(2008). Cellular Phone in Malaysia*. Retrieved October 23, 2008 from http://www.skmm.gov.my/facts_figures/stats/ViewStatistic.asp?cc=77141977&srid=9247989

Manlove, S., Lazonder, A. W., & de Jong, T. (2008). Trends and issues of regulative support use during inquiry learning: Patterns from three studies. *Computers in Human Behavior, 25*(4), 795–803. doi:10.1016/j.chb.2008.07.010

Manning, S. (2005). The promise of podcasting. *Pointers and Clickers, 6*(2). Retrieved December 10, 2009 from http://www.ion.illinois.edu/resources/pointersclickers/2005_03/Podcasting2005.pdf

March, T. (2003). *Webquests for learning.* Retrieved July 10, 2009, from http://bestwebquests.com/

Markett, C., Sanchez, I. A., Weber, S., & Tangney, B. (2006). Using short message service to encourage interactivity in the classroom. *Computers & Education, 46*(3), 280–293. doi:10.1016/j.compedu.2005.11.014

Masten, D., & Plowman, T. (2003). Digital Ethnography: The next wave in understanding the consumer experience. *Design Management Journal, 14*(2), 74–81.

Matthee, M., & Liebenberg, J. (2007). Mathematics on the Move: Supporting Mathematics Learners through Mobile Technology in South Africa. In *Proceeding of the 6th conference on Mobile Learning Mlearn.* Melbourne, Australia. (pp. 156-161).

McFarlane, A., Triggs, P., & Yee, W. (2008). *Mobile learning: Research findings.* Coventry, UK: Becta. Retrieved May 2, 2009 from http://partners.becta.org.uk/uploaddir/downloads/page_documents/research/mobile_learning.pdf.

McKenzie, J. (1998). Grazing the Net. [from http://www.fno.org/text/grazing.html.]. *Phi Delta Kappan, 79*(September), 26–31. Retrieved July 9, 2009.

McLellan, H. (1993). Evaluation in a situated learning environment. *Educational Technology, 33*(3), 39–45.

McMahon, M. (1997). Social Constructivism and the World Wide Web - A Paradigm for Learning. *Paper presented at the ASCILITE conference*, Perth, Australia. Retrieved October 15, 2008 from http://www.ascilite.ord.au/conference/perth97/paper/Mcmahon/Mcmahon.html

McMichael, H. (1999). An Activity based perspectives for Information Systems research, In Proceedings of the *10th Australian Conference on Information Systems*, Wellington, New Zealand, pp. 610-621.

Merchand, J. E. (1995). Problem-based learning in the business curriculum: An alterative to traditional approaches. In Gijselaers, W., Tempelaar, D., Keizer, P., Bernard, E., & Kasper, H. (Eds.), *Educational innovation in economics and business administration: The case of problem-based learning* (pp. 261–267). Dordrecht: Kluwer.

Metiri Group. (2006). Technology in Schools, What the Research Says, *Cisco Systems*. Retrieved July 20, 2009 from http://www.cisco.com/web/strategy/docs/education/TechnologyinSchoolsReport.pdf

Mevarech, Z. A. (1993). Who benefits from cooperative computer-assisted instruction? *Journal of Educational Computing Research, 9*(40), 451–464.

Miniwatts Marketing Group. (2009). *Internet World Statistics. Usage and Population* Statistics. Retrieved October 21, 2009, from http://www.internetworldstats.com/stats.htm

Mitchell, K., & Race, N. J. P. (2005). uLearn: Facilitating ubiquitous learning through camera equipped mobile phones. In H. Ogata, M. Sharples, Kinshuk & Y. Yano (Eds.). In *Proceedings of the Third IEEE International Workshop on Wireless and Mobile Technologies in Education 2005,* Tokushima, Japan, pp.274-281.

Mobi Math Project. (n.d.). *Mobi Math Project*. Retrieved July 8h, 2009 from http://www.mymobi.co.za/mobi_signin.php

Mobil, E. D. (2009). *MobilED Initiative*. Retrieved on July 10, 2009 from http://mobiled.uiah.fi/

Molebash, P. (2004). Web historical inquiry projects. *Social Education, 68*(3), 226–234.

Montague, W. E., & Knirk, F. G. (1993). What works in adult instruction: the management, design and delivery of instruction? *International Journal of Educational Research, 19*, 329–331.

Moraveji, N., & Kim, T. Ge, J., Pawar, U., Inkpen, K., & Mulcahy, K. (2008). Mischief: Supporting Remote Teaching in Developing Regions. In *Proceedings of the Computer Human Interaction Conference CHI.* Florence, Italy.

Moraveji, N., Inkpen, K., Cutrell, E., & Balakrishnan, R. (2009). A Mischief of Mice: Examining Children's Performance in Single Display groupware Systems with 1 to 32 Mice. In *Proceedings of the International Conference on Computer Human Interaction CHI.* Boston.

Mount, N., & Chambers, C. (2008). Podcasts and Practicals. In Salmon, G., & Ediringsha, P. (Eds.), *Podcasting for Learning in Universities*. Glasgow: McGraw Hill.

MXit. (2009). *MXit*. Retrieved July 8, 2009 from http://www.mxit.co.za

Nah, K. C., White, P., & Sussex, R. (2008). The potential of using a mobile phone to access the Internet for learning EFL listening skills within a Korean context. *ReCALL*, *20*(3), 331–347. doi:10.1017/S0958344008000633

Naismith, L., Lonsdale, P., Vavoula, G., & Sharples, M. (2006). *Literature review in mobile technologies and learning*. Bristol, UK: Futurelab.

Naismith, L., Lonsdale, P., Vavoula, G., & Sharples, M. (2004). *Literature Review in Mobile Technologies and Learning*. Birmingham UK: Futurelab & University of Birmingham. Retrieve May 1, 2009, from www.futurelab.org.uk/research/lit_reviews.htm

Nardi, B. (1996). Studying context: A comparison of activity theory, situated action models, and distributed cognition. In Nardi, B. (Ed.), *Context and consciousness: Activity theory and human computer interaction*. Cambridge, MA: MIT Press.

Nardi, B., Whittaker, S., & Bradner, E. (2000). Interaction and outeraction: instant messaging in action. *Proceedings of CSCW '00*. Philadelphia, PA, pp. 79-88.

Nataatmadja, I. & Dyson, L.E. (2008) The role of podcasts in student's learning. *iJIM, 2*(3), 17-21

National Council for Curriculum and Assessment. (2008). *Primary Curriculum Review, Phase Two: Final report with recommendations*. Research Report No. 7. Dublin: Author.

National Council for the Social Studies. (1994). *Expectations of excellence: Curriculum standards for social studies*. Washington, DC: National Council for the Social Studies.

National Research Council. (1999). *How people learn: Brain, mind, experience and school*. Washington, DC: National Academy Press.

Newmann, F. (1996). *Authentic achievement: Restructuring schools for intellectual quality*. San Francisco: Jossey-Bass.

Newmann, F. (2000). Authentic intellectual work: What and why? University of Minnesota. Retrieved November 21, 2009, from http://education.umn.edu/CAREI/Reports/Rpractice/Fall2000/newmann.html

News Straits Times. NST (2008). Times are changing. *The News Straits Times Publication*. Nov 1, pp. 4.

News, B. B. C. (2005). *Wordsmiths hail podcast success*. Retrieved May 20, 2009 from: http://news.bbc.co.uk/1/hi/technology/4504256.stm

Ng, W., & Nicholas, H. (2009). Introduction of pocket PCs in schools: attitudes and beliefs in the first year. *Computers & Education*, *52*(2), 470–480. doi:10.1016/j.compedu.2008.10.001

Ng, W. (2008). Self-directed learning with web-based sites: How well do students' perceptions and thinking match with their teachers? *Teaching Science*, *54*(2), 26–30.

Ng, W., & Hanewald, R. (2009). Concept maps as a tool for promoting online collaborative learning in virtual Mapping. In Marriott, R., & Torres, P. (Eds.), *Handbook of Research on Collaborative Learning Using Concept Mapping* (pp. 81–99). Hershey, PA: IGI Global Publisher.

Ng, W., & Nicholas, H. (2009b). Self-directed learning with web-based technologies. In Rogers, P., Berg, G., Boettcher, J., Howard, C., Justice, L., & Schenk, K. (Eds.), *Encyclopedia for online and distance Learning* (pp. 1847–1852). Hershey, PA: IGI Global Publishers.

Ng, W. (2006). Web-based Technologies, Technology Literacy and Learning in Tan, L.W.H. and Subramaniam, R. (Eds). *Handbook of Research on Literacy in Technology at the K-12 Level* (pp. 94-117) Hershey, PA: Idea Group Publishing.

Ng, W. (in press, accepted 4 March 2009). Empowering students to be scientifically literate through digital literacy in Rodrigues, S. (Ed). *Multiple Literacy and Science Education: ICTs in Formal and Informal Learning Environments.* Hershey, PA: IGI Global Publishing.

Ng, W., & Nicholas, H. (2007). *Pocket PCs for mGeneration learners*. Unpublished research report. Bundoora: La Trobe University.

Ng, W., & Nicholas, H. (2007). Ubiquitous learning with handhelds in schools. In Norman, A. & Pearce, J. (Eds) *Proceedings of the 6th International Conference on Mobile Learning*, Melbourne, Australia. 16-19 October, 2007, p 186-193.

Ng, W., & Nicholas, H. (2008). Capturing individual thinking and through collaboration with handheld computers. *British Educational Research Association*, Edinburgh 3-6 September, 2008

Ng, W., Nicholas, H., Loke, S., & Torabi, T. (2010). Designing effective pedagogical systems for teaching and learning with mobile and ubiquitous devices in Goh, T. T. (Ed). *Multiplatform E-Learning Systems and Technologies: Mobile Devices for Ubiquitous ICT-Based Education*. (pp. 42 - 56). Hershey, PA: IGI Global Publishing.

Ngwenyama, O. K., & Lee, A. (1997). Communication richness in electronic mail: critical social theory and the contextuality of meaning. *Management Information Systems Quarterly*, *21*(2), 145–167. doi:10.2307/249417

Nicholas, H., & Ng, W. (2009). Ubiquitous learning and handhelds: An overview of theory and pedagogy. In Rogers, P., Berg, G., Boettcher, J., Howard, C., Justice, L., & Schenk, K. (Eds.), *Encyclopedia for online and distance learning* (*Vol. IV*, pp. 2171–2176). Hershey, PA: IGI Global Publishers.

NMC. (2004). *Standards of Proficiency for Pre-registration Midwifery Education*. London: NMC Publishing.

NMC. (2008). *The Code Standards of conduct, performance and ethics for nurses and midwives*. London NMC Publishing.

Nokia (2009a). M4Girls. *Exploring the Use of Mobile Phones for Learning among High School Girl Learners; Improving the Learning Condition of 2 High Schools*. Retrieved July 10, 2009 from http://www.nokia.com/NOKIA_COM_1/Corporate_Responsibility/World_map/pdf/MEA/CI_OnePager_South_Africa_m4girls.pdf

Nokia (2009b). *Education – Youth Development – Community – Corporate responsibility*. Retrieved July 13, 2009 from http://www.nokia.com/corporate-responsibility/society/youth-development/education

Nonaka, I. (1994). A dynamic theory of organizational knowledge creation. *Organization Science*, *5*(1), 14–37. doi:10.1287/orsc.5.1.14

NSW Department of Education and Training (NSWDET). (2003a). *Quality teaching in NSW public schools: Discussion paper*. Sydney: NSWDET.

NSWDET. (2003b). *Quality teaching in NSW public schools: An annotated bibliography*. Sydney: NSWDET.

Nuthall, G. (1999). The way students learn. *The Elementary School Journal*, *99*, 303. doi:10.1086/461928

Nyiri, K. (2002). Towards a philosophy of m-learning. In *Proceedings of the 1st IEEE International Workshop on Wireless and Mobile Technologies in Education*, Växjö, Sweden, pp. 121-124.

O'Malley, C. (1994). *Computer supported collaborative learning* (*Vol. 128*). NATO ASI Series, Computer and Systems Sciences.

O'Malley, C., Vavoula, G., Glew, J., Taylor, J., Sharples, M., & Lefrere, P. (2003) *Guidelines for learning/teaching/tutoring in a mobile environment. Mobilearn project deliverable*. Retrieved April 21, 2009 from http://www.mobilearn.org/download/results/guidelines.pdf

O'Reilly, T. (2009). *Web Squared: Web 2.0 Five Years On*. Retrieved October 28, 2009, from http://www.web2summit.com/web2009/public/schedule/detail/10194

Oblinger, D. (2003). Understanding the new students. *Educause*. Retrieved October 15, 2008 from http://www.educause.edu/ir/library/pdf/ERM0032.pdf

Ogata, H., Matsuura, K., & Yano, Y. (1996). Knowledge awareness: Bridging between shared knowledge space and collaboration in Sharlok. *Proceeding of Educational Telecommunications*, *96*, 232–237.

Ogata, H., & Yano, Y. (2000). Combining Knowledge Awareness and Information Filtering in an Open ended Collaborative Learning Environment. *International Journal of Artificial Intelligence in Education, 11*, 33–46.

Ogata, H., & Yano, Y. (2004). Context-Aware Support for Computer Supported Ubiquitous Learning, *Proceedings of the 2nd IEEE International Workshop on Wireless and Mobile Technologies in Education*, (pp.27-34).

Ogata, H., & Yano, Y. (2004). Knowledge Awareness Map for Computer-Supported Ubiquitous Language-Learning. *Proceeding of the International Workshop on Wireless and Mobile Technologies in Education*, Taiwan, IEEE Computer Society, pp. 19-26.

Ogata, H., Imai, K., Matsuura, K., & Yano, Y. (1999). Knowledge Awareness Map for Open-ended and Collaborative Learning on World Wide Web. In *Proceeding of the International Conference on Computers in Education ICCE99*, pp.319-326.

Ohashi, Y., Ogawa, H., & Arisawa, M. (2008). Making new learning environment in zoo by adopting mobile devices. In *Proceedings of the 10th International Conference on Human Computer Interaction with Mobile Devices and Services*, (pp. 489-490).

Pachler, N. (2007). *Mobile learning towards a research agenda*, WLE Centre, Institute of Education. Retrieved April 21, 2009 from http://www.wlecentre.ac.uk/cms/files/occasionalpapers/mobilelearning_pachler2007.pdfb

Palm Inc. (2002). *Palm Education Pioneers Program Final Evaluation Report*. Resource document. Retrieved 21 October 2008 from http://www.palmgrants.sri.com/PEP_Final_Report.pdf.

Papert, S. (1980). *Mindstorms: Children, Computers, and Powerful Ideas*. Brighton: Harvester Press.

Pask, G. (1976). *Conversation Theory: Applications in Education and Epistemology*. Amsterdam, New York: Elsevier.

Pastore, S. (2004). Experiences of mobile learning in science: technological solutions for wireless network and content delivery. Paper presented in *CUC 2005: Internet Users' Conference*, Dubrovnik, Croatia. Retrieved December 15, 2008 from http://cuc.carnet.hr/cuc2005/program/papers/abs/h1_pastore_abs.pdf

Pawar, S., Pal, J., & Toyama, K. (2006). Multiple Mice for computers in Education in Developing Countries. In *Proceedings of International Conference on Information and Communication Technologies and Development ICTD.* Berkley, CA. (pp. 64-71).

Pawar, S., Pal, J., Gupta, R., & Toyama, K. (2007). Multiple Mice for Retention Tasks in Disadvataged Schools. *Proceedings of the International Conference on Computer Human Interaction.* San José, CA. (pp. 1581-1590).

Pęcherzewska, A., & Knot, S. (2007). Review of existing EU projects dedicated to dyslexia, gaming in education and m-learning. WR08 Report to CallDysc project. June 2007. Retrieved June3, 2009 from http://www.calldysc.info/attachments/114_WR08_existingprojectreview.doc/

Perry, D. (2003). *Handheld computers (PDAs) in schools*. Coventry, UK: Becta.

Perry, D. (2003). *Handheld computers (PDAs) in schools*. Coventry: Becta (British Educational Communications and Technology Agency). Retrieved May 12, 2008 from http://partners.becta.org.uk/uploaddir/downloads/page_documents/research/handhelds.pdf

Perry, D. (2006). *Learning2Go Wolverhampton Mobile Learning: Report - end of Phase II*. Retrieved March 15, 2009 from http://www.learning2go.org/media/download_gallery/dpa%20L2G%20Ph2%20FINAL%20Report.pdf

Piaget, J. (1955). *The construction of reality in the child*. London: Routledge & Keegan Paul.

Piaget, J. (1972). *Psychology and epistemology: Towards a theory of knowledge*. London: Penguin University Books.

Pierroux, P. (2008) Extending Meaning from Museum Visits through the Use of Wikis and Mobile Blogging. *Paper presented at ICLS 2008: Internation Perspectives in the Learning Sciences, Utrecht, the Netherlands.*

Pimsleur Language Learning System. (n.d.). *Pimsleur Language Learning System*. Retrieved July 8[h], 2009 from, http://en.wikipedia.org/wiki/Pimsleur_language_learning_system

Pink, S. (2001). *Doing Ethnography: Images, Media and Representation in Research*. London: Sage.

Pinkwart, N., Hope, H., Milrad, M., & Perez, J. (2003). Educational scenarios for cooperative use of Personal Digital Assistants. *Journal of Computer Assisted Learning*, *19*, 383–391. doi:10.1046/j.0266-4909.2003.00039.x

Polanyi, M. (1966). *The tacit dimension*. London: Routledge and Kegan Paul.

Polanyi, M. (1958). *Personal Knowledge*. Chicago: The University of Chicago press.

Pontecorvo, C. (1991). *Introduzione. La scuola come contesto sociale naturale. Discutendo si impara. Interazione sociale e conoscenza a scuola*. Berlin, Germany: NIS.

Porto, M. (2001). Cooperative writing response groups and self-evaluation. *ELT Journal*, *55*(1), 38–46. doi:10.1093/elt/55.1.38

Prain, V., & Waldrip, B. (2006). An exploratory study of teachers' and students' use of multi-modal representation of concepts in primary science. *International Journal of Science Education*, *28*(15), 1843–1866. doi:10.1080/09500690600718294

Prensky, M. (2004). *What can you learn from a cell phone? – Almost anything!* Retrieved June 3, 2009 from http://www.marcprensky.com/writing/

Prinz, W. (1999). NESSIE: An awareness environment for cooperative settings. In S. Bødker, M. Kyng, and K. Schmidt (eds.), *ECSCW"99: Proceedings of the Sixth European Conference on Computer-Supported Cooperative Work*. (pp.391–410). Copenhagen. Dordrecht: Kluwer Academic Publishers.

Proctor, N., & Burton, J. (2003). Tate modern multimedia tour pilots 2002-2003. In Attewell, J., Da Bormida, G., Sharples, M., & Savill-Smith, C. (Eds.), *M Learn 2003: Learning with mobile devices* (pp. 54–55). London: Learning and Skills Development Agency.

Prosser, J. (2008). *Introducing Visual Methods*. ESRC National Centre for Research Methods Review Paper. Retrieved 12 October 2009 from http://www.reallifemethods.ac.uk/publications/articles/

Quinn, C. (2000). *mLearning: mobile, wireless and in-your-pocket learning*. Line Zine Magazine. Retrieved June 3, 2009, from http://linezine.com/2.1/features/cqmmwiyp.htm

Rachel, H. L., & Irit, B. N. (2002). Writing development of Arab and Jewish students using cooperative learning (CL) and computer-mediated communication (CMC). *Computers & Education*, *39*, 19–36. doi:10.1016/S0360-1315(02)00019-2

Ramsay, A., Hanlon, D., & Smith, D. (2000). The association between cognitive style and accounting students' preference for cooperative learning: an empirical investigation. *Journal of Accounting Education*, *18*, 215–228. doi:10.1016/S0748-5751(00)00018-X

Raven, M. E., & Flanders, A. (1996). Using contextual inquiry to learn about your audiences. *ACM SIGDOC Asterisk Journal of Computer Documentation*, *20*(1), 1–13. doi:10.1145/227614.227615

Renda, M., & Straccia, U. (2005). A personalized collaborative Digital Library environment: a model and an application. *Journal of Information Processing and Management*, *41*(1), 5–21. doi:10.1016/j.ipm.2004.04.007

Rennie, L. J., Feher, E., Dierking, L. D., & Falk, J. H. (2003). Toward an agenda for advancing research on science learning in out-of-school settings. *Journal of Research in Science Teaching*, *40*(2), 112–120. doi:10.1002/tea.10067

Resnick, L. (1987). Learning in school and out. *Educational Researcher*, *16*, 13–20.

Resource document. Retrived May 12, 2008 from http://www.learning2go.org/media/download_gallery/dpa%20L2G%20Ph2%20FINAL%20Report.pdf.

Rettie, R. 2003. Connectedness, Awareness and Social Presence, 6th Annual International Workshop on Presence 2003.

Ribble, M., & Bailey, G. (2007). *Digital citizenship in schools*. Washington, DC: ISTE.

Rickinson, M., Dillon, J., Teamey, K., Morris, M., Choi, M. Y., Sanders, D., & Benefield, P. (2004). *A review of research on outdoor learning*. London: National Foundation for Educational Research and King's College London.

Rogers, A. (1997). Learning: Can we change the discourse? *Adults Learning*, *8*(5), 116–117.

Rogers, Y., Price, S., Fitzpatrick, G., Fleck, R., Harris, E., & Smith, H. (2004). Designing new forms of digital augmentation for learning outdoors. In *Proc. Interaction Design and Children, ACM* (pp. 3–10). New York: Ambient Wood.

Rogers, Y., Price, S., Randell, C., Fraser, D. S., Weal, M., & Fitzpatrick, G. (2005). Ubi-learning Integrates Indoor and Outdoor Experiences. *Communications of the ACM*, *48*(1), 55–59. doi:10.1145/1039539.1039570

Roschelle, J. (2003). Unlocking the learning value of wireless mobile devices. *Journal of Computer Assisted Learning*, *19*(3), 260–272. doi:10.1046/j.0266-4909.2003.00028.x

Roschelle, J., Patton, C., & Tatar, D. (2007). Designing networked handheld devices to enhance school learning. *Advances in Computers*, *70*, 1–60.

Roschelle, J. (2003). Unlocking the learning value of wireless mobile devices. *Journal of Computer Assisted Learning*, *19*(3), 260–272. doi:10.1046/j.0266-4909.2003.00028.x

Roschelle, J., Patton, C., & Tatar, D. (2007). Designing networked handheld devices to enhance school learning. In *Advances in Computers, 70,* 1-60.

Roth, W. M., & Roychoudhury, A. (1992). The social construction of scientific concepts or the concept map as conscription device and tool for social thinking in high school science. *Science Education*, *76*, 531–557. doi:10.1002/sce.3730760507

Rothwell, L. (2008) in Salmon, G., Ediringsha, P.,(Eds), *Podcasting for Learning in Universities*. Glasgow: McGraw Hill

Russell, T., & McGuigan, L. (2001). Promoting understanding through representational redescription: an illustration referring to young pupils' ideas about gravity. In D. Psillos, P. Kariotoglou, V. Tselfes, G. Bisdikian, G. Fassoulopoulos, E. Hatzikraniotis, & E. Kallery (Eds.), *Science education research in the knowledge-based society. Proceedings of the Third International Conference of the ESERA* (pp. 600–602). Thessaloniki, Greece: Aristotle University of Thessaloniki.

Ruuska, S., & K. Vaananen-VainioMattila, (1998). User needs for mobile communication devices: Requirements gathering and analysis through contextual inquiry. In C. Johnson (ed.) *Proceedings of the first Workshop on Human Computer Interaction with Mobile Devices.*

Ryder, M., & Wilson, B. (1997. From center to periphery: shifting agency in a complex technical learning environments. Paper presented at the *Meeting of the American Educational Research Association*, Chicago, IL.

Sakamura, K., & Koshizuka, N. (2005). Ubiquitous Computing Technologies for Ubiquitous Learning. *Proceeding of the International Workshop on Wireless and Mobile Technologies in Education*, Japan, pp.11-18.

Salmon, G., & Ediringsha, P. (Eds.). (2008). *Podcasting for Learning in Universities*. Glasgow: McGraw Hill.

Sánchez, J. Flores, H., & Sáenz, M. (2008). Mobile science learning for the blind. *CHI '08 extended abstracts on Human factors in computing systems,* Florence, Italy.

Sánchez, J., & Salinas, A. (2008). Science problem solving learning through mobile gaming. In *Proceedings of the 12th international Conference on Entertainment and Media in the Ubiquitous Era.* Tampere, Finland, (pp. 49-53).

Sariola, J. (2001). What are the Limits of Academic Teaching? In *Search of the Opportunities of Mobile Learning 2001*. Vancouver, Canada: Telelearning.

Savill-Smith, C., & Kent, P. (2003). *The Use of Palmtop Computers for Learning: A Review of the Literature*. Learning and Skills Development Agency. Retrieved June 25, 2009 from http://www.m-earning.org/docs/the_use_of_palmtop_computers_for_learning_sept03.pdf

Scanlon, E., Jones, A., & Waycott, J. (2005). Mobile technologies: prospects for their use in learning in informal science settings. *Journal of Interactive Media in Education*, 21(5), Retrieved December 10, 2009 from: http://jime.open.ac.uk/2005/25/scanlon-2005-25.html

Schank, C. (1995). What We Learn When We Learn by Doing. *(Technical Report No. 60)*, Northwestern University, Institute for Learning Sciences.

Schenker, J., Kratcoski, A., Lin, Y., Swan, K., & van 't Hooft, M. (2007). Researching ubiquity: ways to capture it all. In van 't Hooft, M., & Swan, M. (Eds.), *Ubiquitous computing in education: invisible technology, visible impact* (pp. 167–186). Mahwah, NJ: Lawrence Erlbaum Associates.

Schroyen, J., Gabriëls, K., Luyten, K., Teunkens, D., Robert, K., Coninx, K., et al. (2008). Training social learning skills by collaborative mobile gaming in museums. In *Proceedings of the 2008 international Conference on Advances in Computer Entertainment Technology*, pp. 46-49.

Schrum, W., Duque, R., & Brown, T. (2005). Digital Video as Research Practice: Methodology for the Millennium. *Journal of Research Practice. 1*(1), Article M4.

Schwab, J. J. (1962). The teaching of science as enquiry. In Schwab, J. J., & Brandwein, P. F. (Eds.), *The teaching of science*. Cambridge, MA: Harvard University Press.

Schwartz, M. (2008, April). Maths via mobile is South African pupil's to learning. *Developing Telecoms*, Retrieved July 20, 2009, from http://www.developingtelecoms.com

Seppälä, P., & Alamäki, H. (2003). Mobile learning in teacher training. *Journal of Computer Assisted Learning, 19*, 330–335. doi:10.1046/j.0266-4909.2003.00034.x

Shannon, V. (2006). *A 'more revolutionary' Web*. International Herald Tribune. Retrieved October 22, 2009 from http://www.iht.com/articles/2006/05/23/business/web.php.

Shariq, Z. S. (1999). How does knowledge transform as it is transferred? Speculations on the possibility of a cognitive theory of knowledgescapes. *Journal of Knowledge Management, 3*(4), 243–251. doi:10.1108/13673279910303998

Sharples, M. (2003). Disruptive devices: mobile technology for conversational learning. *International Journal of Continuing Engineering Education and Lifelong Learning, 12*(5/6), 504–520. doi:10.1504/IJCEELL.2002.002148

Sharples, M. (2000). The design of personal mobile technologies for lifelong learning. *Computers & Education, 34*, 177–193. doi:10.1016/S0360-1315(99)00044-5

Sharples, M. (2005). Learning as conversation: Transforming education in the mobile age. In *Proceedings of* (pp. 147–152). Seeing, Understanding, Learning in the Mobile Age.

Sharples, M., Corlett, D., & Westmancott, O. (2002). The design and implementation of a mobile learning resource. *Personal and Ubiquitous Computing, 6*, 220–234. doi:10.1007/s007790200021

Sharples, M. (Ed.). (2007). *Big issues in mobile learning*. Nottingham: Learning Sciences Research Institute.

Sharples, M., Graber, R., Harrison, C., & Logan, K. (2009). E-safety and Web 2.0 for children aged 11–16. *Journal of Computer Assisted Learning, 25*, 70–84. doi:10.1111/j.1365-2729.2008.00304.x

Sharples, M., Taylor, J., & Vavoula, G. N. (2007). A Theory of Learning for the Mobile Age. In Andrews, R., & Haythornthwaite, C. (Eds.), *The SAGE Handbook of E-learning Research* (pp. 221–247). London: Sage.

Sharples, M. (Ed.). (2006). *Big Issues in Mobile Learning: Report of a workshop by the Kaleidoscope Network of Excellence Mobile Learning Initiative*. Retrieved November 29, 2009 from http://telearn.noekaleidoscope.org/ warehouse/Sharples-2006.pdf.

Sharples, M., Lonsdale, P., Meek, J., Rudman, P., & Vavoula, G. (2007). An Evaluation of My ArtSpace: A mobile Learning Service for School Museum Trips. Norma, A., & Pearce, J. (eds). In *Proceedings of the 6th Annual Conference on Mobile Learning mlearn 2007*. Melbourne: University of Melbourne, 238-244

Sharples, M., Taylor, J., & Vavoula, (2005). *Towards a Theory of Mobile Learning*. Paper presented at the 4th World Conference on mLearning, Cape Town, South Africa.

Shield, L., & Kukulska-Hulme, A. (2008). Editorial. *ReCALL*, *20*(3), 249–252. doi:10.1017/S095834400800013X

Shin, N., Norris, C., & Soloway, E. (2007). Findings from early research on one-to-one handheld use in K-12 education. In van 't Hooft, M., & Swan, M. (Eds.), *Ubiquitous computing in education: invisible technology, visible impact* (pp. 19–39). Mahwah, NJ: Lawrence Erlbaum Associates.

Short, J., Williams, E., & Christie, B. (1976). *The Social Psychology of Telecommunications*. Toronto: Wiley.

Shudong, W., & Higgins, M. (2006). Limitations of Mobile Phone Learning. *The JALTCALL journal* *2*(1), 3-14.

Shuler, C. (2009) Pockets of Potential: Using mobile technologies to promote children's learning. Retrieved December 10, 2009 from: http://joanganzcooneycenter.org/pdf/pockets_of_potential.pdf

Siemens, G. (2005). Connectivism: A learning theory for a digital age. *International Journal of Instructional Technology and Distance Learning, 2* (1). Retrieved December 10, 2009 from http://www.itdl.org/journal/jan_05/article01.htm

Siemens, G. (2008) Learning and Knowing in Networks: Changing roles for Educators and Designers. Retrieved December 10, 2009 from http://it.coe.uga.edu/itforum/Paper105/Siemens.pdf

Sinclair, J. McH., & Coulthard, R. M. (1975). *Towards an analysis of discourse: The English used by teachers and pupils*. Oxford, UK: Oxford University Press.

Sinitsa, K. (2000). Learning individually: a life-long perspective. *Journal of Educational Technology & Society*, *3*(1), 17–23.

Slavin, R. E. (1989). Research on cooperative learning: consensus and controversy. *Journal of the Educational Leadership*, *47*(4), 52–54.

Smith, R. K. (1995). Emerging tools for team collaboration: desktop videoconferencing experience and insight. *FID News Bulletin*, *45*(4), 121–127.

Smith, R. (2008). *GoMobile! Maximising the potential of mobile technologies for learners with disabilities*. Learning and Skills Network.

Smith, M. (1996). *Self-direction. The encyclopaedia of informal education.* Retrieved July 9, 2009 from http://www.infed.org/biblio/b-selfdr.htm

Smyth, E., Dunne, A., Darmody, M., & McGoy, S. (2007). *Gearing Up for the Exam? The Experiences of Junior Certificate Students*. ESRI: NCCA and the Gender Equality Unit funded research report. Dublin: The Liffey Press in association with the Economic and Social Research Institute.

Smyth, E., Dunne, A., McGoy, S., & Darmody, M. (2006). *Pathways through the Junior Cycle: The Experiences of Second Year Students*. ESRI: NCCA and the Gender Equality Unit funded research report. Dublin: The Liffey Press in association with the Economic and Social Research Institute.

Soloway, E., Norris, C., Blumenfeld, P., Fishman, B. J., & Marx, R. (2001). Hand-held devices are ready at hand. *Communications of the ACM*, *44*(6), 15–20. doi:10.1145/376134.376140

Soloway, E., Norris, C., Blumenfeld, P., Fishman, B., Krajick, J., & Marx, R. (2001). Handheld Devices are Ready-at-Hand. *Communications of the ACM*, *44*, 15–20. doi:10.1145/376134.376140

Song, Y. (2006). What consequences have handheld educational uses brought about. *Proceeding of the International workshop on Mobile and Ubiquitous Learning Environments MULE in conjunction with ICCE2006*, China, pp. 1-7.

Spender, D. (2007). Digi-kids and a new way of learning, in *The Age Newspaper*. Retrieved July 20, 2009 from http://www.smh.com.au/news/technology/digikids-and-a-new-way-of-learning/2007/05/18/1178995417590.html

STAAT (2007) *Measuring Village ICT in Sub- Saharan Africa.* Markets and Statistics Unit STAAT. ITU's Development Sector.

Stanczak, G. C. (2007). *Visual research methods: Image, society, and representation.* Thousand Oaks, CA: Sage Publications Inc.

Stanton, D., & Neale, H. (2002). Designing mobile technologies to support collaboration. *Equator Annual Conference*.

State of Victoria. (2009) *Information Privacy Act 2000.* Victorian Consolidated Legislation Australasian Legal Information Institute. Retrieved August 19, 2009, from http://www.austlii.edu.au/au/legis/vic/consol_act/ipa2000231/

Statistics Newslog, I. C. T. *Mobile Internet Becoming A Daily Activity For Man.* Wednesday, March 18, 2009. Retrieved October 21, 2009 from http://www.itu.int/ITUD/ict/newslog/Mobile+Internet+Becoming+A+Daily+Activity+For+Man.aspx

Stead, G. (2006). Mobile technologies: transforming the future of learning, in *Emerging Technologies for Learning*, BECTA. Retrieved July 20, 2009 from http://partners.becta.org.uk/upload-dir/downloads/page_documents/research/emerging_technologies.pdf

Stenhouse, L. (1997). An Appeal for Evidence of the Effectiveness of an Alternative Role to that of Neutral Chairman in Promoting Rational Inquiry into Moral Issues: A Reply to John T. Hyland. *Journal of Further and Higher Education, 1*(2), 50–54.

Stewart, K., Thompson, K., Hedberg, J., & Wong, W. (2009). Using technology to support quality learning for school activities involving field studies. In Caballe, S., Xhafa, F., Daradoumis, T., & Juan, A. A. (Eds.), *Architectures for distributed and complex m-learning systems: Applying intelligent technologies*. Hershey, PA: IGI Global.

Stewart, K. (2007). Mobile learning: Designing the learning context. In A. Norman, & J. Pearce (Eds.), *Making the connections*. Proceedings of the 6th Annual International Conference on Mobile Learning, Melbourne (pp. 253–258). Parkville, Vic: University of Melbourne. Available online at http://www.mlearn2007.org/files/mLearn_2007_Conference_Proceedings.pdf

Stockwell, G. (2007). Vocabulary on the Move: Investigating an intelligent mobile phone-based vocabulary tutor. *Computer Assisted Language Learning, 20*(4), 365–383. doi:10.1080/09588220701745817

Stockwell, G. (2008). Investigating Learner Preparedness for and usage patterns of mobile learning. *ReCALL, 20*(3), 253–270. doi:10.1017/S0958344008000232

Stylianou, D., & Pitta-Pantazi, D. (2002). Visualization and high achievement in mathematics: A critical look at successful visualization strategies. In Hitt, F. (Ed.), *Representations and mathematics visualization* (pp. 31–46). Mexico: Cinvestav-IPN.

Suchman, L. (1987). *Plans and situated actions: The problem of human–machine communication.* Cambridge, UK: Cambridge University Press.

Suchman, J. R. (1961). Inquiry training: Building skills for autonomous discovery. *Merrill-Palmer Quarterly of Behavior and Development, 7*(3), 147–169.

Sun, C. T., & Chou, C. (1996). Experiencing CORAL: design and implementation of distant cooperative learning. *IEEE Transactions on Education, 39*(3), 357–366. doi:10.1109/13.538759

Swain, M. (2001). Integrating language and content teaching through collaborative tasks. *Canadian Modern Language Review, 58*, 44–63. doi:10.3138/cmlr.58.1.44

Swan, K., van't Hooft, M., Kratcoski, A., & Schenker, J. (2007). Ubiquitous computing and changing pedagogical possibilities: Representations, conceptualisations and uses of knowledge. *Journal of Educational Computing Research, 36*(4), 481–515. doi:10.2190/B577-7162-2X11-17N5

Swan, K., van 't Hooft, M., Kratcoski, A., & Unger, D. (2005). Uses and effects of mobile computing devices in K-8 classrooms. *Journal of Research on Technology in Education, 38*(1), 99–113.

Synthesis of Research. 1987–2003. *Review of Educational Research, 76*, 1–62.

Tan, N., Chen, W., Looi, C.-K., Zhang, B., Seow, P., & Chan, A. (2007). Handheld computers as cognitive tools: An environmental learning project in Singapore. In T. Hirashima, U. Hoppe & S. Shwu-Ching Young (Eds.), *Supporting learning flow through integrative technologies* (162, pp. 377–384). Amsterdam: IOS Press.

Tangkuampien, J. (2009). The Potential for Technology-Enabled Connections: Kids, Education and Cellular Handsets. *Interaction, 16*(1).

Taylor, R. P., & Gitsaki, C. (2003). Teaching WELL in a computerless classroom. *Computer Assisted Language Learning, 16*(4), 275–294. doi:10.1076/call.16.4.275.23412

Taylor, J. (2007). Evaluating mobile learning. In Sharples, M. (Ed.), *Big issues in mobile learning* (pp. 26–28). Nottingham: Learning Sciences Research Institute.

Taylor, J. (2006). Evaluating Mobile Learning: What are appropriate methods for evaluating learning in mobile environments? In Sharples, M (ed) *Big Issues in Mobile Learning: Report of a workshop by the Kaleidoscope Network of Excellence Mobile Learning Initiative:* University of Nottingham. Retrieved November 29, 2009 from http://telearn.noe-kaleidoscope.org/ warehouse/Sharples-2006.pdf.

Tectonic (2007). South Africa High School Math Syllabus goes Mobile. *Tectonic*. Retrieved July 8, 2009, from http://www.tectonic.co.za/wordpress/?p=1589

Tétard, F., Patokorpi, E., & Carlsson, J. (2008). A conceptual framework for mobile learning. *Research Report 3/2008*. Retrieved October 15, 2008 from http://iamsr.abo.fi/publications/openFile.php?pub_id=464

The New London Group. (1996). A Pedagogy of Multiliteracies: Designing Social Futures. *Harvard Educational Review, 66*(1), 60–92.

The New Media Consortium. (2009). 2009 Horizon Report ANZ Edition. *COSN*. Retrieved October 30, 2009, from http://www.nmc.org/pdf/2009-Horizon-Report-ANZ-Edition.pdf

The Parliament of Queensland. (2009) Information *Privacy Act 2009*. Act No. 14 of 2009. Retrieved August 19, 2009, from www.legislation.qld.gov.au/LEGISLTN/ACTS/2009/09AC014.pdf

Thirumeni, S., & Yusoff, Y. (2009). A conceptual framework for OUM m-learning portal. In *Proceedings of IADIS International Conference Mobile Learning* (pp. 250-254).

Thompson, K., & Stewart, K. (2007). *The mobile jigsaw: A collaborative strategy for mlearning about the environment*. Proceedings of the 6th Annual International Conference on Mobile Learning, Melbourne (pp. 248–252). Parkville, Vic: University of Melbourne. Available online at http://www.mlearn2007.org/files/mLearn_2007_Conference_Proceedings.pdf

Thornton, P., & Houser, C. (2005). Using mobile phones in English education in Japan. *Journal of Computer Assisted Language Learning, 21*, 217–228. doi:10.1111/j.1365-2729.2005.00129.x

Thornton, P., & Houser, C. (2002). M-learning in transit. In Lewis, P. (Ed.), *The changing face of CALL* (pp. 229–243). Lisse, The Netherlands: Swets and Zeitlinger.

Thornton, P., & Houser, C. (2003). Using mobile web and video phones in English language teaching: Projects with Japanese college students. In Morrison, B., Green, C., & Motteram, G. (Eds.), *Directions in CALL: Experience, experiments and evaluation* (pp. 207–224). Hong Kong: English Language Centre, Hong Kong Polytechnic University.

Thornton, P., & Houser, C. (2004). Using Mobile Phones in Education. In *Proceedings of the 2nd IEEE International Workshop on Wireless and Mobile Technologies in Education, pp. 3-10.*

Tollmar, K., Sandor, O., & Shömer, A. (1996). Supporting social awareness at Work, design and experience. In *Proceedings of CSCW '96* (pp. 298–307). New York: ACM Press.

Tough, A. M. (1967). *Learning Without a Teacher. A study of tasks and assistance during adult self teaching projects*. Toronto: Ontario Institute for Studies in Education.

Traxler, J. (2008). Mobile learning: Moving through philosophical space. [Algarve] [Portugal] [: ACM.]. *Proceedings of IADIS International Conference Mobile Learning, 2008*, 235–239.

Traxler, J. (2009). Current State of Mobile Learning. In Ally, M. (Ed.), *Mobile Learning: Transforming the Delivery of Education and Training* (pp. 9–24). AU Press, Athabasca University.

Traxler, J. (2005). Mobile learning: it's here but what is it? *Interactions*, 9,1. University of Warwick. Retrieved December 10, 2009 from http://www2.warwick.ac.uk/services/ldc/resource/interactions/archive/issue25/traxler

Traxler, J. (2005). Strategic aspects of wireless and mobile learning. *Landscape Study in Wireless and Mobile Learning in the Post-16 Sector.* Retrieved August 19, 2008 from http://www.jisc.ac.uk/eli_outcomes.html

Trifonova, A., & Ronchetti, M. (2003, November). Where is Mobile Learning Going? *Proceedings of the World Conference on E-learning in Corporate, Government, Healthcare, & Higher Education*, Phoenix, AZ, 1794-1801.

Twarog, L., & Pereszlenyi-Pinter. (1988). Telephone-assisted language study and Ohio University: A report. *Modern Language Journal*, *72*, 426–434.

Twining, P., & Evans, D. with Cook, D., Ralston, J., Selwood, I., Jones, A., et al. (2005). Should there be a future for tablet PCs in schools? *Journal of Interactive Media in Education*, 20, 1–13. Retrieved July 7, 2009, from http://jime.open.ac.uk/2005/20/twining-2005-20-paper.html

UN. (2008). *The Global Information Society. Partnership on Measuring ICT for Development*. Santiago, Chile: United Nation UN Publications.

Unsworth, L. (2001). *Teaching multiliteracies across the curriculum: Changing contexts of text and image in classroom practice*. Buckingham, UK: Open University Press.

Vahey, P., & Crawford, V. (2002). *Palm Education Pioneers Program Final Evaluation Report*. Retrieved on July 7, 2009, from http://www.palmgrants.sri.com/PEP_Final_Report.pdf.

van 't Hooft, M. (2006). Tapping into digital literacy: Handheld computers in the K-12 classroom. In Tan, L. W. H., & Subramaniam, R. (Eds.), *Handbook of Research on Literacy in Technology at the K-12 Level* (pp. 287–307). Hershey, USA: Idea Group Publishing.

Vavoula, G., & Sharples, M. (2002). KLeOS: A personal, mobile, knowledge and learning organisation system. In Milrad, M, Hoppe, U and Kinshuk (eds) *Proceedings of the IEEE International Workshop on Mobile and Wireless Technologies in Education* (WMTE2002) (pp.152-156), Aug 29-30, Vaxjo, Sweden.

Vavoula, G., Sharples, M., Rudman, P., Meek, J., & Lonsdale, P. (2009). Myartspace: Design and evaluation of support for learning with multimedia phones between classrooms and museums. *Computers & Education*, *53*(2), 286–299. doi:10.1016/j.compedu.2009.02.007

Veenman, S., Benthum, N. V., Bootsma, D., Dieren, J. V., & Kemp, N. (2002). Cooperative learning and teacher education. *Teaching and Teacher Education*, *18*, 87–103. doi:10.1016/S0742-051X(01)00052-X

Von Glasersfeld, E. (1989). Cognition, construction of knowledge, and teaching. *Synthese*, *80*(1), 121–140. doi:10.1007/BF00869951

Vygotsky, L. S. (1962). *Thought and language*. Cambridge, MA: MIT Press. doi:10.1037/11193-000

Vygotsky, L. S. (1978). *Mind in Society: the Development of Higher Psychological Processes*. Cambridge Mass, London: Harvard University Press.

Wagner, J. (2006). Visible materials, visual theory and images of social research. *Visual Studies, 21*(1), 56–61. doi:10.1080/14725860600613238

Waldrip, B., Prain, V., & Carolan, J. (2006). Learning junior secondary science through multi-modal representations. *Electronic Journal of Science Education, 11*(1), 87–107.

Wang, C., & Burris, M. A. (1997). Photovoice: Concept, Methodology, and Use for Participatory Needs Assessment. *Health Education & Behavior, 24*(3), 369–387. doi:10.1177/109019819702400309

Waycott, J., Jones, A., & Scanlon, E. (2005). PDAs as lifelong learning tools: an activity theory based analysis. *Learning, Media and Technology, 30*, 107–130. doi:10.1080/17439880500093513

Weber, S. (2008). Visual Images in Research. In Knowles, J. G., & Cole, A. L. (Eds.), *Handbook of the Arts in Qualitative Research* (pp. 44–45). London: Sage.

Weiser, M. (1991). The Computer for the Twenty-First Century. *Scientific American*, (September issue), 94–10. doi:10.1038/scientificamerican0991-94

Weiser, M., Gold, R., & Brown, J. S. (1999). The origins of ubiquitous computing research at PARC in the late 1980s. *IBM Systems Journal, 38*, 693–696. doi:10.1147/sj.384.0693

Weiser, M. (1996, March). *Open house. ITP Review 2.0*. Retrieved March 15, 2008 from http://www.itp.tsoa.nyu.edu/~review.

Weiser, M., & Brown, J. S. (1996). The coming of age of calm technology. Retrieved March 15, 2008 from http://www.ubiq.com/hypertext/weiser/acmfuture2endnote.htm.

Wells, G. (2001). *Action, Talk, and Text: Learning and Teaching Through Inquiry*. Teachers College Press.

Wetherell, M., Taylor, S., & Yates, S. J. (2001). *Discourse theory and practice*. A Reader London: SAGE.

Williamson, B. (2005). *What are multimodal, multisemiotics and multiliteracies?* Retrieved November 15, 2009, from http://www.futurelab.org.uk/resources/publications_reports_articles/web_articles/Web_Article532

Willis, J. (2007). The Neuroscience of joyful education. *Educational Leadership 64*. Retrieved May 30, 2009 from http://www.ascd.org

Winter, N. (2006). What is Mobile Learning? In Sharples, M. (Ed.). *Big Issues in Mobile Learning: Report of a workshop by the Kaleidoscope Network of Excellence Mobile Learning Initiative*. Retrieved 29 Nobember from http://telearn.noekaleidoscope.org/warehouse/Sharples-2006.pdf.

Wishart, J., Ramsden, A., & McFarlane, A. (2007). PDAs and Handhelds: ICT at your side and not in your face. *Technology, Pedagogy and Education, 16*(1), 95–110. doi:10.1080/14759390601168106

Wolfson, L., & Willinsky, J. (1998). What service-learning can learn from situated learning. *Michigan Journal of Community Service Learning, 5*, 22–31.

Yatani, K., Onuma, M., Sugimoto, M., & Kusunoki, F. (2004). Musex: A system for supporting children's collaborative learning in a museum with PDAs. *Systems and Computers in Japan, 35*(14), 54–63. doi:10.1002/scj.10696

Yin, R. K. (2002). *Case Inquiry Research, Design and Methods* (3rd ed.). Newbury Park, London: SAGE Publications.

Yoo, Y., & Lyytinen, K. (2005). Social Impacts of Ubiquitous Computing: Exploring Critical Interactions Between Mobility, Context and Technology. *Journal of Information and Organization, 15*(2), 91–94. doi:10.1016/j.infoandorg.2005.02.006

Zurita, G., & Nussbaum, M. (2004). Computer supported collaborative learning using wirelessly interconnected handheld computers. *Computers & Education, 42*(3), 289–314. doi:10.1016/j.compedu.2003.08.005

About the Contributors

Wan Ng (Ph.D.) is Professor in Technology Enhanced Teaching and Learning & Science Education in the School of Education at the University of New South Wales, Australia. Her major research interest is in the use of ICT, including Web 2.0 and mobile technologies, in education. Wan has published widely in the areas of science education, gifted education, online learning at both school and tertiary levels and the use of mobile technologies for learning. Wan has co-ordinated several nationally-funded science and technology initiatives and works extensively with primary and secondary science teachers in professional learning courses.

* * *

Stamatina Anastopoulou holds a Ph.D. in the area of Learning Technologies, a M.Sc. in Cognitive Science and a B.Sc. in Industrial Management. She is currently a Research Fellow at the Learning Sciences Research Institute, University of Nottingham, UK. She has participated in research projects funded by the European Union, EPSRC and the Greek Ministry of Education and published several papers in international journals and conferences. Her research interests lie in science education and pupils attitudes to learning with technologies. She is currently exploring how mobile technologies can facilitate scientific inquiries within the classroom, the home and beyond.

André H. Caron, a graduate of Harvard University (Ed.D., 1976), is presently Full professor in the Communication Department of the Université de Montréal and Bell Canada Chair in interdisciplinary research on emerging technologies. He is the founding director of the Center for Youth and Media Studies (1987) and of CITÉ (2000) an Interdisciplinary Research Centre on Emerging A specialist in mass media and new technologies, his research interests include broadcasting policy, political and cultural appropriations of media, and the influences of new technologies on society. He has carried out numerous quantitative investigations on the diffusion and adoption by families of interactive technologies available in Canada, the U.S. and in the United Kingdom. Some of his more recent work has been published in the academic journal Convergence and discusses data gathered on the appropriation of "Mobile Culture" in society. He his co-author of a recently published book titled "Moving Cultures : Mobile Communication in everyday life" (2007) by McGill/Queens University Press.

Letizia Caronia (Ph.D.) is senior researcher at the Department of Education, University of Bologna (Italy). She is also Professor of Qualitative Methodologies in Educational Research. She has been a Visiting Researcher at the Centre for Youth and Media Studies, Department of Communication, University of Montréal. She is a member of the editorial board of the international journal "Encyclopaideia". For the past 15 years she has carried out research on the role of television and other information and communication technologies in children's everyday life and in a variety of intercultural communication settings. Her research focuses principally on the use of the media as a situated activity and a cultural practice whose sense and meanings are constructed in and through people's everyday interactions and conversations. Dr Letizia Caronia has published a number of books, articles and essays on the relationship between everyday language, media uses and educational practices, she has recently co-authored with Dr. A.H. Caron a book titled "Moving Cultures: Mobile Communication in everyday life" (2007) by McGill/Queens University Press.

Samantha Chenery-Morris is a Lecturer in Midwifery in the School of Nursing and Midwifery at University Campus Suffolk. Samantha is a Registered Midwife, Registered Sick Children's Nurse and a Registered General Nurse, and currently practices Midwifery clinically. Samantha lecturers across the Midwifery provision, and also takes a role within Interprofessional course at University Campus Suffolk. Sam has published in nursing journals, and has conducted research into online discussion forums and their use within the health curricula. Sam has recently commenced her doctoral studies in Midwifery.

Loi Peng Chng is a lecturer in the Faculty of Science and Technology, Open University Malaysia since 2003. He graduated with a Bachelor of Science in Bioindustry in 1999 from University Putra Malaysia (UPM) specializing in microbial technology and bioprocessing. He later obtained his Master of Agricultural Science from UPM specializing in land use and land management. His research interests include the application of mobile phones in teaching and learning, e-learning, carbon sequestration processes and modeling, and pest management.

Chien-Wen Chuang is currently a doctoral student in the Department of Information and Learning Technology at National University of Tainan in Taiwan. Her research interests include mobile and ubiquitous learning, web-based learning, and instructional design. The focus domain is social education development, especially in local culture pedagogy.

Elba del Carmen Valderrama Bahamóndez is a Ph.D. student at University of Duisburg-Essen, Germany, since 2008. In 2002 she received her B.Sc. degree in Computer Science Engineering from the Technological University of Panama (UTP). From 2003 to 2005 she has worked as an adjunct lecturer in the departments of Computer Programming, and Computing and Simulation of the faculty of Computer Science Engineering at UTP. In the same year, Elba won a DAAD scholarship to study a M.Sc. in Media Informatics in RWTH-Aachen University, Germany, which she finished in 2008. Elba research interests are in the field of ICT for development in particular for learning, mobile learning, user interfaces for mobile devices, user centered design and mobile phones interactive systems.

Moushir El-Bishouty is a researcher at the Informatics Research Institute, Mubarak City for Science and Technology, Egypt. He received his B.Sc. and M.Sc. degrees in Computer Science from Alexandria University, Egypt. He obtained his Ph.D. from the Information Science and Intelligent Systems Department, The University of Tokushima, Japan. His research interests include ubiquitous learning, knowledge awareness, software agents and personalization. He received the Best Paper Award at ICALT(IEEE), 2006 and CollabTech(IPSJ), 2008, and International Exchange Research Award from the University of Tokushima, 2009. He also received the Outstanding Contribution Award at ICCE(APSCE), 2008 and EJISST, 2008. Currently, he is a visiting researcher at The University of Tokushima, Japan.

Pascal Gagné is a graduate student in communication at the University of Montreal. Pascal Gagné works presently with the Research Group Language, Organization and Governance (LOG). He has also worked with the *Interdisciplinary Research Centre on Emerging Technologies* (CITÉ), the Centre for Youth and Media Studies (GRJM) and the Information and Communication Technologies Use and Design Laboratory (LUDTIC). Interested in the rhetoric of the Art in organizational settings, his thesis consists of an ethnographic look at argumentative practices expressed in political controversies about artistic works.

Tim Goodchild is a Senior Lecturer in Health in the School of Nursing and Midwifery at University Campus Suffolk. He is a Registered Nurse and has clinical experiences in acute medicine, head injuries and stroke care. He lectures across the nursing provision, taking a lead role in the support and development of anatomy & physiology teaching. Tim also takes a lead role in the interprofessional course and runs modules in the health foundation degree courses. He also lectures in aspects of educational technology and runs modules at Masters level in educational technology and research methods. Tim has conducted research in the application of educational technology in the Health curricula in various areas. He has presented at several national and international conferences based upon developments in educational technology.

Norlia Goolamally is a senior lecturer in the Faculty of Science and Technology, Open University Malaysia. She graduated with a Bachelor of Science in Education specializing in Mathematics and Chemistry from the University of Malaya in 1981. She obtained her Master in Education specializing in Measurement and Evaluation from the University of Malaya in 1998. She has 22 years of experience as a mathematics and chemistry teacher and has served as an assistant director at the Malaysian Ministry of Education (MOE). She was actively involved in educational policy research and programme evaluation while at the MOE.

Ria Hanewald (Ph.D.) is a Research Fellow in the Graduate School of Education at The University of Melbourne, Australia. Her expertise is in the use of ICT in education at all levels to improve teaching and learning, online teaching and learning, cyber safety and emerging technologies. The current focus of her work is on using digital technologies as tools for research. Dr Hanewald is state councillor for ICT in Education Victoria and editor of the perennial first semester issue of ICT in Education. Her background comprises teaching at early childhood, primary, secondary and tertiary levels in Australia, Europe and Asia. She has presented the findings of her research at conferences, seminars and workshops in Australia, Asia, Europe and America. Dr Hanewald is interested in conducting further research on an international level and can be contacted via her email: r.hanewald@yahoo.com.

John Hedberg (Professor) holds the Millennium Innovations Chair in ICT and Education and is Head of the Department of Education at Macquarie University, Sydney, Australia. He is well known for his research into learning environments that engage students and employ problem-based learning. Professor Hedberg has worked in several countries exploring how interactive and digital media can be employed as part of innovative pedagogies. His current work is on the role of digital technologies in new approaches to discipline knowledge and how the technologies can be employed to represent ideas that cannot be represented in any other ways.

Gwo-Jen Hwang is currently Professor of the Department of Information and Learning Technology, as well as Dean of the College of Science and Engineering, at National University of Tainan in Taiwan. He has published nearly 300 academic papers, including over 100 papers in professional journals such as Educational Technology and Society, Innovations in Education and Teaching International, Computers & Education, IEEE Transactions on Education, IEEE Transactions on Systems Man and Cybernetics, IEEE Transactions on Mobile Computing among others.

Harvinder Kaur Dharam Singh is an academic staff at Open University Malaysia. Having worked in the open and distance learning institution for the past seven years, Harvinder Kaur has presented and published numerous research papers locally and internationally in the area of e-learning, open and distance learning, lifelong learning and recently, in mobile learning. As well, Harvinder Kaur has published a chapter in an online learning book, and has co-authored a self-learning module which has received the Commonwealth of Learning (COL) Excellence in Distance Education Award for Distance Education Materials. She is also a member of the Malaysia Mobile Learning Society. She can be reached at harvinder@oum.edu.my.

Ray M. Kekwaletswe (Ph.D.) is currently a Senior Lecturer of Information Systems at the Tshwane University of Technology, South Africa. Previously, he was the Manager of Multimedia and Digital Computing at the American University, Washington D.C., USA where he attained his B.A., BSc. and the MSc in Computer Information Systems and Technology. He has a Ph.D. in Information Systems from the University of Cape Town. His research areas include ubiquitous mobile computing, context and presence awareness, learning and knowledge technologies and business informatics.

Katrina Keogh is an education officer with the National Council for Curriculum and Assessment (NCCA). Her areas of work at NCCA include modern languages at primary level, ICT, English as an additional language and the Mobile-Assisted Language Learning / FÓN project. Katrina has a B.Sc. (Hns.) in Applied Computational Linguistics from Dublin City University, where she specialised in Computer-Assisted Language Learning. Katrina taught German at primary level as a visiting teacher for over five years. She is currently working on her Ph.D. thesis in the area of Intelligent Computer-Assisted Language Learning.

Cecilie Murray specialises in learning innovation in the K-12 sector. She is Director of Delphian eLearning, an education consultancy company and App-titude Learning, developing applications for the iPhone/iPod Touch. She is an Australian leader in the use of ICT in education, particularly in relation to emerging technologies and school transformation. She is also well known for her work in mobile learning research, strategic planning, online teaching and digital content development. She was previously Senior Manager of eLearning at the Department of Education and Early Childhood Development, Victoria and at Emagine, the Centre of Excellence in Online Learning for the Department of Education, Tasmania where she established two virtual schools. She managed the first digital content development in Australia for The Learning Federation, and has worked on international projects with the UK, USA, Singapore, Japan and Indonesia. http://www.delphian.com.au and http://www.app-titude.com.au.

Howard Nicholas (Ph.D.) is Senior Lecturer in Language Education at La Trobe University. He has worked with mobile technologies in school contexts to explore the potential of these tools to support creative interactions between students and ways in which new technologies can support students to gain their own control of knowledge that they engage with. His work with mobile technologies builds on his long interest in the relationship between language, interaction and learning in both children and adults. He has published extensively in the areas of second language acquisition, language teaching and language policy, with a particular focus on relationships between interaction, feedback and views of learning that involve the restructuring of knowledge or behaviour.

Hiroaki Ogata is an Associate Professor in the Department of Information Science and Intelligent Systems, at Tokushima University, Japan. He has been the recipient of numerous Best Papers awards, among which are "Best Paper Award of Japanese Society for Information and Systems in Education (1998), WebNet 99 Top Paper Award (1999), Paper Award of SIG GN, IPSJ (2004), CSCL 2005 Contribution Award (2005), ICALT 2006 Best Paper Award (2006), MULE 2007 Best Paper Award (2007), and CollabTech 2008 Best Paper Award (2008). His research areas cover Computer Supported Ubiquitous Learning and Mobile Learning, CSCL (Computer Supported Collaborative Learning) and Human Computer Interaction.

Albrecht Schmidt is a professor for Pervasive Computing and User Interface Engineering at the University of Duisburg-Essen in Germany. Previously he was head of department at the Fraunhofer Institute for Intelligent Information and Analysis Systems. From 2003 to 2006 he headed the Embedded Interaction Research Group at the University of Munich. Albrecht studied in Ulm, Karlsruhe and Lancaster, where he completed his PhD on the topic "ubiquitous computing – computing in context". His teaching and research interests are in media informatics and ubiquitous computing, and in particular in the area of user interface engineering. Albrecht enjoys creating new interaction techniques and interfaces technologies for specific environments such as the home or the car. In recent years he has organized several workshops and conferences and served on various committees in the pervasive computing community.

Ju-Ling Shih is currently an Associate Professor of the Department of Information and Learning Technology in National University of Tainan, Taiwan. Dr. Shih earned her Ed.D. in Communication and Education from Teachers College, Columbia University, specializing in distance education and instructional technology. She has two master degrees; Ed.M. in Communication and Education and M.S. in Broadcasting Production from Columbia University and Boston University respectively. Her research interests include instructional design and qualitative research in mobile learning, digital games, and technology-mediated education at various levels and fields. She is involved in several individual and collaborative researches, ranging from extending digital archive use in online education, designing instructional model, and digital system to support qualitative studies in the fields, as well as searching for possible application of mobile devices in natural and social sciences education.

Kathy Stewart lectures in science and environmental education at Macquarie University, Sydney, Australia and researches in the area of mobile learning. Her academic interest is focussed on pedagogies and technologies that support learner inquiry in local environments. Dr Kathy Stewart has significant experience in teaching and researching in learning environments beyond the classroom and has developed a wide range of resources to support the informal learning sector. She has held elected positions on boards of community environmental education centres and organisations and is currently involved with community-led projects to monitor and restore natural environments in urban areas.

Zoraini Wati Abas (Ed.D).As a professor at the Faculty of Education and Languages and Director of the Institute of Quality, Research & Innovation (IQRI) at the Open University Malaysia (OUM), Zoraini is entrusted to help the university achieve its vision of becoming a leader of flexible learning as well as to drive research and innovation activities among the academicians. She currently leads the mobile learning project at OUM. She graduated with a Bachelors of Science in Biological Sciences in 1978, specialized in Instructional Technology at the Masters and doctoral levels, graduating with the latter from Northern Illinois University in 1985. She has authored several books and computer dictionaries and published in journals. She was also a weekly columnist in the national newspapers promoting the use and benefits of computers in education and how to benefit from Internet and e-Learning. She has written over 800 articles spanning more than a decade. Today, she actively promotes Learning 2.0.

Yoneo Yano is currently a full professor at the Faculty of Engineering, the University of Tokushima, Japan. He received his Ph. D. in communication engineering from Osaka University, Japan, in 1974. He was a visiting Research Associate at the Computer-Based Education Research Lab, University of Illinois, USA. He is the president of APSCE (Asia-Pacific Society for Computers in Education) from 2008. His current interests are in technology enhanced learning and e/u-Learning. He is a member of the IEICE, JSET, JSiSE, AACE and a senior member of IEEE.

Yuzery Yusoff is a lecturer in the Faculty of Information Technology and Multimedia Communication, Open University Malaysia and has been a member of the faculty since 2005. He received his B.I.T. degree in Science and Management System and M.I.T degree in Information Science from National University of Malaysia (UKM). Prior to joining the faculty he was a lecturer at University Tun Abdul Razak (UNITAR) from 1998 and a member of Multimedia Unit for the Faculty of Information Technology. His research interests are in the areas of multimedia application development, multimedia programming, mobile learning and open source systems initiatives.

Index

H

I

K

L

M

N

O

P

R

Breinigsville, PA USA
05 January 2011
252757BV00005B/89-158/P

9 781616 928490